Voltaire's Riddle

Micromégas and the measure of all things

Illustration on p. 222 from *Le Petit Prince* by Antoine de Saint-Exupéry, copyright 1943 by Harcourt, Inc. and renewed 1971 by Consuelo de Saint-Exupery, reprinted by permission of the publisher.

Storming of the Bastille by Jean-Pierre Louis Laurent Houel on p. 73. Reprinted with permission ©Musée Carnavalet / Roger-Viollet / The Image Works.

Folios 48 and 53 from Holkam Manuscript 48 on p. 42. Reprinted with permission of the Bodleian Library, University of Oxford.

Engraving of Maupertuis on p. 192. Reprinted with permission from the Owen Gingerich collection.

© 2010 by
The Mathematical Association of America (Incorporated)
Library of Congress Catalog Card Number 2009938662
ISBN 978-0-88385-345-0
Printed in the United States of America
Current Printing (last digit):
10 9 8 7 6 5 4 3 2 1

The Dolciani Mathematical Expositions

NUMBER THIRTY-NINE

Voltaire's Riddle

Micromégas and the measure of all things

Andrew J. Simoson
King College

ELMHURST COLLEGE LIBRARY

Published and Distributed by
The Mathematical Association of America

DOLCIANI MATHEMATICAL EXPOSITIONS

Committee on Books
Paul Zorn, *Chair*

Dolciani Mathematical Expositions Editorial Board
Underwood Dudley, *Editor*
Jeremy S. Case
Rosalie A. Dance
Tevian Dray
Patricia B. Humphrey
Virginia E. Knight
Mark A. Peterson
Jonathan Rogness
Thomas Q. Sibley
Joe Alyn Stickles

The DOLCIANI MATHEMATICAL EXPOSITIONS series of the Mathematical Association of America was established through a generous gift to the Association from Mary P. Dolciani, Professor of Mathematics at Hunter College of the City University of New York. In making the gift, Professor Dolciani, herself an exceptionally talented and successful expositor of mathematics, had the purpose of furthering the ideal of excellence in mathematical exposition.

The Association, for its part, was delighted to accept the gracious gesture initiating the revolving fund for this series from one who has served the Association with distinction, both as a member of the Committee on Publications and as a member of the Board of Governors. It was with genuine pleasure that the Board chose to name the series in her honor.

The books in the series are selected for their lucid expository style and stimulating mathematical content. Typically, they contain an ample supply of exercises, many with accompanying solutions. They are intended to be sufficiently elementary for the undergraduate and even the mathematically inclined high-school student to understand and enjoy, but also to be interesting and sometimes challenging to the more advanced mathematician.

1. *Mathematical Gems,* Ross Honsberger
2. *Mathematical Gems II,* Ross Honsberger
3. *Mathematical Morsels,* Ross Honsberger
4. *Mathematical Plums,* Ross Honsberger (ed.)
5. *Great Moments in Mathematics (Before 1650),* Howard Eves
6. *Maxima and Minima without Calculus,* Ivan Niven
7. *Great Moments in Mathematics (After 1650),* Howard Eves
8. *Map Coloring, Polyhedra, and the Four-Color Problem,* David Barnette
9. *Mathematical Gems III,* Ross Honsberger
10. *More Mathematical Morsels,* Ross Honsberger
11. *Old and New Unsolved Problems in Plane Geometry and Number Theory,* Victor Klee and Stan Wagon
12. *Problems for Mathematicians, Young and Old,* Paul R. Halmos
13. *Excursions in Calculus: An Interplay of the Continuous and the Discrete,* Robert M. Young
14. *The Wohascum County Problem Book,* George T. Gilbert, Mark Krusemeyer, and Loren C. Larson
15. *Lion Hunting and Other Mathematical Pursuits: A Collection of Mathematics, Verse, and Stories by Ralph P. Boas, Jr.,* edited by Gerald L. Alexanderson and Dale H. Mugler
16. *Linear Algebra Problem Book,* Paul R. Halmos
17. *From Erdős to Kiev: Problems of Olympiad Caliber,* Ross Honsberger

18. *Which Way Did the Bicycle Go? . . . and Other Intriguing Mathematical Mysteries,* Joseph D. E. Konhauser, Dan Velleman, and Stan Wagon
19. *In Pólya's Footsteps: Miscellaneous Problems and Essays,* Ross Honsberger
20. *Diophantus and Diophantine Equations,* I. G. Bashmakova (Updated by Joseph Silverman and translated by Abe Shenitzer)
21. *Logic as Algebra,* Paul Halmos and Steven Givant
22. *Euler: The Master of Us All,* William Dunham
23. *The Beginnings and Evolution of Algebra,* I. G. Bashmakova and G. S. Smirnova (Translated by Abe Shenitzer)
24. *Mathematical Chestnuts from Around the World,* Ross Honsberger
25. *Counting on Frameworks: Mathematics to Aid the Design of Rigid Structures,* Jack E. Graver
26. *Mathematical Diamonds,* Ross Honsberger
27. *Proofs that Really Count: The Art of Combinatorial Proof,* Arthur T. Benjamin and Jennifer J. Quinn
28. *Mathematical Delights,* Ross Honsberger
29. *Conics,* Keith Kendig
30. *Hesiod's Anvil: falling and spinning through heaven and earth,* Andrew J. Simoson
31. *A Garden of Integrals,* Frank E. Burk
32. *A Guide to Complex Variables* (MAA Guides #1), Steven G. Krantz
33. *Sink or Float? Thought Problems in Math and Physics*, Keith Kendig
34. *Biscuits of Number Theory*, Arthur T. Benjamin and Ezra Brown
35. *Uncommon Mathematical Excursions: Polynomia and Related Realms*, Dan Kalman
36. *When Less is More: Visualizing Basic Inequalities*, Claudi Alsina and Roger B. Nelsen
37. *A Guide to Advanced Real Analysis* (MAA Guides #2), Gerald B. Folland
38. *A Guide to Real Variables* (MAA Guides #3), Steven G. Krantz
39. *Voltaire's Riddle: Micromégas and the measure of all things*, Andrew J. Simoson
40. *A Guide to Topology,* (MAA Guides #4), Steven G. Krantz
41. *A Guide to Elementary Number Theory*, (MAA Guides #5), Underwood Dudley

MAA Service Center
P.O. Box 91112
Washington, DC 20090-1112
1-800-331-1MAA   FAX: 1-301-206-9789

Contents

Introduction xi

Vignette I: A Dinner Invitation 1
I The Annotated Micromégas 7

Vignette II: Here be Giants! 41
II The Micro and the Mega 51

Vignette III: The Bastille 71
III Fragments from Flatland 75

Vignette IV: A want-to-be mathematician 95
IV Newton's Polar Ellipse 101

Vignette V: A Bourgeois Poet in the Temple of Taste 117
V A Mandarin Orange or a Lemon? 121

Vignette VI: The Zodiac 139
VI Hipparchus's Twist 145

Vignette VII: Love Triangles 159
VII Dürer's Hypocycloid 163

Vignette VIII: Maupertuis's Hole 189
VIII Newton's Other Ellipse 197

Vignette IX: The Man in the Moon 217
IX Maupertuis's Pursuit Problem 221

Vignette X: Voltaire and the Almighty 237
X Solomon's π 241

Vignette XI: Laputa and Gargantua 257
XI Moon Pie 263

Vignette XII: A Last Curtain Call 279
XII Riddle Resolutions 281

Appendix 303
Cast of Characters 321
Comments on Selected Exercises 333
References 359
Index 367
About the Author 377

in memory of my father
Harold Lester Simoson, Jr.
(1920–1982)

who taught me
early on
to love books

My earliest image of him:

Dad in an overstuffed chair,
In one hand a pipe,
In the other, a burgundy volume—
Onto his lap I climb
Asking,
"Teach me to read."

Introduction

Voltaire?

A mathematics book featuring Voltaire?

Did you know that Voltaire was the first to publish the legend of Isaac Newton being inspired about gravity upon seeing an apple fall from a tree? That Voltaire tried for about eight years to be a mathematician? That he was responsible in no small way for Newton's *Principia* being translated from Latin into French? That he wrote a story whose backdrop is the French expedition to the polar regions in 1736–37 whose mission was to test Newton's theories about gravity?

This book is about that story, *Micromégas*. Micromégas, whose name means small-large, is a giant alien from Sirius who visits earth. He stumbles across a shipload of mathematicians, led by Pierre-Louis Moreau de Maupertuis, returning from a year-long expedition to measure the length of a degree of arc along a line of longitude in Lapland. Nearly fifty years had passed since Newton had predicted that the earth was flattened at the poles and bulged at the equator. Some simple but tedious and careful work by a scientific expedition could verify Newton's claims once and for all. In 1735, the French decided to do just that. Like the giant's name, Voltaire's story is about quantifying the very large and the very small, from gnats and worms to planets and stars, from the penchant for man to make war, to dreams of understanding the soul.

After interviewing the French savants about human knowledge and about humanity, Micromégas gives them a parting gift: a book with the answers to all the questions in the universe. Upon opening the book, it is found to be empty.

What does such an ending mean?

This is the question we call *Voltaire's riddle*. As well as suggesting some resolutions in the last chapter, the book presents the adventure, the people, and the related mathematics behind the polar expedition.

This book is a series of vignettes and chapters. The vignettes present episodes in the life of Voltaire or expand upon the literary traditions he used in his story. No special mathematical background is needed to read them. The chapters focus on mathematics, and much of the material is accessible to anyone interested in mathematics. The student who has studied undergraduate linear algebra, vector calculus, and differential equations will be able to understand almost all of the details.

Chapter I is an annotated edition, newly translated for this book, of Voltaire's *Micromégas*. Chapter II explores the measurement of small, medium, and large, and offers a case study of how the ancients could determine the size of the earth, sun, and moon, and the distances between them. Chapter III visits the world of Flatland, wherein a hero named A. Square writes his memoirs from prison, and—to pass the time—solves the equations of motion for planets moving about a sun in dimensions two and four, which in turn solves a troublesome riddle that A. Square once posed to his three-dimensional guide, the Sphere. Chapter IV reviews Newton's solution of planetary motion in three dimensions, with a case study analyzing the comet trajectory along which Micromégas toured the solar system. Chapter V shows why the earth is flattened at the poles by about seven kilometers, and bulges at the equator by about fourteen kilometers. Chapter VI explains why a flattened earth forces the earth's pole to rotate with a period of about twenty-six thousand years. Chapter VII is a look at Johann Bernoulli's riddle for Newton, and its solution. Chapter VIII extends the argument of Chapter VII to a rotating earth, and demonstrates, among other things, that a hole drilled by a pebble falling freely through the rotating earth, when dropped at the equator, is a familiar curve. Chapter IX considers a pursuit problem posed by a bishop of Copernican persuasion who imagines a wedge of swans flying from the earth to the moon at constant speed, always flying toward the moon—a problem studied by Maupertuis much later. Chapter X, in deference to Voltaire's habit of questioning everything, is a light-hearted survey of various explanations of how a round object could have circumference to diameter ratio of three. Chapter XI is a story about a giant who goes to the moon, written in the tradition of Voltaire and his precursors, namely, Jonathan Swift and Rabelais. Chapter XII gives possible resolutions to Voltaire's riddle of why the book is blank.

Some of the material in this book has appeared as my articles over the years: Some exercises of Chapter II appeared in *The American Mathematical Monthly* [119] and [120]. A case study of Chapter IV and Chapter VI appeared in *Primus* [123]. Parts of Chapter VII appeared in *The Mathematics Magazine* [121] and in *Primus* [125]. A version of Chapter VIII appears in *The Mathematics Magazine* [128]. Versions of Chapters IX and X appeared in

Introduction   xiii

the *College Mathematics Journal* [**124**] and [**126**]. Parts of Chapters VII and VIII of this volume include solutions to Exercises 9 and 10 of Chapter VII in [**122**, p. 107]. Some of the fractals appearing in Chapters II and XII appeared in *Primus* [**127**].

By and large, the chapters are independent of each other. However, Chapter IV is a follow-up chapter to Chapter III. Chapter V uses the relationship between polar angle and time for elliptical trajectories as developed in Chapter IV. Chapter VIII is a follow-up chapter to Chapter VII. And Chapter XII will make sense only to those who read the Micromégas story of Chapter I.

At the end of each chapter are exercises. Some are routine, others are conversation starters, and others are open-ended and could serve as the beginnings of honors projects for the interested student. Some of the theoretical details of the chapters are posed as exercises; complete solutions for these are given in the appendix.

I confess to some liberties taken in this book. For the translation of *Micromégas* itself in Chapter I, in the fall of 2006, I commissioned the King College advanced French conversation class to produce a literal translation from the French version [**145**] and [**148**]. I then compared that text with other translations, and pruned Voltaire's verbiage, updated terminology, and simplified action. Hopefully the text is lively in today's ear while preserving Voltaire's laughter and spirit. In Chapter II, even though Archimedes and Eratosthenes were friends, and Eratosthenes's nickname was Beta, I trust that the shade of Archimedes will not be offended with the nickname Alpha. Furthermore, there is an extensive literary tradition in the mathematical community in fleshing out the universe of Flatland. I have added a little to that number in Chapter III, offering fragments from A. Square's notebooks. Although A. Square was a lawyer [**1**, p. 73], he dabbled in mathematics much like Pierre Fermat. Finally, although no one but Rabelais can do justice to his Pantagruel, the temptation to try resulted in Chapter XI.

As I am neither a historian nor a linguist, I cite many secondary sources in the text. I was often struck by the wide disparity among the experts on many issues. Even on items as simple as physical appearance, the experts disagree. For example, one biographer estimates Voltaire's height at five feet three inches, presumably on the basis of the extant likenesses of Voltaire [**157**, p. 65], whereas another source [**101**, pp. 37, 422] gives Voltaire's height as five feet ten inches in stockinged feet, even in old age, citing three eyewitness accounts. As a more global example of disagreement, the biographies [**87**] and [**157**] of Emily du Châtelet, a friend to Voltaire, present her as two entirely different personalities, both sympathetic, the former as human and the latter as perfection. I have tried to strike a middle ground throughout this book; hopefully I have neither over-simplified issues nor stereotyped personalities in my presentation of the volatile writings and the controversial life of Voltaire. For translations of various phrases into English, I have gone with my best guesses. For example, should I translate Bernoulli's Latin phrase about Newton, "ex ungue leonem," as *claw* or *paw*? A more seasoned mathematical expositor than I suggested, "*Claw* has too many connotations of violence and the ripping of flesh while *paw* is altogether softer and gentler and captures Newton's stealthy behavior." I went with *paw*.

How does one read this book? The intended audience includes anyone interested in how scientific knowledge and perspective, including mathematics, becomes part of common knowledge and perspective. However, a (possibly false) publishing rule says that the relationship between a book's audience size and the abundance of formulas therein is an inverse one, prompting one reviewer of an early version of this book to predict that "its audience will be close to the empty set." With this rule in mind, here is what I suggest. Read Voltaire's *Micromégas*, which is Chapter I of this book. If it appeals to you, then read the Vignettes (and perhaps the first few paragraphs in each chapter) and the final chapter, which is an attempt to answer the riddle of a blank book. After all is said and done about scientific inquiry (not just in this book), what does such a discussion imply, if anything, about how to look at the universe, or, more grandly, about how to live? If any of the technical questions strike your fancy, then read the chapters. For those readers who have had a number of undergraduate mathematics courses, some chapters are easy to read, while others may take careful study, working through one line after another. Be forewarned, Chapter VI, while yet a very intuitive explanation of the earth's 26,000 year polar precession period, is the most technical of the chapters; however, compared to rigorous studies of this phenomenon the chapter is child's play.

As to the style of this book, I have violated a second rule, a Voltaireian axiom. In his entry on style in *The Philosophical Dictionary*, he admonishes the would-be writer [**144**, vol. xiv, p. 5],

Avoid pleasantry in mathematics.

This entire volume is a counterexample to this rule.

As some reason for ignoring his advice, Voltaire's perspective was flawed about many mathematical ideas. Of course, he merely reflected the informed viewpoints of his day, so he should not be blamed. For example, he missed non-Euclidean geometry [**144**, vol. vii, *Cato*, pp. 53–54]:

Introduction

> A young man, beginning to study geometry, comes to me. "Are you not certain," said I, "that the three angles of a triangle are equal to two right angles? I demonstrated it to him. He then became very certain of it, and will remain so all his life.

He missed the butterfly effect and chaos theory [**144**, vol. vii, *Chain of Events*, p. 61]:

> The motion produced by Magog in spitting into a well cannot have influenced what is now passing in Moldavia.

He missed the gravitational bending of light (unless of course he viewed nearby masses as impediments) [**144**, vol. ix, *Fire*, p. 92]:

> Why does light never move out of a right line when it is unimpeded in its rapid course?

Even though Voltaire's logic and facts were not always right, his word magic—at least in his romances, his letters, and his philosophical dictionary—endures through the centuries and the translation from French into English. Beauty, truth, justice: those were his themes. Such notions are ever in season.

However, it may be that I have misinterpreted Voltaire in these last few quotes. After all, I am a simple man. Sometimes—in fact, quite often—Voltaire means something altogether different than what he appears to say. For example, when talking of Plato's insight about the way the universe is constructed, Voltaire says,

> An author should always be explained in the most favorable sense.

To which my initial response was disbelief. He continues [**144**, vol. xii, pp. 209–210],

> [Plato] demonstratively proves that there can be but five worlds, because there are but five regular solid bodies. I confess that no philosopher in Bedlam has ever reasoned so powerfully.

At first reading, my impression was that Voltaire had complimented Plato. But then—wait a minute!—is not Bedlam an insane asylum?

My favorite example of Voltaire's craft at fashioning verbal time bombs is this selection on liberty of the press from *The Philosophical Dictionary* [**144**, vol. xi, p. 134]:

> You fear books, as certain small cantons fear violins. Let us read and let us dance—these two amusements will never do any harm to the world.

My first reaction was, "No Voltaire, you have it all wrong. Dancing is the very reason to keep on living. And reading will change the world." As I thought a little more and saw what he did not say—like looking at the background in a painting for meaning rather than its foreground—I understood and sheepishly confessed, "Oh, that was Voltaire's point exactly."

Finally, I wish to thank a number of people:

- The many biographers and translators whose works I used in this book. Thanks, too, to those who translated upon request: Stelio Cro, language professor and a native Italian, for translating selections from Galileo's first public lecture; Joseph Fitsinakis, a free-lance writer of political science and a native Grecian, for help with a sentence of Aristotle; Natalie Walters, a double major in French and mathematics, and Chris Thron, a former colleague who lectured in French for a year in Chad, for help with Maupertuis's *Lettre sur le progrès des sciences* [84].

- Professor Tracy Parkinson's 2006–07 advanced French conversation class: Kathyrn Baley, Rachel Barker, Ashley Caire, Heather Flowers, Amber Vandivort, and Anna Waldbart, for translating the *Micromégas* story from French to English. Thanks again to Chris Thron for critiquing my translation of *Micromégas*, many of whose suggestions I adopted. Thanks to Paul Zorn for improving upon the translation of a particularly thorny passage of the story.

- Emeritus physicist Dan Cross and astronomer Ray Bloomer for discussions concerning torque and angular momentum, and emeritus astronomer Edward Burke, jr., for discussions on comets.

- English professor Craig McDonald for inviting me to lead a cross-disciplinary honors seminar on measuring value, during which we showcased the *Micromégas* story, the discussion from which snow-balled into this book. Thanks to him also for some Middle English translations.

- Three of my former students for working on senior honor projects related to this book: Shane Morrison and Jared Newton [91] for floating pyrolithic graphite in a magnetic field to simulate Swift's floating island of Laputa, and Jessica Miller Waterman [150] for a discussion of Flatland mechanics.

- My fall 2008 electricity and magnetism class for staging one of Dante's clues about Nimrod's height, as shown in Figure 10 of Chapter II: Morgan "Taylor" Gillie (left bottom), Elizabeth Gillenwaters (right bottom),

Introduction

William Edwards (middle), and Timothy "Jordan" Smith (top). Thanks also to Taylor and Elizabeth for spotting some typos in the MS.

- Joanna Hopkins, picture curator of the Royal Society, for rummaging in the archives for some images for me.

- Stephen Simoson for the gift of a ten franc Voltaire bill and a Finnish Maupertuis stamp used in Chapter V. Jack Boyles and Trudie Simoson for commissioned sketches appearing in Chapter XI.

- Rudy Rucker for an exchange of letters on *Flatland*, including a *Mathematica* file [113] involving background gravity model calculations for *Spaceland*, a romance on life in a four dimensional universe. Rick Mabry for help with some measure theory exercises.

- Professor of mathematics Donald Teets of the South Dakota School of Mines for pointing out various standard celestial mechanics approaches to problems discussed in Chapters IV, V, and VI.

- Professor of religious studies James Bowley of Millsaps College for conversations about the history and structure of the *Talmud* for Chapter X.

- Don Albers for being willing to consider a mathematics book featuring Voltaire for the Mathematical Association of America.

- Underwood Dudley and the Dolciani Committee for critiquing the manuscript. Elaine Pedreira and Beverly Ruedi, MAA production editors, for transforming the manuscript into a real book.

- My wife Connie—her presence ever reminds me that life is in the dance.

October 2009
King College
Bristol, Tennessee

Vignette I: A Dinner Invitation

Imagine being invited to dinner at the court of Frederick the Great in Potsdam of old Prussia. The year is 1751. It's a wintry evening. Lantern lights twinkle orange and yellow. Inside the palace, fireplaces are warm. Lively, yet mellow chamber music of violin, cello, flute, and harpsichord melodies directed by Carl Philip Emmanuel Bach filter through the halls and antechambers. Bewigged and powdered man-servants direct seating assignments at the long candelabra-lit table with smooth, self-effacing charm. Despite the music, guests converse in familiar tones. They know one another well. Quiet laughter and nods, a touch of hands, passing of salt: seated around the table are the elite intellects of the day, all assembled at the behest and enticement of a philosopher king, who enjoys collecting minds as some collect rare coins or stamps.

Everyone at the table is a male of distinction with two exceptions, the queen-mother Sophia and her six-year-old grandniece Charlotte, who, wide-eyed, follows much of the French conversation passing back and forth across the table. Sophia insists on visits to these chambers because nothing delights her more than hearing about novel discoveries and prospective expeditions. A few seats from her is Leonhard Euler, a regular at this table for the past ten years, the queen's favorite savant. She often tells the story of their first dinners when he had recently arrived from St. Petersburg. She tells it again this night.

"Many were my attempts at drawing Monsieur Euler into conversation, but he would only reply in monosyllables. One day I asked him directly, 'Why do you not speak with me?'" All eyes were on her. Most everyone had heard the story before. They loved it, especially the Russians. She continued, "'Madam,' he replied, 'it is because I have just come from a country where every person who speaks is hung'" [**42**, p. 19].

Euler bowed to Sophia, and nodded to the princess—who, nine years later, he would tutor for several years.

Frederick laughed quietly, tapping his glass with silver in suitable applause. "Our president has some announcements."

By this time dessert was being served, a light, lemon-iced sherbert.

Pierre-Louis Moreau de Maupertuis, president of the newly reformulated Berlin Academy, arose and bowed. In doing so, some papers crinkled in his coat pocket, and he was reminded of an errand.

"Some day the postal service between Paris and Berlin will be no more than a day. But before I make it any longer than what it is now...," he said, whereupon he retrieved two letters from his pocket, leaned across the table, and set them before Euler.

Euler inspected them, turning his head slightly to the right so as to see with his one good eye, and pronounced the names Bouguer and Clairaut slowly. He held the second, tapping one corner upon the table.

"I have been expecting this. If it is what I think," and the tapping of the letter slowed, "it will be the most important and profound discovery that has ever been made in mathematics" [60, p. 35]. He paused for dramatic effect, "No less profound than your expedition to the arctic," he said, making eye contact with the president.

Maupertuis took this signal as his cue to continue the dialogue. Perhaps he was inspired by the princess sitting high atop a pillow in a chair across from him, for an attentive child in clean clothes will melt the heart of almost anyone. Whatever the reason, the president began describing a scientific expedition that had begun sixteen years ago.

"As you may know, the most-Catholic King Louis XV commissioned two expeditions in 1735 and 1736." His gaze circled the table, resting a moment on the queen. "We were to determine once and for all whether Monsieur Descartes or Monsieur Newton was right. The one argued that the earth is shaped like a lemon, flattened at the equator; the other argued that its shape was like a mandarin orange, flattened at the poles. Who was right?

"My friend, Pierre Bouguer who, by the way," he looked directly at Charlotte, "was first a professor when but fifteen years of age . . . " He abandoned his broken sentence, starting afresh, "Pierre sailed south to the equator, to the high Andes. They measured three degrees of arc near Quito."

He coughed and spat into a crisp handkerchief. Blood and pus. He had been battling a lung infection for years [16, p. 264]. He refolded the linen carefully and returned it to a pocket. "Some of us sailed to the arctic circle. With me was my friend, Alexis Claude de Clairaut, who," he again faced the child princess, "read his first mathematics paper to the French Academy at age 13."

"And what was the verdict?" asked the king's physician, Julien Offray de la Mettrie, who knew the answer. Everyone did. It was an achievement for the ages. But it is ever fun to tease, especially someone like Maupertuis, who was quick to detect even mild criticism and who subsequently turned red in the face.

"Sir Isaac. It is always Sir Isaac." He raised his hands, and bowed. His face was red.

Euler broke in, as he had scanned Clairaut's letter during the president's remarks, "And it is Sir Isaac again, despite ourselves. Five years ago Clairaut and I were convinced that the sun, earth, and moon as a system violated the inverse square law. But now, Clairaut, I believe, has shown that Newton was right after all."

The president thereafter updated everyone about current and upcoming business of the academy. "At present," he said as he finished his report, "we are considering a number of projects." Now his gaze went beyond the dinner guests as he saw the list in his mind. "I will tell you one of them. Another time for the others. We propose to dig a hole."

Silence greeted his plan. The words were clear. The audience was receptive. They simply needed more information.

"A deep hole."

He outlined the enterprise:

Rather than constructing the pyramids, it would have been far better had the pharoahs of Egypt demonstrated their greatness by using their huge labor force to dig deep holes in the earth. At present, we know nothing about the earth's interior. Our deepest mines barely dent the crust. If we could reach toward the earth's core, it is quite possible that we would encounter strange situations and singular phenomena. This controversial force [gravity] which extends to all objects and explains nature so well, is only known by its effects on the surface of the earth. It would be desirable to see how things are deep underground. [**84**, p. 157]

He coughed again, and continued, "Today we imagine a homogeneous earth. Is this true? We shall see."

Maupertuis bowed, and sat down. All clapped. Some chuckled.

Meanwhile, the dessert crystal had been cleared. Now was the time for after-dinner entertainment. Sometimes Frederick played a minuet on his flute. But, for his niece—a story would be pleasant. Frederick raised a hand.

"Thank you, Herr Doktor," he said, proud of his eccentric president. *Not every idea,* thought Frederick, *is a good one. But in order to have some good ones, one must have many ideas. The good ones will rise to the top.* At least that was his approach in monitoring his salon. "Now let's hear a story. Monsieur Voltaire has many."

The dinner party murmured their agreement.

Roused by the heightened conversation intensity, a young greyhound, who had been napping at Frederick's feet during the meal, yawned, stretched, turned about in a circle, and lay back down.

"Yes," Sophia concurred, "please do."

Attention shifted to the poet at Frederick's right hand.

Sophia continued, "How about the one with the giant, the one you wrote for my son so many years ago when he was only just grown?"

The historian and playwright rose to his feet. As he described himself, he was "thin, tall, dried-up, and bony" [**101**, p. 36]. He loved performances, and had a fine voice. He often substituted on-stage for actors in his own plays, as he knew the parts from memory. He bowed to his audience.

"Thank you, kind gentlemen," he added, "and ladies. That story has already been told in part tonight." He gestured to Maupertuis. "But let me tell you more of that story, a story that the gazettes failed to report, and a story that my former tutors, and that includes both Monsieur Clairaut and the president, have kept to themselves—primarily because no one would have believed them had they done otherwise."

Voltaire began the story. His voice rose and fell. His hands swooped this way and that. He stooped, he reached high. For an hour, as the wind jostled the windows and as snow fell in great drifts, he told them the story, the text of which, translated to English, is Chapter I of this book.

After dinner notes on the proceedings

The dinner party portrayal at Frederick's court in this vignette is largely accurate. The suppers began "at half-past eight (concert just over) and last[ed] till towards midnight." Of them, Voltaire said [**21**, p. 78],

> At these suppers there was a great deal of esprit going. The King had it, and made others have; and, what is extraordinary, I never felt myself so free at any table.

The anecdote told by the queen mother is a reminiscence of Euler taken from his *Letters to a German Princess* (1768). Although Euler's right eye was deformed and useless from an infection in 1738, and although his left eye developed a cataract rendering him blind by 1771, "his left eye worked well enough in 1751" [**37**]. The second of Euler's cited quotes was written by him in April of 1751 to Clairaut; as can be seen by this quote, Euler was free with praise for others. Euler maintained an extensive correspondence.

Series IV, volume I, of his *Opera Omnia* documents 2829 letters written to and from the great mathematician; 18 are with Bouguer, and 61 are with Clairaut. Maupertuis published in an open letter his proposal about digging a hole, *Lettre sur le progrès des sciences* (1752), along with a number of other proposals, many of which are listed in Vignette VIII.

CHAPTER I

The Annotated Micromégas

Preface

Voltaire probably told this tale at the table of Frederick the Great. It was published in 1752. An earlier version, entitled *The Voyage of Baron Gangan*, now lost, had been written for Frederick in 1739, a year before he assumed the throne. The story involves the French expedition of 1736 whose mission was to test Newton's claim that the earth is flattened at the poles. However, Voltaire embellishes the adventure by adding an alien giant, who questions the French savants—including Maupertuis and Clairaut—to their very core, indeed, to the core of everyman. The tale ends abruptly when the giant gives man a book with the answers to everything. But the book is blank. Why? That is the open-ended riddle with which this book wrestles. In doing so, we present the mathematics, and some related mathematics and stories, behind the tale. In the last chapter we give some possible resolutions to the riddle.

Scene I: A Voyage from Sirius to Saturn

On a planet circling Sirius[1] lived a young man of great intelligence whom I[2] had the honor to meet on his last[3] voyage to our little ant-hill. His name was Micromégas,[4] an apt name for a great person. He was eight leagues[5] tall; that is, 24,000 paces of five feet each.

---

[1] Besides the sun, Sirius is the brightest star in the sky as seen from earth, of magnitude -1.43. Sirius could be seen in the summer skies of ancient Egypt in a constellation referred to as the Dog, hence came the expression the *dog days of summer*.
[2] The narrator is a Voltaire figure.
[3] This construction suggests that Micromégas visits earth more than once.
[4] The name *Micromégas* means *small-big*.
[5] A league is about 3 miles long.

Figure 1. Voltaire, age 41, portrait by Maurice Quentin de La Tour (1735)

Some algebraists—people ever useful to the public—at once used their pens[6] and, since Monsieur Micromégas is 120,000 feet from head to foot while we earthlings are barely five feet and since earth is 9,000 leagues around, estimated his world to be 216 million leagues around.[7] Nothing is more ordinary in nature than such wide variety. The ability to traverse the periphery of some German or Italian monarchy within a half hour in comparison to the same for the empires of Turkey, Russia, and China gives but a weak picture of the vast range that nature displays.

---

[6]This comment is probably Voltaire's acknowledgment of Euler in the audience whom he describes as a "mathematician who in a given time fills more sheets of paper with calculation than any other" [36, p. xxvi].

[7]With this circumference, the radius is about one *astronomical unit* (a.u.), the distance of the earth from the sun. In the 1717 edition of the *Memoires of the Royal Academy of Paris*, Jacques Cassini (1677–1756) estimated that the diameter of Sirius was 100 diameters of the sun [148, p. 6], which would make its circumference about 100 million leagues, about half that of Micromégas's world. This value would be the star's measure, not that of any of its planets. Today's astronomers put Sirius's radius as 1.75 that of the sun. As we will see from these footnote remarks, Voltaire borrowed some imagery for his story from Bernard de Fontenelle's 1686 *Conversations on the Plurality of Worlds*. In this passage about the size of the Sirian's world, Voltaire may have adopted de Fontenelle's opinion that "I have a peculiar notion that every star could well be a world" [32, p. 10].

From his Excellency's height, any sculptor or painter would agree that his belt would be 50,000 feet around: a most ideal proportion. His nose was one third of his face,[8] which in turn was one seventh of his height,[9] which means that his nose was 6333 feet.[10]

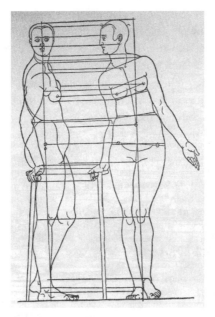

Figure 2. Dürer woodcut of the human body (1528), [**38**, cut 346]

His mind is the most cultivated we have encountered. He knows many things, some of which were his own discoveries. When he was 250 years old

---

[8] Figure 2 is an Albrecht Dürer woodcut of a woman showing this classic one third ratio of nose to face.

[9] Figure 3 is Leonardo da Vinci's 1492 *Vitruvian Man* sketch showing this one seventh ratio of head to body. However, Figure 2 gives a one ninth ratio of head to body. See also Figure 10(a) of Chapter II showing a similar Dürer sketch for the male body.

[10] Should it be 120,000 divided by 21, namely, 5714 feet rather than 6333? In fact, Voltaire seems to make a number of arithmetic errors throughout the story. Some editors change his original numbers to be in accordance with a natural interpretation. However, much of Voltaire's work is satirical. Therein lies a dilemma. After hearing a compliment from his lips or pen, should he be thanked, or should the statement be more carefully examined for hidden satire? We err on the side of keeping his numbers as he wrote them.

while studying, according to custom, at the Jesuit[11] school[12] on his planet, he proved more than 50 Euclidean propositions. This achievement far exceeds that of Blaise Pascal,[13] who, according to his sister,[14] proved 32 of the propositions for his amusement, and who has since[15] became a rather decent geometer and a very poor metaphysician.

At 450 years old—on the verge of manhood—Micromégas dissected small insects measuring nearly 100 feet in diameter, a size invisible to ordinary Sirian microscopes. He wrote a very curious treatise based upon these dissections which caused him much trouble.[16] Being of little understanding and great ignorance, the mufti[17] of his country considered his book offensive, reckless, and heretical in that Micromégas wondered in print whether fleas and snails shared a common morphology.[18]

---

[11] The Jesuits, a religious order within the Catholic church, was founded by Ignatius Loyola (1491–1556) in 1540. They are especially known for their emphasis on education. Voltaire, too, was schooled by the Jesuits. Many of his classmates were life-long friends. Although he lampooned many religious orders, including the Jesuits, he often praised them.

[12] Voltaire's first edition reads "most famous school" rather than "Jesuit school." But as the Jesuits literally went everywhere in the world, the satire of having them establish a school on Sirius was rich enough for Voltaire or a later editor to make the school a Jesuit one.

[13] Pascal (1623–1662) had a conversion experience in 1654, after which he mostly abandoned mathematical pursuits. His wager that God exists is famous: "If you win, you win all; if you lose, you lose nothing." Whereas, if you wager that God fails to exist, "If you win, you win nothing at all." With respect to this wager, Voltaire said, "I do not say to you with Pascal, 'Choose the safest.' There is no safety in uncertainty" [**144**, vol. ix, *Gods*, p. 241]. Pascal was a champion of the Jansenists, a religious sect that vied in power with the Jesuits. In Voltaire's day, as he saw it, the Jansenists more or less controlled the religious tone—and hence the political tone—in France. Very simply, the Jansenists emphasized faith whereas the Jesuits emphasized reason. Of the two, faith tends to institutionalize superstition and intolerance, whereas reason tends to reform arbitrary and unjust rules. Voltaire's prime focus in his life—if one is allowed to ferret out a life principle from Voltaire's 84 years—was to promote the latter. Hence arises Voltaire's lasting irritation with Pascal. However, Voltaire does concede that Pascal "contributed much to the improvement of our language and eloquence" [**144**, vol. xxxviii, *Notes on writers,* p. 296].

[14] Jaqueline Pascal entered a Jansenist convent in 1651.

[15] Voltaire, in this instance and throughout the text, refers to many people, who had long since died when he wrote the story, as if they were yet alive—so giving an air of timelessness to the narrative.

[16] Voltaire was forever getting into trouble with the authorities for what he wrote. See Vignette III.

[17] A mufti is a Mohammedan cleric. Voltaire uses it as a pejorative term for an official censor of the state. One of the unforgiveable sins of Voltaire—according to his critics—was that he served as historiographer for Louis XV for a season. Since such service included the possible censoring of new works, Voltaire sometimes muzzled free-thinkers like himself, an activity that branded him as a hypocrite among his peers. Perhaps Voltaire is poking fun at himself here.

[18] "Essential Nature" might be a better eighteenth-century translation. Apparently, a bishop of Mirepoix, Jean-François Boyer (1675–1755), had considered Voltaire's assertion that the human body and animal bodies developed in similar ways as offensive, reckless, and heretical,

The Annotated Micromégas   11

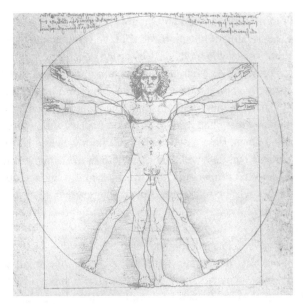

Figure 3. Da Vinci sketch of the human body, c. 1492 [53, Figure 48]

Micromégas defended himself with verve. All the women supported him.[19] His trial dragged on for 220 years after which the book was condemned by a panel of judges who had not read it.[20]

The author was exiled from court for 800 years.[21] Micromégas was scarcely bothered at such a rebuke because he considered the court rife with empty protocol[22] and pettiness. He composed a song lampooning[23] the

---

and therefore as advisor to Louis XV, he recommended that Voltaire not be given a vacant seat in the French Academy as "it would be an offence against God should a profane person like [Voltaire] succeed a cardinal" [147, pp. 75–80].

[19] Voltaire said that the most dangerous books were those "read by the idle of the court and by the ladies," which more or less included his own writings [144, vol. ix, *Gods,* p. 231].

[20] This event is much like Galileo's (1564–1642) trial, about which Galileo said that his judges had condemned his book "without understanding it, without hearing it, without even having seen it" [114, p. 97]. Voltaire echoes this image throughout his writing: "We must a hundred times remind mankind that the holy conclave condemned Galileo" [144, vol. xii, *Philosophy,* p. 188].

[21] When Voltaire was faced with another stint in the Bastille, he suggested exile. The powers that be agreed, and Voltaire spent the next two years in England.

[22] Here is a sample of Voltaire's sentiments on courtly protocol: "Which is more useful to a state—a well-powdered lord, who knows precisely at what hour the king rises and retires, playing the part of a slave in the antechamber; or a merchant who enriches the country, and contributes to the happiness of the world?" [144, vol. vii, *Commerce,* pp. 213–214].

[23] This is classic Voltaire behavior. As can be seen, at this point in the story, Micromégas is a Voltaire incarnation. However, in a moment, and throughout much of the remainder of the

mufti, who in turn failed to find any humor therein. So, as the saying goes, to complete the training of his mind and soul, he ventured out to travel from planet to planet. Those who travel only by coach or carriage will be surprised by his method of travel. As Micromégas was familiar with the laws of gravity and all the attractive and repulsive forces in nature, he applied that knowledge so well that, sometimes with the help of a ray of sun, sometimes by the convenience of a comet,[24] he went from planet to planet like a bird fluttering from branch to branch. He crossed the Milky Way[25] quickly. However, I am obliged to confess that he never saw anywhere in the canopy of these stars that beautiful celestial Empyrean[26] that the illustrious vicar Derham[27] claims to have seen through the telescope. Not to suggest that Monsieur Derham was mistaken, God forbid! But Micromégas was there, at the place, and he is a good observer. Let me not contradict anyone.

After many adventures, Micromégas arrived on Saturn. Despite being accustomed to novelty, he initially failed, when seeing the small size of the planet and its inhabitants, to smother those smiles of superiority that sometime escape from the wise.[28] Saturn is barely 900 times larger than the earth,[29] and its citizens are dwarves of about one thousand toises[30] in height. At first, he laughed a little at the people, somewhat in the way that Italian musicians visiting France may disregard Lully's[31] music. But the good-hearted Sirian

---

story, Micromégas becomes a Frederick the Great figure. In his stories, Voltaire often interjected his own voice through his characters, even though the characters might thereby speak out-of-character. Perhaps this penchant arose from his habit of substituting on stage for various actors in his plays, as he knew the dialogues from memory.

[24] In 1686, De Fontentelle suggested that "the neighboring [stars] occasionally send us comets" [32, p. 68].

[25] Our galaxy has a diameter of about 200,000 light years.

[26] The Empyrean Heaven is suppposedly the highest heaven, the place where God lives.

[27] William Derham (1657–1735), an astronomer/cleric and member of the Royal Society, suggested that nebulae were windows into a spiritual heaven. In his book *Astro-Theology*, he imagines that angels inhabited planets and stars. This particular passage from *Micromégas* is somewhat like a Voltaire version of the apocryphal quote by cosmonaut Yuri Gagarin (1934–1968) on his first flight into space, "I do not see any god up here."

[28] Voltaire is poking fun at himself here. Many portraits and busts of Voltaire capture the expression of the superior smile, as in Figure 1.

[29] Saturn has radius 57,550 km, making it nine times the size of the earth. Its mass is 95 times that of earth's.

[30] One toise is about six feet. More precisely, one toise was 1.949 meters in France before the year 1812.

[31] Jean-Baptiste Lulli (1632–1687) was a musician in the court of Louis XIV (1638–1715). Apparently, Voltaire is espousing the old proverb, "Can anything good come out of Nazareth [versus Jerusalem]?" Ironically, Lulli was an Italian although he spent much of his professional career in Paris.

understood that no one should be ridiculed merely for being only 6,000 feet tall. After their astonishment subsided, he made many Saturnian acquaintances. He formed a close friendship with the secretary[32] of the Saturn Academy, a man of great intelligence,[33] who, it is true, invented nothing.[34] Instead the secretary excelled at describing the inventions and discoveries of others[35] as well as writing poetry[36] and performing calculations.[37] For the satisfaction of the readers, I report a conversation between Micromégas and this secretary.

Scene II: Dialogue on Saturn

His Excellency[38] reclined himself so that the secretary could speak to him face to face.

"Must it always be," asked Micromégas, "that nature be so greatly varied?"

"Yes," said the Saturnian, "nature is like a flowerbed full of color."

"Humph!" said the other, "forget your flowerbed."[39]

"She is," retorted the secretary, "like a bevy of blondes and brunettes adorned with much jewelry."

"What is it with you and your brunettes?" asked Micromégas.[40]

---

[32] The Saturnian secretary is a Voltaire figure. Just as Micromégas and the secretary are friends, so too are Frederick and Voltaire. Of course, the size differential helps make the analogy more politically correct. Wade [**148**] and other Voltaire specialists suggest that in this passage Voltaire is mocking Bernard de Fontenelle, the secretary of the French Academy of Sciences from 1697–1739. See footnote 39.

[33] Yes, Voltaire is complimenting himself here.

[34] Voltaire describes himself as a philosopher. However, his critics have more or less concluded that Voltaire failed to introduce a single new philosophical idea; instead he embellished what he found in the marketplace of ideas. In short, his talent lay in being able to say most anything better than most anybody.

[35] Voltaire may be referring to his 1738 popularized *Elements* account of Newton's *Principia*.

[36] Voltaire is credited with being the first epic French poet due to his versification of Henry IV's life (1553–1610) in the *Henriade*.

[37] If this skill is a Voltaireian trait, perhaps he refers to his ability to manage finances—Voltaire died a very wealthy man largely by dint of investing his resources. It is also possible that in this passage, Voltaire is mocking his mathematical skills, which might mean that some of the obvious arithmetic errors in this story are deliberate.

[38] This address supports the idea that Micromégas is a Frederick the Great figure.

[39] As evidence that the Saturnian figure may also be a de Fontenelle figure, here is how Voltaire characterized the secretary's style: "If de Fontenelle is in some places too flowery, we should consider them as rich harvests wherein flowers naturally grow up with the corn" [**144**, *Notes on writers*, vol. xxxviii, p. 286].

[40] Apparently, Frederick the Great was a homosexual [**21**, pp. 81–81], [**101**, p. 142, 168–169]. Voltaire may be making light of his employer's sexual proclivities in this passage. If so, it is easy to see why Voltaire's wit was so controversial.

Figure 4. Saturn, as viewed from earth over a five year period, courtesy of NASA

"Then it is like an art gallery full of various paintings."

"Eh! No!" said the traveler, "I will say it again: nature is like nature. Why look for comparisons?"

"To please you," said the secretary.

"In this matter I do not wish to be titillated," replied the traveler, "I want to be informed. You may begin by informing me about the number of senses you have on this planet."

"Seventy-two," said the savant, "and everyday we complain about having so few. Our desire exceeds our need. Despite having seventy-two senses, a planetary ring, and five moons,[41] we are often blind and witless. In spite of our curiosity and the passion aroused by our senses, we are ever restless and dissatisfied with the depth of our understanding."

"I believe you," said Micromégas. "Although in our world we have about one thousand senses, we also are afflicted with a vague and inscrutable restlessness[42]—I am not sure what it is. We seem to be nothing compared to those beings who surpass us. In my travels, I have seen mortals very inferior

---

[41] Saturn has at least ten named satellites.

[42] Augustine's (354–430) comment would be, "Our hearts are restless, O Lord, until they find their rest in thee." However, Voltaire is probably referring to the general human condition. Here is how Pliny the Elder (23–79) describes us [**106**, book vii, p. 509]:

to us; I have seen mortals very superior to us, but each of them always have more desire than need and more need than satisfaction. Maybe one day I will discover a country where people truly lack nothing, but so far I have heard no credible report that such a place exists."[43]

The Saturnian and the Sirian continued in these conjectures,[44] but after many very ingenious arguments and tenuous hypotheticals, Micromégas returned to facts.[45]

"How long do you live?" asked the Sirian.

"All too short," replied the dwarf.

"So it is with us," said the Sirian; "we ever rue the brevity of life.[46] Such a perspective must be a universal one."

"Alas! We live no longer," sighed the Saturnian, "than 500 revolutions of the sun (about fifteen thousand earth years). It seems as if we begin dying almost at the moment of our birth. Our existence is a mathematical point, a single instant on a tiny atom in space. Having scarcely begun to learn, we die before we can use that knowledge. For this reason, I refrain from starting open-ended projects. I am only a drop of water in an ocean. I am ashamed, especially before you, of how ridiculous I appear to be."

Micromégas replied, "If you were not a philosopher, it would sadden you to learn that our average life span is seven hundred times longer than yours. But you well know that at death your body is rendered to the elements, and thereafter returns to nature in other forms. When the moment of

---

On man alone of living creatures is bestowed grief, on him alone luxury, and that in countless forms; he alone has ambition, avarice, immeasurable appetite for life, superstition, anxiety about burial, and even about what will happen after life is no more. No creature's life is more precarious, none has a greater lust for all enjoyments, a more confused timidity, a fiercer rage.

[43] This country may be an allusion to Thomas More's *Utopia*, a word meaning *nowhere*. Once, upon arriving in Brussels in 1739, Voltaire hosted a grand evening party replete with fireworks; the dinner invitations announced his status as *Ambassador of Utopia* [**101**, p. 163].

[44] Of the value of this kind of talk, Voltaire was of many minds. Here's one of his perspectives: "Metaphysical arguments seem to me like balloons filled with air used between disputants. The bladders burst, and nothing remains" [**144**, *The Atheist and the Sage*, vol. ii, p. 163]. But, of course, he loved to dispute anyway.

[45] One reason Voltaire may have decided to bring a Saturnian into his story is that de Fontentelle, in his book, had characterized Saturnians as "quite phlegmatic. They are people who don't know what it is to laugh and who always take a day to answer the slightest question asked them" [**32**, p. 60].

[46] Here's an amplification of this observation from *The Huron*, one of Voltaire's romances: The indolent "languish in bed till the sun has performed half its daily journey, unable to sleep, but not disposed to rise, and lose so many precious hours in that doubtful state between life and death, and who nevertheless complain that life is too short" [**144**, *The Huron*, vol. iii, p. 73].

transformation comes, to have lived an eternity or a day is precisely the same. I have been among those who live a thousand times longer than we do, and still they murmured. But everywhere are sensible men who accept their lot and give thanks to the Author of nature. Though the universe has much diversity, that variety is graced with an admirable commonality. For example, every person is unique, yet we all resemble one another at the core in thought and desire. Matter is everywhere; but it appears differently in each world. From your perspective, how many properties does matter have?"

"If you speak of essential properties," said the Saturnian "we count three hundred, among which are extension, impenetrability, mobility, gravitation, and divisibility."[47]

"Apparently," replied the voyager, "this small number is sufficient for appreciating the phenomena present in your world. Everywhere I see great variance, yet in proper proportion. Your planet is small, as are your fellow men; you have few senses; you perceive a small number of physical properties. Yet all is the work of Providence. So tell me, what color is your sun when you examine it closely?"

"Yellowish white," answered the Saturnian. "When its rays are divided, seven colors appear."[48]

"Our sun is tinted red,[49] and we have thirty-nine primary colors," said the Sirian. "Among all of the suns I've observed, none is like another. Just as among your kind, every person's face is different."

After many like questions, the Saturnian listed the thirty various substances to be encountered on Saturn such as God, space, matter, beings occupying space which feel and think, thinking beings which do not occupy space, those which possess penetrability, and others which do not.[50] The Sirian whose people had systematically categorized three hundred such levels and who had discovered three thousand more in his voyages, listened carefully to his friend's philosophy. At length, after explaining to each other a little of what they knew and much about what they did not—the dialogue having

---

[47] *Extension* includes the notions of length, surface area, and volume. Euler defines *impenetrability* as the inability of two bodies to occupy the same place simultaneously [42, p. 237].

[48] Newton developed a theory of light, and explained the colors of the spectrum.

[49] Sirius is a blue star of type A. Stars are classified by temperature from hot to cold as types O, B, A, F, G, K, and M. Our sun is a type G star. Curiously, the ancients classified Sirius as a red star; certain sects of old Rome sacrificed red dogs to the star. It is possible, but unlikely, that the star has changed its color.

[50] These phrases are technical terms from eighteenth century Cartesian ontology. A more simple-minded list might be God, arch-angels, angels, man, brutes, worms and insects, bacteria, inanimate matter, and empty space.

lasted an entire revolution of the sun[51]—they resolved to embark on a brief philosophical expedition.

Scene III: The voyage from Saturn to earth

As the two philosophers were preparing to soar into the Saturnian stratosphere, equipped with a splendid assortment of mathematical instruments, the Saturnian's mistress[52] approached them in tears. She was a petite brunette not even 660 toises tall. What she lacked in stature was more than compensated by charm.

"Such cruelty!" she cried. "After having resisted you for 1500 years, and now after a mere century in your arms, you leave me to go on a voyage with an alien giant. Go! You never loved me. If you were a true Saturnian, you would be faithful. Do you know where you are going? Do you even know for what you are looking? Saturn's five moons are less erratic than you,[53] her ring less unstable. This is it. I will never love another."

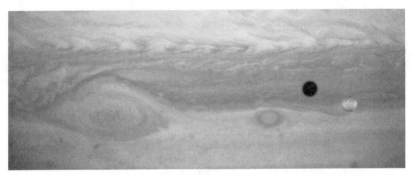

Figure 5. Io and its shadow on Jupiter's surface, courtesy of NASA

The philosopher embraced her, and wept with her, philosopher though he was.

The woman, after having swooned, soon consoled herself with a fashionable young consort of the country.[54]

---

[51] Saturn's orbital period is 29.5 earth years.

[52] This is a figure of Emily, Mme. du Châtelet, who upon the occasion of Voltaire's extended visit to Frederick's court in 1740, wrote to Maupertuis, 'I do hope [Frederick] will soon send me back the man with whom I intend to spend the rest of my life. I only lent him for a few days'" [101, p. 167]. We shall have more to say about Emily.

[53] A reference to the unpredictability of trajectories in the n-body problem as raised by Euler and Clairaut in Vignette I, where n is an integer greater than 2.

[54] Emily had romantic liasons with others, including Maupertuis and Clairaut [54], [87, p. 244], [101, pp. 118–119].

Meanwhile the two curious adventurers left. They jumped onto Saturn's ring, which they found to be nearly flat as was conjectured by a renowned inhabitant[55] of our little world. From there they hopped from moon to moon. A comet[56] passed close to the last one, and they hurled themselves aboard, together with their servants and instruments. About 150 million leagues further on, they reached the moons of Jupiter[57] and passed to Jupiter proper. There they stayed for a year[58] learning wonderful secrets which would actually be printed if not for the mufti,[59] who found the proposals too hard to understand. But I have read the manuscript[60] in the library[61] of an illustrious archbishop[62] who allowed me to look at his books at great length with such generosity and goodness that I cannot praise him enough. In the next edition of the *Moréri*,[63] I shall include a long article on him and his sons.[64]

Leaving Jupiter, they traveled 100 million leagues, and neared Mars, which, as we know, is five times smaller[65] than our globe. They saw two moons[66] orbiting Mars, orbs that have escaped notice from our astronomers.

---

[55] Christiaan Huygens (1629–1695).

[56] A well-known comet appeared in the skies at this time; the comet was first seen in 1737 by the Jesuit missionary Ignatius Kegler (c. 1680–1746), from Beijing, China—now identified as Comet 109P: Swift-Tuttle, after Lewis Swift (1820–1913) and Horace Tuttle's (1837–1923) observations in 1862.

[57] Figure 5 is a photo of Io, a moon of Jupiter, and its shadow on the Jovian surface taken by the Cassini spacecraft in 2002.

[58] Jupiter's period is almost twelve earth years.

[59] These mufti could be Jovian ones. Perhaps every society has them.

[60] How did this manuscript wind up in this library? Voltaire might be satirizing the Jesuits' habit of going everywhere and returning with curious artifacts.

[61] Voltaire had a large library himself consisting of over seven thousand volumes [**157**, p. 119]. After his death, Catherine the Great (1729–1796) of Russia bought it.

[62] Wade [**148**, p. 26] speculates that this archbishop was the Jesuit Pierre-Guérin de Tencin (1680–1758). See footnote 64.

[63] The *Moréri* was a precursor of the encyclopedia: the "Grand Dictionnaire Historique" of 1670 was compiled by Louis Moréri (1643–1680).

[64] This apparent compliment is vintage Voltaire satire. Catholic priests take vows of celibacy, and evidently the archbishop (who became a cardinal in 1739) had several children. Even though he himself was an anti-Jansenist, Voltaire was upset with the archbishop for suspending eighty-year-old Bishop Jean Soanen of Senez who had appealed the censoring of a certain Jansenist work.

[65] Mars has radius 3394 km whereas earth has radius 6378 km, making it 0.53 the size of the earth. Its mass is about 10% that of earth.

[66] Voltaire borrows this remarkable prediction from *Gulliver's Travels*, Part III: Laputa, Chapter III, first printed in 1726. More than one hundred years later, in 1877, the American astronomer Asaph Hall discovered two Martian moons. Swift may have gotten the idea from a congratulatory letter from Kepler to Galileo upon discovering four Jovian moons: "I long for a telescope to anticipate you in discovering two [moons] around Mars" [**55**, p. 93].

Figure 6. Surfing the aurora borealis

I know very well that Father Castel[67] will write quite convincingly against the existence of these two moons, but I rely on those who reason by analogy. Such philosophers know that Mars, which is so far from the sun, could little endure with less than two moons. However, the voyagers deemed the planet so small that they feared that they would not find room enough to sleep. So they continued on their way like two travelers declining a flophouse in a

---

[67]Charles-Irénée Castel L'Abbé de Saint-Pierre (1658–1743) was a member of the French Academy who, in Voltaire's words, "proposed for the most part things impractical" [**144**, *Notes on writers*, vol. xxxviii, p. 302]. One of these ideas was a universal parliament, a Diet of Europe, a kind of European Common market or a United Nations. Another reason for Voltaire mentioning Castel in his story at this point is that Maupertuis is in the audience—and on Castel's death, Maupertuis assumed Castel's seat (seat number 8) in the French Academy. So, in a back-handed way, what Voltaire writes about Castel can really be directed towards Maupertuis.

village in favor of what might be found in the next town.[68] The Sirian and his traveling companion soon regretted their decision for they traveled a very long time and found nothing. At last they noticed the faint glimmering of the earth. Compared to Jupiter, it was pitiful. Nevertheless lest they squander a second chance, they disembarked. Slipping off the tail of their comet and falling through the aurora borealis, they landed on the shores of the Baltic Sea, July 5, 1737.[69]

Scene IV: A walkabout on earth

After resting for a time, they each ate a mountain for breakfast[70] which their servants had properly prepared. Next, wanting to explore the countryside, they went first from north to south. As the normal step of the Sirian was around 30,000 feet, the Saturnian dwarf lagged far behind, panting. He took twelve steps to the Sirian's one.[71] To illustrate (if the reader will permit me such comparisons), the two were like a small greyhound following the King[72] of Prussia's captain of the guard.

These strangers strode along, circling the globe in 36 hours. The sun (the earth really) does the same in a day, but turning on an axis is considerably easier than walking. The hikers eventually returned to their starting point after having passed the Mediterranean Pond and the larger pond called the *Great Ocean*[73] which surrounds our molehill. When wading through the waters, the dwarf never sank below his calf while the other barely wetted his heel.[74] As they went, they looked for signs of habitation. They stooped.

---

[68] In *Conversations on the Plurality of Worlds* de Fontenelle uses the mind-experiment technique of imagining what it would be like to live on each of the planets. When he came to Mars after visiting the moon, Venus, Mercury, and the sun, he passed it by, saying "Mars isn't worth the trouble of stopping there" primarily because of its small size and its similarity to the planets already visited [**32**, p. 52].

[69] Maupertuis's research team left Tornio on the north shore of the Baltic for France on June 10, 1737. They encountered a storm which drove the ship onto the Bothinian coast. After repairs, they set sail again on July 18, and arrived in Paris on August 20. In the winter photo of Figure 7, the white region near Tornio is ice.

[70] Although the outward physique of the alien giants is comparable in proportion to humans, apparently their digestive systems and other internal organs follow a completely different paradigm.

[71] Should it be 20 rather than 12?

[72] The King is Frederick the Great. One of his passions was raising greyhounds. He often allowed a favorite to follow him around, even on the battlefield. An oft-repeated saying of his, "The more I see of men, the better I like my dog" [**131**, p. 99].

[73] The Pacific Ocean.

[74] As Voltaire describes the hike, the travellers basically follow lines of longitudes 20° E and 160° W, which means that they would pass near Hawaii and encounter water at least three miles

They lay prone. They poked into everything. However they were incapable of seeing or feeling any of earth's creatures.

The dwarf, prone to jump to conclusions, decided, due to lack of evidence to the contrary, that the planet was devoid of life. Micromégas politely debunked his reasoning.

"With your little eyes," he said, "you cannot see stars of the fiftieth magnitude[75] that I can see clearly. Can you conclude that they do not exist?"

"Look at how poorly constructed, how irregular, and how jumbled everthing is," said the dwarf. "It is ridiculous. The streams are crooked. The ponds lack any regular shape. The earth's surface is like sandpaper. Its sharp grains have rubbed my feet raw." (He was speaking of the mountains.[76]) "Remember how the globe appeared from space? It is flat at the poles.[77] Its rotation axis is skewed compared to its orbit around the sun.[78] The arctic climate is brutal.[79] No one in his right mind would choose to live here. That is my argument."

"That may be," said Micromégas. "But this land was not made for nothing. You view this place as irregular because you are comparing it to Jupiter and Saturn. That is why you are a little confused. Remember the diversity which I have met in all my travels?"

The Saturnian rebutted with weighty arguments of his own. Their dispute might never have ended except that in the excitement of their dialogue, the clasp to Micromégas's necklace broke, and diamonds were scattered everywhere. They were pretty little stones ranging from fifty to four hundred pounds.[80] The dwarf gathered a few, and happened to notice as he looked through them that they made excellent microscopes.[81] He chose one of

---

deep, well above Micromégas's ankle; the secretary would be swimming. Even according to the ancients, the secretary would be swimming, for Pliny the Elder cites Fabianus, saying that "the deepest sea has a depth of nearly two miles" [**106**, book ii, section cv, p. 351].

[75] The unaided human eye on a clear night can see stars down to magnitude 6. The north star, Polaris, has magnitude about 2.12. A star of magnitude n is 2.512 times as bright as a star of magnitude $n + 1$. This system dates to Ptolemy (60–168) who categorized stars into six levels, with the brightest stars as level one and the faintest stars as level six.

[76] Voltaire seems to be forgetting that a short while ago the two giants breakfasted on two mountains.

[77] Measuring this phenomenon was the primary focus of Maupertuis's polar expedition.

[78] Saturn's *obliquity* is 26.7°, exceeding earth's obliquity of 23.5°.

[79] The average surface temperature on Saturn is −130° C. Here is de Fontenelle imagining how a Saturnian would fare on earth: "If we placed [Saturnians] in our coldest countries, in Greenland or Lapland, we'd see them sweat huge drops and die of the heat [**32**, p. 59]."

[80] If Micromégas could see a 50 pound diamond then he would be able to see a five foot man.

[81] Voltaire could be describing a compound microscope here, an invention attributed to lens maker Hans Janssen and his son Zacharias (1580–1638) in 1590, when Zacharias was ten

160 feet in diameter while Micromégas chose one of two thousand five hundred feet.[82] They were excellent magnifiers, but it took practice to focus them properly.

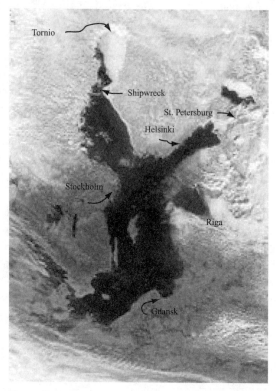

Figure 7. The Baltic Sea in winter, courtesy of NASA

Through these crystal lenses, the Saturnian glimpsed a whale on the waves of the Baltic Sea. He deftly snagged it with the little finger of his right hand, and placed in on his left thumbnail. He showed it to the Sirian who laughed again at the miniscule size of the natives. The Saturnian, now convinced that the place was inhabited, rashly decided that only whales lived there: and since he was a great thinker, he wondered about many things. How do such small creatures move? Do they have ideas, a will, freedom? Micromégas was

---

years old [12]. Perhaps Zacharias discovered the compound microscope by playing with his father's lenses, somewhat in the same way as Voltaire's dwarf was fiddling with Micromégas's diamonds.

[82] A moment ago, Voltaire estimated that the heaviest diamond was 400 pounds. A diamond thousands of feet in diameter would weigh many, many times this weight.

stumped; he made a patient examination, but could find nothing to make him believe that the whale had a soul.[83] But then they saw something else the same size as the whale floating nearby. As it happened, a flock of philosophers[84] were returning by ship from the arctic circle where they had made some unprecedented observations. The newspapers later reported they ran aground in the Gulf of Bothnia and barely survived. But one rarely knows the truth of stories in this world. Here I will tell the real unembellished truth—which is no small feat for a historian.[85]

Scene V: A shipload of savants

Micromégas gently and with great caution grasped the vessel with two fingers and deposited it on his thumbnail.

"Here is a creature very different from the first," said the dwarf from Saturn.

Micromégas transferred the supposed creature to the palm of his hand.

The frantic passengers and crew of the ship, believing themselves lifted by a hurricane, and foundering upon a large boulder, began abandoning ship.[86] For life-rafts, the sailors threw barrels of wine into Micromégas's hand and then jumped in after them. The geometers took their quadrants, their sectors, and their two Lapland ladies,[87] and descended onto the fingers of the Sirian. They scurried with such energy that he felt them on his fingers. Indeed, one sailor thrust an ice axe[88] a foot deep into the giant's index finger. Micromégas judged, from the pin prick, that something had poked him, but beyond that he had no idea what was happening. The crystal with which he could scarcely see the whale and the ship failed to resolve the mystery. For all practical purposes, the sailors were invisible. I do not want to patronize anyone, but

---

[83] A related, modern-day question for computer scientists is, *How do we determine whether a machine is intelligent?*

[84] Maupertuis's expedition.

[85] Voltaire wrote lengthy histories of France, Russia, and Sweden. He makes light of his profession here.

[86] As mentioned in footnote 69, Maupertuis's team really did experience a shipwreck on their way home.

[87] During the long winter nights in Lapland, the Frenchmen socialized with the locals. Apparently Maupertuis was given to accompanying himself with guitar and to reciting romantic poetry. Two young Lapland ladies, Elisabeth and Christine Planström, fell in love with the dashing French mathematicians and followed them to Paris.

[88] Voltaire uses the term "bâton ferré" which was a long staff with a sharp iron ferrule at the end; this tool was a precursor to the ice axe.

I am obliged to stress again the relative sizes of the giant and the sailors. A five foot tall man walking on earth is no more impressive than a dust mite 1/600,000 of an inch crawling on a ball ten feet around.[89] Think of a being who could hold the earth in his hand and whose body is proportionate to our own. How would they view the interminable battles on earth where villages are lost, and won, and lost again?

Were a captain of tall grenadiers to read this story, he would immediately issue hats[90] two feet taller than those now worn by his troop. But I warn him that it will be a useless exercise. He and his men can never be more than infinitely small.[91]

Figure 8. ... immediately issue hats two feet taller than now worn ...

---

[89] Yes, Voltaire has his ratios more or less correct in this analogy. However a better ratio match might be for the dust mite to be 1/200,000 of an inch.

[90] Figure 8 shows a tall hat on a guard at Buckingham Palace.

[91] Frederick Wilhelm I (1688–1740), the emperor's father, took pride in a special troupe of his grenadiers, the Potsdam Giants, comprised of soldiers from 5 feet 11 inches to beyond seven feet tall—the possible target of Voltaire's wit. Apparently, the king's agents kidnapped tall monks and innkeepers as recruits from all over Europe. Upon Frederick the Great's accession, the giants were dismissed, and, it is said, thereafter the roads home were littered with lost half-wits.

When the two travellers looked more closely, they perceived at last the living atoms. The excitement felt by Leeuwenhoek[92] and Hartsoeker[93] when they first saw what they believed to be the seeds that formed us was nothing in comparison to the jubilation of Micromégas's discovery. What pleasure it gave Micromégas to observe the little machines[94] and their activities. He whooped in exultation and eagerly handed a better microscope to his fellow traveler.

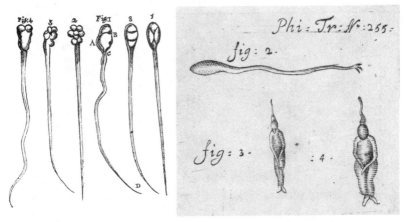

(a) Various animal sperm [78]   (b) Human sperm vs some fantasies [79]

Figure 9. van Leeuwenhoek sketches of "the seeds that formed us"

"I see them!" they said as one.

"Do you see how they carry burdens? They bend down and then rise up."

As they spoke their hands trembled both with the pleasure of seeing these new things and with fear lest they lose sight of them. The Saturnian, converted instantly from rank skepticism to boundless credulity, claimed to see them in the process of reproduction.

"Aha," he said, "I have surprised nature in the act."

---

[92] Anton van Leeuwenhoek (1623–1723), the "father of microbiology," was the first to observe single celled organisms, which he called *animacules*, through microscopes he had made, some of which magnified 270 times. Figure 9 shows some of his sketches of spermatozoa. See also Footnote 95.

[93] Nicolas Hartsoeker (1656–1725) invented the screw-barrel simple microscope.

[94] The king's physician, La Mettrie (1709–1751), had written *Man-Machine* in 1748, and this phrase is Voltaire's acknowledgment of him in his audience.

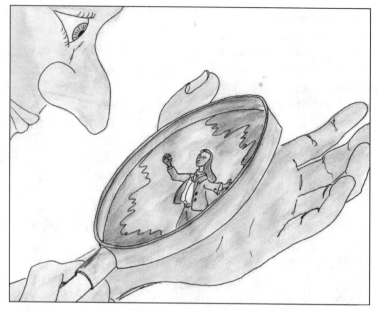

Figure 10. Man and Micromégas
Some story versions replace the crystals with a magnifying glass

But he was deceived by appearances,[95] which is all too common, even when microscopes are not involved.

Scene VI: Learning French

Micromégas, a better observer than the dwarf, saw clearly that the atoms spoke, and so remarked to his companion, who, red-faced over his mistake concerning procreation, was reluctant to believe such a species could communicate ideas. The dwarf had a facility with languages equal to that of the Sirian, and had not heard anything from the atoms that could be construed as speech. Moreover, he had reasons for supposing they could not. How could these imperceptible beings have vocal cords? What could they possibly have

---

[95] As a specific instance of being fooled by appearances, in Figure 9(b), van Leeuwenhoek, in the drawings marked 3 and 4, sketches what other people sometimes imagined they saw when they looked at the human sperm cell, and explains that it is easy for these animalcules "to deceive our sight," and suggests, for those who see a shadowy yet tiny human form in the sperm cell, that possibly "four [cells] may lie together [on the viewing glass] in such a posture that two of their tails might represent the arms, and the other two the legs" [**79**, pp. 307–8].

to say one to another? To speak, it is necessary to have thoughts or at least some fairly close approximation; but in such case, they should have the equivalent of a soul—which was patently absurd.

"But," said the Sirian, "a moment ago you claimed they were making love. Can one make love without thoughts, or words, or at the very least making oneself heard? Which is more difficult: to formulate an argument or conceive a child?"

"I view both as great mysteries," replied the dwarf stiffly. "I will refrain from speculation and remain impartial. Let us first examine these insects, then discuss afterwards."[96]

"Well said," said Micromégas.

At once he took out his nail clippers and snipped a paring from his thumbnail. This he fashioned into a funnel-shaped ear trumpet.[97] He inserted the apex into his ear and positioned the open end over the ship and its crew. The faintest of voices entered the circular fibers of the nail so that, thanks to this ingenious contraption, the philosopher up above could perfectly hear the buzzing of the insects below. After a few hours, he had identified a few words, and finally was able to follow the sense of their spoken French. The dwarf did the same, although it came harder to him. The interplanetary voyagers' amazement grew stronger every instant. The dust mites were speaking intelligibly. This variety in nature awed them. You can well imagine that the Sirian and the dwarf burned with impatience to converse with the atoms. The dwarf feared that his own thunderous voice, let alone that of Micromégas, would deafen the mites. Somehow they needed to turn down the volume. So they placed in their mouths a kind of small toothpick, positioning the sharp end near the ship. With the dwarf perched on the Sirian's knees, and with the ship and crew poised on his fingernail, Micromégas lowered his head and, with infinite caution, began his announcement in the softest possible tones.

"Invisible insects, who by the hand and good pleasure of the Creator have been given birth in the depths of the infinitely small—I thank Him for allowing me to discover such seemingly impenetrable secrets. Though no one at my court might deign to recognize you, still I despise no one, and I offer you my protection."

Never was anyone more astonished than the men who heard this speech. They had no clue as to who, what, or where the sound had came. The

---

[96] This paragraph summarizes the scientific method.
[97] Figure 11 shows some hearing aids from Voltaire's era.

Figure 11. Eighteenth century hearing-aid trumpets

ship's priest[98] exorcized. The sailors swore. The philosophers theorized, but to no avail—their theories could not determine who spoke to them. The softer-voiced dwarf of Saturn took his turn, briefly describing the expedition from Saturn, Micromégas's background, and so on. After commiserating with them for their puniness, he peppered them with questions such as whether they had always been in this miserable state verging on oblivion, how they got along on this planet that seemed to belong to whales, whether they were happy, whether they reproduced, whether they had a soul, and so on. One reasoner of the group, bolder than the others,[99] and shocked that anyone could doubt the existence of his soul, took a quadrant[100] and made three sightings of the Saturnian.

At the third station he spoke, "So you believe thus, sir, since you are one thousand fathoms[101] from head to foot, that you are ... "

---

[98] Réginald Outhier was also an astronomer. Maupertuis's expedition had stopped in Stockholm before proceeding to the arctic circle, as that territory was controlled by Sweden, a protestant country. They successfully petitioned for the liberty of conducting Catholic mass during their stay. However, the Swedish king imposed a fine on any but French souls from attending these ceremonies.

[99] Maupertuis. Voltaire can give compliments.

[100] Figure 12 shows a typical mid-eighteenth-century quadrant, a navigational instrument that measured the angle of inclination of a heavenly object with respect to the vertical.

[101] One fathom is a length of six feet.

The Annotated Micromégas   29

"A thousand fathoms!" cried the dwarf. "Good heavens! But how can he know my height? He is not an inch off the mark. What! This atom has measured me! A geometer. He can ascertain my height, while I, who can see him only by microscope, cannot begin to measure his."

"True, I have measured you," said the physicist, "and I will also measure the height of your taller companion."

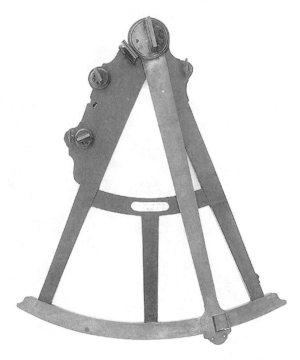

Figure 12. Eighteenth-century Hadley quadrant, courtesy of the West Sea Company

The proposal was accepted; Micromégas lay down at full length, for, had he stood upright, his head would have been too far above the clouds to be seen. To facilitate this measurement, the philosophers had him stuck with a great tree in a location that Dr. Swift[102] would readily call by name, but I refrain from doing so out of consideration for the ladies. Then, by using

---

[102] Jonathan Swift (1667–1745) wrote the Gulliver stories, and Voltaire refers to him as the English Rabelais. Not being able to resist their ribald tradition of off-color humor, Voltaire leaves the reader to imagine what place in the recumbant giant's body cannot be mentioned in mixed company.

similar triangles,[103] they concluded that what they saw was, in effect, a young man of 120,000 feet.

Then Micromégas declaimed, "Now more than ever it is clear to me that one must judge nothing based on its apparent size. O God, Who has given intelligence to substances that seem so insignificant, You esteem the infinitely small equally as the infinitely great; and, if it is possible that smaller beings exist, they may have intellects superior to those of the magnificent animals I have seen on other planets, whose feet alone would cover this globe."

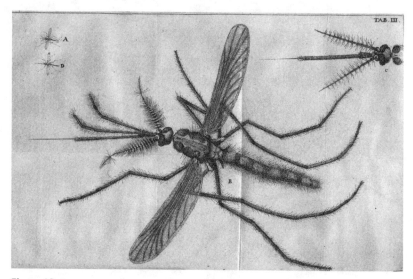

Figure 13. Swammerdam's sketch of a mosquito, from *Historia Insectorum Generalis*, 1669

One of the philosophers answered him that he could verily believe that intelligent beings smaller than man exist. He mentioned Virgil[104] and his fabulous bees, Swammerdam[105] and his discoveries, and Réamur[106] and his dissections. He told them at last that there are some animals which are to bees as bees are to men—that the Sirian, in the same way, was dwarfed by the massive animals of which he spoke, and that these grand animals in turn are dwarfed by others, appearing in turn to be no more than

---

[103] Presumably the tree forms an altitude or one of the sides of these similar triangles.
[104] Virgil (70–19 BC) devotes his entire fourth *Georgic* to the society and care of bees.
[105] Jan Swammerdam (1637–1680) was a noted entomologist and an early pioneer of the microscope. Figure 13 is his sketch of a small insect.
[106] René Réamur (1683–1757) wrote a classic study, *Natural History of the Bees*.

atoms.[107] This information increasingly excited Micromégas's interest, and he responded as follows.

Scene VII: A Parting Gift

"O intelligent atoms, in whom the Eternal Being pleased himself to manifest his skill and power, you must without doubt taste joys here on your planet: because, being so small, yet showing such great spirit, you come to pass your life in love and contemplation; this is the true life of spirits. Nowhere before have I seen true happiness; but here it is, without a doubt."

At this, the philosophers all shook their heads. One of them, more forthcoming than the others, admitted candidly that except for an infinitesimal minority, the rest of their kind were a motley assortment of fools, villains, and unfortunates.[108]

"We lack not the means, either material or intellectual, to wreak all kinds of evil. For example, at this very moment[109] a hundred thousand fools of our kind, wearing helmets and a hundred thousand others wearing turbans are annihilating each other. And it has been this way virtually throughout the world since the dawn of history."

The Sirian shuddered, and inquired what could cause such horrible disputes between such puny animals.

"The argument," said the man, "is over a mudpile called Palestine[110] the size of your heel. Not that any of this gathering of cutthroats has an actual claim to the mudpile. Rather, it is a matter of loyalty to a man called a Sultan, or to another called, I don't know why,[111] Caesar.[112] Neither of these men have ever seen, nor will they ever see, this small corner of earth. And virtually

---

[107] This never-ending sequence of scaling is characteristic of what is now called a *fractal*, a topic explored in Chapter II.

[108] There is an old story of two philosopher friends conversing. One says to the other, "Everyone except us is crazy." The other replied, "Yes, but sometimes I wonder about you."

[109] Wade [**148**, p. 29–30] claims that this conflict refers to the 1736 conflict between the Turks and the allies Austria and Russia, concluded by the 1739 Treaty of Belgrade. To me, the reference is to the Crusades as well as to wars in general.

[110] This reference to Palestine does not appear in the first edition. Whether Voltaire or another added it, its poignancy unfortunately is ever timely.

[111] *Caesar* was the family name of Julius Caesar. After his death, it was the title of Rome's emperors from Augustus to Hadrian. Pliny the Elder says that this family name originated because the first of the Caesars was born by caesarian section, from the Latin word to cut, *caedere* [**106**, book vii, section ix, p. 537]. When translated into German and Russian, *caesar* becomes *keiser* and *tsar* or *czar*, respectively. Alternatively, by this phrase, Voltaire is incredulous that society bestows honorable titles to people of great power who nevertheless choose to belittle the value of human life.

[112] Some translations render *Caesar* as *Pope*, a title that literally means *Papa*.

none of the soldiers participating in the slaughter have ever seen the leaders for whom they fight!"

"How horrible!" exclaimed the Sirian with indignation, "What inconceivable madness! I have a mind to take three steps and demolish this anthill of absurd assassins with three swift kicks."[113]

"Don't bother," the philosopher answered. "They will destroy themselves soon enough, without your help. Within ten years, less than one percent of these men will remain. Even if they never drew their swords, hunger and fatigue and intemperance would consume them anyway. These are not the ones who deserve punishment. Rather, it is those lazy barbarians who, while using the water closet,[114] order the massacre of millions of men and then solemnly thank God for being able to perform His work."[115]

The traveler felt moved with pity for the small human race in which ranged such astonishing contrasts.

"Since you are among those who are wise," he said to the philosophers, "and apparently aren't mercenaries, pray tell me how you occupy your time."

"We dissect flies," the bolder philosopher responded. "We measure lines and collect data. We agree on two or three points that we happen to understand, and we argue about two or three thousand points we do not understand at all."

Immediately the Sirian and Saturnian began questioning these thinking atoms more deeply, so as to learn general information about them and their mind set.

Micromégas posed the first question, "Tell me, do you know how far it is from Sirius to the great star[116] in Gemini?"

They responded as a chorus, "Thirty-two and one half degrees."[117]

"What is the distance from here to the moon?"

They all answered again, "Sixty earth radii, in round numbers."

---

[113] Voltaire initially hoped that Frederick might adopt an attitude of peacemaker in Europe such as Micromégas espouses here. However as Frederick's mania for militarization increased over the years so did Voltaire's disenchantment with Frederick, although the two maintained a cordial correspondence until Voltaire's death in 1778.

[114] An eighteenth century term for loo, john, outhouse, bathroom, toilet.

[115] The fictional Frederick points blame at the real Frederick. This kind of writing is what tended to get Voltaire in trouble. Indeed, Wade [**148**, p. 30] says that, "It is inconceivable that [Voltaire] would have dared write [or read] this passage while at the court of Frederick."

[116] Pollux (magnitude 1.16) is brighter than Castor (magnitude 1.97), the twin stars in Gemini. In Roman mythology, Castor and Pollux were the patrons of sailors.

[117] Sirius is 2.7 parsecs from earth and Pollux is 12 parsecs from earth. By the law of cosines, the two stars are 9.8 parsecs apart. A parsec is 3.26 light years, where a light year is the distance traveled by light through a vacuum in a year. Figure 14(b) is a mirror image of parts of two Dürer woodcuts; he printed the stars as God would see them, looking in on the cosmos.

The Annotated Micromégas

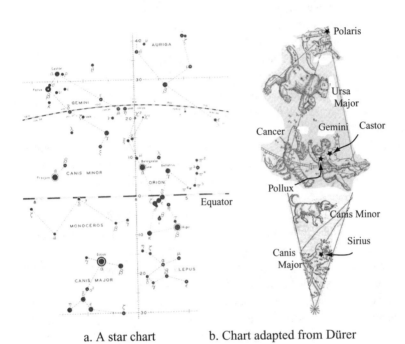

a. A star chart   b. Chart adapted from Dürer

Figure 14. Sirius and the Twins

"How much does your air weigh?"[118] The Sirian had hoped to trick them with this question, but they replied that air weighs around nine hundred times less than the same volume of pure water, and nineteen hundred times less than gold. The midget from Saturn, surprised by their responses, was tempted to take them for sorcerers—the very same people to whom he had refused to grant the possibility of a soul only a quarter of an hour before.

Finally Micromégas said to them, "Since you know your physical environs so well, you must surely be acquainted as well with inward things. Tell me about the soul, where it is, and how you form ideas and thoughts."

To this, the philosophers spoke as before, but all were of different opinions. The eldest among them quoted Aristotle, another pronounced the name of

---

[118] In checking Voltaire's computations here, use one cubic centimeter of air at sea level at 0° C. Incidentally, Anders Celsius (1701–1744), who developed his temperature scale in 1742, was a part of Maupertuis's Lapland expedition.

Descartes;[119] this one of Malebranche; the other of Leibniz and still another of Locke.

An old peripatetic said with great confidence, "The soul is a fact, and as such it has the power to be what it is. Aristotle specifically declares this on page 633 in the edition at the Louvre:[120] Εντελεχεια τις εστι, χαι λογος του δυναμιν εχοντος τοιουδι ει ται."[121]

"I do not understand Greek, I'm afraid," said Micromégas.

"Nor do I," responded the philosophic mite.

"Why then do you quote Aristotle in Greek?"

"It is very necessary to quote what one does not comprehend in a language that one rarely hears."[122]

The Cartesian spoke next, saying, "The soul is a pure spirit that receives in its mother's womb all of the metaphysical ideas of the universe, but once it leaves there must go to school to relearn anew a small portion of what it once knew so well."

"Does it not stretch credibility," asked the giant of eight leagues, "that your soul should be so wise in the womb and so ignorant as a bearded adult? And what do you mean by *spirit?*"

"What are you asking?" responded the reasoner. "I haven't the slightest idea, though we say that it does not consist of matter."

"But do you at least know what matter[123] is?"

"We know it very well. For example this stone is grey and has a certain shape. It has three dimensions, and is heavy and capable of being divided into smaller parts."

"Good!" said the Sirian. "This thing that appears to you to be divisible and heavy, can you tell me what it is? You mentioned its attributes, but do you know its essence?"

---

[119] René Descartes (1596–1650) is the "father of modern philosophy." He believed that everything could be explained rationally.

[120] The Louvre dates to 1190 when it was a fortified palace in the center of Paris. It became a public museum in 1793, during the French Revolution.

[121] This passage is from Aristotle's *De Anima (On the Soul)*, II 414a, 15. Professor of philosophy Harry Platanakis of Cambridge University, whose speciality is Aristotle, says that there are eight known primary manuscripts containing this sentence of Aristotle, and all eight are different; the most widely accepted translation is "The soul's force is the word (reason) inside (behind) the power [that] things (beings) have." He interprets this sentence, "That is to say that life is not what it is by virtue of living beings, but by virtue of the force of the soul that resides in them." The manuscript mentioned by Voltaire is still at the Louvre. **[104]**

[122] Voltaire's caricature of scholasticism.

[123] An old philosophical jingle about the undefinability of matter and mind: *What is matter? Never mind. What is mind? No matter.*

"No," said the other.

"Then you do not know what matter is at all."

Then Monsieur Micromégas addressed the same question to another wise man whom he held on his thumb, "What is your soul and what does it do?"

"My soul does nothing at all," responded the disciple of Malebranche.[124] "It is God who does all for me. Through Him I see all. Through Him I do all. It is He that does all without my interference."

"Then what is the point in existing at all?" asked the wise man from Sirius. "And you, my friend," said he to a follower of Leibniz[125] who was standing there, "what is your soul?"

"It is," he responded, "a hand that shows the hours while my body rings them, or if you prefer, my soul rings the hour while my body shows the hour; or rather my soul is the mirror of the universe, and my body is the border of the mirror: that much is clear."

A supporter of Locke[126] was there, and when it was finally his turn to answer the question he said, "I do not know how I think, but I know that my thoughts only arise as occasioned by my senses. That there exist substances which are immaterial and intelligent, I do not doubt—neither do I doubt at all God's ability to communicate thought to matter. I revere the Lord's power, and it is not my place to limit it. I assert nothing, but I am content to believe that more things are possible than man is able to imagine.[127]

---

[124] Nicholas Malebranche (1638–1715) contends that we "see all things in God"—that every idea is a shadow of an archetype in the mind of God. Here is Voltaire's assessment of a Malebranchean argument: "Father Malebranche proves resurrection by the caterpillars becoming butterflies. This proof is no more weighty than the wings of the insects from which he borrows it" [**144**, vol. xiii, *Resurrection*, p. 97].

[125] Gottfried Wilhelm Leibniz (1646–1716) was the co-discoverer with Newton of calculus. Of Leibniz's flowery style, Voltaire says, "Leibniz made great books, in which he did not even understand himself" [**144**, vol. xii, *Optimism*, p. 82].

[126] John Locke (1632–1704) was an empiricist and utilitarian. At birth, man's mind is a *tabula rasa*, a blank page. Voltaire calls Locke "the Plato of England, so superior to the Plato of Greece" [**144**, vol. vii, *Church*, p. 167]. "There never was a more solid and more methodical understanding, nor a more acute and accurate logician, than Locke, though he was far from being an excellent mathematician" [**144**, vol. xxxix, *Locke*, pp. 33–34]. As can be seen, Voltaire has a difficult time giving unadulterated praise.

[127] Geneticist J. B. S. Haldane (1892–1964) said the same thing [**59**, p. 286]: "The universe is not only queerer than we suppose, but queerer than we *can* suppose." Haldane acknowledges in turn that he borrowed this thought from Shakespeare's Hamlet: "There are more things in heaven and earth, Horatio, than are dreamt of in your philosophy" (Act I, Scene V). This thought probably goes back to the first man who opened his eyes.

The creature from Sirius smiled for he himself could give no wiser answer. The dwarf from Saturn would have embraced Locke's pupil[128] but for the size differential.

By mischance, the discussion was interrupted at this point by a small creature in a square hat who happened to be found among the philosophers. According to him, all answers were in the *Summa* of St. Thomas.[129] Looking the two visitors up and down, he announced that their own persons, their worlds, suns, stars, and everything else were made exclusively for man's use alone.[130] This assertion sent the two travelers off into paroxysms of laughter which, according to Homer, is the gift of the gods.[131] Their shoulders and stomachs heaved in and out. As they doubled over in uncontrollable convulsions the ship, which had been perched upon the Sirian's finger, fell into the pocket of the Saturnian's knickers. When they finally regained control of themselves, the two extraterrestrial gentlemen searched long and hard before finally locating the vessel and its crew. Returning the ship to its former place, the Sirian resumed the conversation with the small mites and spoke to them with great kindness, although inwardly he was somewhat piqued that the infinitely small had a pride that was infinitely large.

Micromégas promised to give them a book of philosophy, in miniature writing[132] for their use, and that, in this book, they would see the answer and end of all things. Indeed he gave them the volume before his departure.

It was brought to the Academy of Science[133] in Paris. But when the secretary[134] opened it he saw nothing but blank pages in an empty book.

"Ah!" said he, "just as I thought."

---

[128] Another Voltaire figure.

[129] Thomas Aquinas (1225–1274) wrote of natural theology in the *Summa Theologica*, championed scholasticism, and embraced Aristotelianism.

[130] The Greek philosopher Protagoras's (circa 484–414 BC) most famous quote is "Man is the measure of all things," a phrase somewhat descriptive of the Renaissance spirit.

[131] Book I of the *Illiad* and Book VIII of the *Odyssey*.

[132] Very small script.

[133] The Academy was founded in 1635 by Cardinal de Richelieu (1585–1642). The Academy of Science was founded in 1666 by Louis XIV.

[134] The actual secretary of the French Academy of Science at this time was Bernard le Bovier de Fontenelle (1657–1757), who was permanent secretary from 1697–1739. Here is his perspective on how our world, and perhaps our knowledge of it, contrasts to the universe [**61**, p. 101]: "Behold a universe so immense that I am lost in it. I no longer know where I am. I am just nothing at all. Our world is terrifying in its insignificance."

Exercises

1. (a) Can you explain how Voltaire may have arrived at the length of Micromégas's nose as 6333 feet on p. 9?
   (b) Check Voltaire's figures involved in Micromégas's trick question, footnote 118, on the relative masses of air, water, and gold.
   (c) Find instances in the story where Voltaire uses proportional reasoning correctly.
   (d) How tall must a supergiant G be for his feet alone to cover the globe of our earth, as described on p. 30? How big must G's home globe be, using proportional reasoning? Are there supergiant stars this large?

2. (a) Micromégas walks around the globe in 36 hours. Is this a realistic number for the giants? To help answer this question here are a few records: Pliny the Elder tells us that a boy of eight ran 68 miles between noon and evening, and that Tiberius Nero travelled 182 miles in 24 hours by carriage [**106**, book vii, section xx, p. 561]. Magellan's expedition needed three years to sail around the world in 1522, and Jules Verne, a hundred years after Voltaire, has a hero who goes *Around the World in 80 Days* using a variety of transport.
   (b) Elephants and whales have long vocal cords; some of the sounds they make are below the threshold of human hearing which ranges from about 20 hertz to 20,000 hertz. With what sound frequency might Micromégas speak?

3. (a) On p. 29, Micromégas needs to lie down so as to be measured. On a clear day, could the philosophers see Micromégas's head while he was standing? That is, could one see the top of a 23 mile tall skyscraper from the ground at noon?
   (b) On the basis of his height and using proportional reasoning, estimate Micromégas's life-span. How does this compare with Voltaire's guess on p. 15?
   (c) Why do you think Voltaire created Micromégas to be 23 miles tall? Why not 100 feet, or merely 10 feet?

4. (a) Voltaire quantifies the distances traveled between the planets Saturn, Jupiter, and Mars. Assume that the orbits of these planets are circles with the sun at their common center. Are the distances Voltaire gives reasonable? If not, what should they be?

(b) Halley predicted his comet to return in 1756. Imagine that such a comet flies by Saturn, Jupiter, Mars, and earth. Ignore the gravitational influence of these planets on the comet. Determine the arclength along the comet's trajectory through these points—enumerating the distances and time lapses between these points, and compare these results with the numbers in Exercise 4a.

(c) Imagine that when the travelers land on Jupiter, they find that the earth and Mars are in ideal positions for the comet they have been riding to fly by both Mars and the earth. Determine how many years later for a similar arrangement to occur (assuming they can conjure up a Halley's Comet clone at any given moment).

5. (a) Explain how the factor 2.512 from footnote 75 arises in Ptolemy's system of star brightness. (Hint: Imagine that the only stars that exist are stars of brightness rated 1, 2, 3, 4, 5, or 6. A total range of 5 units separates the brightest from the dimmest. Now use a geometric progression to go from one brightness level to the next. Think in terms of base ten. Here is another hint from Voltaire: "For want of defining terms, almost all laws, that should be as arithmetic and geometry, are as obscure as logarithms" [**144**, vol. 5, *Ambiguity*, p. 141].)

(b) Explain why Descartes' system of vortices predicts an earth shaped like a lemon.

(c) How did the ancients determine the distance between the earth and the moon?

6. (a) Read some of Thomas Aquinas's proofs for the existence of God, such as is found in [**73**]. Are any more compelling for you than others?

(b) On p. 23, Micromégas disparages the idea of testing a creature for the existence of a soul. A similar dilemma arises with artificial intelligence. Briefly, how might we determine whether a machine is intelligent?

7. (a) What does Voltaire mean by the word "variety" in nature as it appears on pp. 13, 21, and 27?

(b) Explain the rationale, from the dwarf's perspective, of his comment about open-ended projects on p. 15.

8. (a) Choose one of the philosopher-savants: Aristotle, Descartes, Malebranche, Leibniz, Locke, or any other philosopher-savant. Write an

essay, being true to their philosophy, on how they would answer Micromégas's question, "What is the use in existing at all?"

    (b) What is scholasticism? Voltaire ridicules some of its flaws on p. 34. What were its strengths? Contrast scholasticism with modern day scholarly activity.

9. (a) What are the essential properties of matter? Voltaire lists five on p. 16.

    (b) If a Micromégas figure were to visit the earth today, what questions might he ask of a group of scientists? Contrast these questions with those voiced by Voltaire in this story.

10. Can you give any explanation as to why the book at the end of the story is blank?

Vignette II: Here be Giants!

Voltaire's Micromégas is a giant 23 miles tall. Where did Voltaire get such an idea for his character?

Giants appear often in folklore throughout the ages. They seem to come in three sizes: ordinary, big, and bigger.

Ordinary giants are simply very tall men, such as professional basketball centers or wrestlers like André the Giant. Given the chance, Frederick the Great's father, Frederick Wilhelm I, would surely have enlisted such gentlemen into his grenadier troupe, the Potsdam Giants, whose enlistment requirement was a minimum height of five feet eleven inches. He probably drooled over the report of Moses's vanguard that reconnoitered the promised land: "There we saw giants, and we were as grasshoppers in their sight" (Numbers 13:33). A well-known descendant of these larger-than-life strongmen was Goliath of Gath whose height was six cubits and a span, about ten feet (I Samuel 17:4). In 1752, as another item on his list of prospective scientific endeavors besides the deep hole of Vignette I, Maupertuis proposed looking for giants in Patagonia, the southernmost region in South America, as he had remembered that the Royal Society had a skull from there which, arguing by proportion, should belong to a human ten to twelve feet tall [**84**, p. 152]. Grendel of *Beowulf* fame illustrates (in part xi of the poem) what an ordinary giant can do to an ordinary man:

> Straightway [Grendel] seized a sleeping warrior
> as the first [victim], and tore him fiercely asunder,
> the bone-frame bit, drank blood in streams,
> swallowed him piecemeal: swiftly thus
> the lifeless body was clear devoured,
> even feet and hands.

Although ordinary giants are roughly twice the size of an ordinary man, big giants are an order of magnitude larger: about ten times the size. Pliny the Elder (23–79), some of whose *Natural History* read like *Ripley's Believe It or Not* files, says that [**106**, book vii, section xvi, p. 553],

When a mountain in Crete was cleft by an earthquake a body 69 feet in height was found.

Homer's Ulysses encounters the one-eyed Cyclops when sheep-stealing while finding his way home from the victory at Troy. An eyewitness castaway from Ulysses's crew describes the giant (*Aeneid,* book iii, lines 802–810):

> The towering Cyclops
> is tall enough to strike the high stars—gods,
> keep such a plague away from earth.
> I myself have seen [Cyclops]
> snatch up a pair of us in his huge paw,
> then, bash both of them against a boulder.

(a) The Titans aproned in the earth (b) Lucifer at earth's center

Figure 1. Giants from the Holkham manuscript, circa 1375

A group of big giants, the Titans, were sired by the gods, according to Greek mythology. Eventually most of them were incarcerated in the underworld lest they destroy the gods themselves. Dante picks up on this story in the *Inferno,* [**28**, canto xxxi, lines 58–60], and describes their height, albeit he uses Nimrod, a giant sprung from the union of an angel and a woman, who is somewhat shorter than the Titans.

> As large and long [Nimrod's] face seemed, to my sense,
> As Peter's Pine at Rome, and every bone
> Appeared to be proportionately immense.

From an old illuminated manuscript of the *Inferno,* Figure 1(a) shows the Titans ensconced in the earth to their hips; Nimrod is the giant wielding the horn.

Peter's Pine, shown in Figure 2, is a bronze casting made in the first century BC of a pine-cone that once adorned a fountain in the Baths of Agrippa in Rome, and was later moved to a fountain in front of St. Peter's Basilica in the early Middle Ages; it now stands in the Cortile della Pigna, a courtyard within the Vatican Museums.

Figure 2. Peter's pine at Rome, courtesy of Linda Legg

When Galileo was twenty-four, he gave his first public lecture, which included a commentary on Dante's riddle of canto xxxi [**112**, pp. 26–27]. Galileo took the height of Peter's Pine as $5\frac{1}{2}$ armlengths, and said that

> Because men normally are eight heads tall, even though painters and sculptors, among them Albrecht Dürer in his book on human proportions, hold that the most proportioned bodies must be nine heads tall, although due to the very rare number of bodies found with these

proportions [of 9 heads], we will suppose that the height of the giant [Nimrod] must be eight times that of his head. [**49**, p. 42]

Thus Galileo's estimate for Nimrod's height is $(8)(5.5) = 44$ armlengths, or about 75 feet, since Galileo assumed that a body was three armlengths long, and since a man was about five feet tall in his day. See Exercise 1 for insight about Galileo's reference to Dürer's head to body ratio as 1/9.

However, the height of the bronze part of the pine-cone is 9.5 feet and the classic head to body ratio used by artists, as discussed in footnote 9 of Chapter I, is one-seventh, which would mean that Nimrod and his Titan friends were 67 feet tall, which is a little less than Galileo's estimate.

Of course, the classification of a big giant as being ten times the size of an ordinary person is a relative measure. What happens if the ordinary people are other than five feet tall? For example, in Jonathan Swift's *Gulliver's Travels* of 1726, the Lilliputians are a race whose height is but six inches. Thus Gulliver would be classified as a big giant in their lore. However, Gulliver in turn is dwarfed by the Brobdingnagian race who are as tall as "an ordinary spire-steeple" [**139**, part ii, p. 97]. Incidentally, Voltaire read the Gulliver stories hot-off-the-press, as, for three months in 1727, he and Swift were guests at the country home of Charles Mordaunt, Earl of Peterborough near London [**75**, p. 38]. Swift and Voltaire probably talked at length about giants and savants.

In an essay "On being the right size," biologist J. B. S. Haldane points out a physical problem for a human body scaled upwards by a factor of ten. In particular, of the Giant Pope and the Giant Pagan from an illustrated version of John Bunyan's *Pilgrim's Progress* (1678), he says that while the weight of such a giant would be a thousand times that of a man, the cross-sectional area of the giant's thigh bone would increase only a hundred times.

> As the human thigh bone breaks under about ten times the human weight, Pope and Pagan would have broken their thighs every time they took a step. This was doubtless why they were sitting down in the picture I remember. [**59**, pp. 18–19]

In this vignette and for tales involving giants in general, we shall ignore these physical impossibilities.

Although intimidatingly large, big giants are in turn dwarfed by bigger giants, who are at least 100 times as large as ordinary men. One such creature appears in Dante's *Inferno*.

Vignette II: Here be Giants!

Dante's Satan is a huge hairy humanoid, frozen in rock at the center of the earth, down whose body Dante and Virgil descend on their way to the antipodes. From a time-worn page, Figure 1(b) depicts a voracious Satan with Dante and Virgil to the lower left. Dante describes this fallen angel's height in terms of the Titans (canto xxxiv, lines 29–33):

> Out of the girding ice [Satan] stood breast-high,
> And to his arm alone the [Titan] giants were
> Less comparable than to a giant I;
> Judge then how huge the stature of the whole
> That to so huge a part bears symmetry.

Let H be the height of a typical human, about 5 feet. Let G be the height of a Titan, 60 to 75 feet or Galileo's 44 armlengths. Let A be the length of Satan's arm. Dante's passage translates to

$$\frac{G}{A} \leq \frac{H}{G}, \tag{1}$$

or $A \geq G^2/H$. Since a body is about three armlengths, Galileo concluded that Satan's height, $3A$, was about 1936 armlengths, and rounded it up to 2000 armlengths [**47**, p. 42], over 3000 feet, a height on the same order of magnitude as Voltaire's Saturnian dwarf. See Exercise 1 for some alternate estimates based upon more of Dante's clues.

Another imposing example of a bigger giant comes from Rabelais who wrote his stories about Gargantua and son Pantagruel starting in 1532. However, these giants are unusual in that they appear to change size according to the convenience of the narrative. That is, at times they seem to be ordinary giants, and at others big giants, or bigger giants. As merely an ordinary giant, Rabelais portrays Gargantua as slightly larger than life, allowing him "to skip over a ditch at one leap, spring over a hedge, climb up trees like a cat, and leap from one to another like a squirrel." As a big giant, Rabelais says that Gargantua rode a steed that "was as big as six elephants," as shown in Figure 3. As a bigger giant, Gargantua grooms his hair using a comb 900 feet long. He once ate a salad in which were hiding six pilgrims, who, upon being ingested, managed to avoid Gargantua's grinding molars until being spat out. In passages wherein Rabelais satirizes the pollution levels of the River Seine, Gargantua relieves himself "in copious measure," producing rivers in which drowned a quarter of a million men, not counting women and children [**110**, pp. 36–37, 51, 77–79]. Therefore, Gargantua is anywhere from being an unusually tall person to being a mile or two tall.

Figure 3. Gargantua goes to Paris [**110**, plate facing p. 36], drawn by Gustave Doré (1832–1883)

How then did Voltaire decide upon the height of his giants?

One obvious reason is that Voltaire wanted to impress upon his readers "the wide variety" that appears in nature, both small and large. He wanted at least the same size ratio between Micromégas and earthlings as between scientists like van Leeuwenhoek and Hartsoeker and the micro-organisms they discovered, such as the sperm cell in Figure I.9, about the smallest body that could be resolved with microscopes of that day. Since the human

Vignette II: Here be Giants!

sperm cell is about 40 microns in length, the proportion of a two meter tall man to such a cell is 50,000, a figure on the same order of magnitude as 24,000, Voltaire's proportion of Micromégas to man. Voltaire also wanted a vivid analogy for the sizes of the stars imagined by Cassini, as discussed in footnote 7 of Chapter I. For example, if his giant had lived on a world the size of our sun, his proportionate height would have been only 550 feet. Instead, Voltaire gave his giant a home world with radius of a little more than one astronomical unit (a.u.), the distance of the earth from the sun. That is, the Sirian's world was about 24,000 times that of earth. If everything in the Sirian world was scaled linearly upward from ours, then a Sirian man would be (5 ft)(24,000)/(5280 ft/mi) $\approx$ 23 miles tall. So Voltaire used a giant 23 miles tall to symbolize the extent of man's knowledge, giving the height ratio of man and the smallest observable body and the radii ratio between what was conjectured to be the largest star and earth. Today's astronomers have discovered supergiant stars whose radii exceed seven a.u. If Voltaire had written his story with this information, he would have been tempted to make his giant in excess of 7(23) = 161 miles tall.

However, Christiaan Huygens says (in a book that Voltaire probably read),

> If the magnitude of the planets was to be the standard of [alien] measure, there would be animals in Jupiter ten or fifteen times larger than elephants, and as much longer than our whales. And then their men must be Goliaths in respect to our pygmyships. Now though I do not see any so great absurdity in this as to make it impossible, yet there is no reason to think it is really so, seeing nature has not always tied herself to those rules which we have thought more convenient for her: for example, the magnitude of the planets is not answerable to their distance from the sun. [**70**, pp. 55–56]

That is, there is no necessary connection between the size of the planet and the size of its inhabitants. So, is there any other reason for Voltaire choosing such great heights for his giants?

Being Voltaire, he preferred not to be beaten by anyone, including Dante of whom he said, "His reputation will last because he is little read" [**144**, vol. viii, *Dante,* p. 55], and least of all by Rabelais whom he considered "a drunken philosopher, writing only when he was unable to stand" [**144**, vol. xxxi, *Dean Swift,* p. 90]. Therefore Voltaire makes Micromégas a biggest giant, at least ten times as large as Dante's Satan or Rabelais's giants—twenty-three miles tall.

But besides being in part the result of Voltaire's ego and a demonstration of the great variety in the universe, his choice for Micromégas's height is the solution to a problem.

One day in 1736, Voltaire received fan mail from an unusual source. It was from the twenty-four year old Frederick before he became king, telling Voltaire how much he admired his writings and how he hoped to implement enlightened principles of governance once he assumed the throne. Charmed by this, Voltaire replied in equally complimentary tones. Voltaire hoped that Frederick would indeed be a "philosophical prince born to make the human species happy" [**144**, vol. xxxviii, *Letters,* p. 166].

How could a poet be of service to a king, especially to a young one?

Figure 4. Frederick's father reviewing the Potsdam Giants

Why not write a philosophical fable raising basic questions with which an enlightened ruler must wrestle? Being a wit, and knowing that the young man had been terrorized by his father, Frederick Wilhelm I, Voltaire seized upon the old king's hobby-horse as a means to tell his tale, namely his obsession with parading the Potsdam Giant grenadiers. Figure 4 is an old political cartoon showing the king reviewing his troops; the smallest figure is probably the king, although it could be Frederick as a boy. Thus both Frederick and Voltaire would be giants in the fable. Voltaire's first version of his story was sent to Frederick in 1739; a revised version, a translation of which is Chapter I of this book, was published in 1752. However, even though Frederick had written to Voltaire after his coronation saying, "By God, write to me only as man to man, and share in my disdain for all titles

Vignette II: Here be Giants!

and outward glory" [**101**, p. 165], Voltaire wisely knew his place among kings. Therefore Micromégas—a Frederick figure—was made to be much larger than the Saturnian secretary—a Voltaire figure.

Why did he use the comet as a vehicle of choice?

Voltaire joined the king's entourage in 1750. By this time, Frederick had surrounded himself with the best minds that his money and prestige could attract. In some ways, he was like the NSF—the National Science Foundation—of the eighteenth century. Talk at his court included news of exciting theories, predictions, and research.

One of these items was Halley's prediction of 1682: the great comet of 1682 was supposedly the same as the one that appeared in 1607, and 1531, and therefore the comet might next appear in 1757. By 1751, anticipation had grown. Would it truly arrive? The two giants, as universal philosophers—Frederick and Voltaire—could ride into prominence in style on a heavenly body that was the talk of the day and for all to see. What an entry! What an opportunity! Would Frederick rise to the occasion? What measure of man would he be? Voltaire dreamt that even as Newton ushered in a new way of perceiving and understanding the physical universe, so Frederick would usher in a new order of enlightened, reasonable, self-evident government among men.

He could truly be a giant of a man.

CHAPTER II

The Micro and the Mega

Quantifying the great and small, the physical forms of matter, the intangible ideas of the mind, the mystery of the soul: such is the theme of the Micromégas story. The giant repeatedly points out the range of self-similarity that exists in the universe—although he calls it great variety. Can natural phenomena be truly measured? Measuring the size of a soul may be difficult, but how about the size of a star? More simply, what about finding the length and width of the page upon which these words are written? Anyone should be able to determine the length of an object, right?

Not exactly. We more or less approximate numbers, and say that the measure of an item is within a certain degree of certainty. For example, builders often use a tolerance of about one-eighth to one-fourth of an inch when constructing houses. Mortgage payments are rounded up to the nearest cent. Age of a person is given in years, and the age of a star is given in billions of years.

Our five senses are ill-equipped to understand smallness. Each sense's range is extensive, but limited. We see stars down to magnitude six and dust particles floating in the air. Wine or tea tasters can differentiate beverages of berries or leaves from adjoining fields. Experienced musicians can identify a symphony from hearing a few bars. With fingers or teeth, we can feel thicknesses to within a thousandth of an inch. But eventually for any of our senses, a range exists beyond which differentiation becomes problematic. When things get small we encounter confusion.

Calculus is all about the infinitely small. George Berkeley (1685–1753), in an early critique of Newton's infinitesimals, called them the ghosts of departed quantities. In his seventeenth *Philosophical Letter*, Voltaire describes Newton's calculus as

> the art of numbering and measuring exactly a thing whose existence cannot be conceived.

It took more than a hundred years for the mathematical community to clarify Newton's notion of limit so as to exorcise Berkeley's ghosts, although many students in the midst of studying calculus for the first time might yet agree with Voltaire's description. Of course, the tools of calculus rely on assuming that the things being measured are smooth whereas Lucretius (99–55 BC) reminded us long ago that the apparent continuity of smooth surfaces we see in nature is an illusion [**82**, book i, lines 585–653, pp. 37–38]:

> Something must be the smallest that there is,
> Otherwise, every possible tiny object
> Will be composed of infinite particles,
> A half can always be in halves divided,
> No limit to all this. So how would they differ,
> The universe from the littlest thing? They wouldn't.
>   But since reason tells us
> This makes no sense, we therefore must acknowledge
> That there are things which have no parts at all,
> The smallest natural objects.

So, what can we do when measuring things? We start simply, and extend, and hope that any attendant difficulties are minimal, humbly remembering that we are applying abstract mathematical ideas in a universe that by its very nature may defy being measured. In this chapter we sketch the standard definition of the length of one-dimensional phenomena and point out some attendant difficulties. We then contrast the multi-level structure of the universe with self-similar fractals, which sometimes provide an abstract approximation to reality within a limited range of levels. Finally we look at measuring the tamest of sets, the interval, albeit at an astronomical level, so as to measure the earth, moon, and sun. When we measure the really small, we encounter the difficulty of Berkelian ghosts, Lucretian atoms, and the bosons of quantum mechanics. When we measure the really large, do we meet difficulty afresh?

The length of an interval

On the real number line, intuition suggests that the *length* of the interval from a to b, whether we include the endpoints or not, is $|b - a|$. If a and b are *extended* real numbers (including $\pm\infty$) with $a \leq b$, we use the following conventions: $(a, b) = \{x \mid a < x < b\}$ is an *open* interval, $[a, b] = \{x \mid a \leq x \leq b\}$ is a *closed* interval, and $(a, b]$ and $[a, b)$ are both *half-open* and *half-closed* intervals. Henri Lebesgue (1875–1941) focused on the first of these natural interval forms so as to determine the length of an arbitrary set of numbers. That is, intuition suggests that the length or *measure* of the union of two disjoint open intervals should be the sum of the lengths of the two intervals. For example, if $B = (0, 1/2) \cup (3/4, 1)$, then $m(B) = 3/4$, where $m(B)$ is the measure of the set B. Lebesgue observed that the measure of any set A of real numbers ought to be no more than the sum of the lengths of open intervals whose union contains A. That is,

$$m(A) \leq \sum_{\alpha \in \mathcal{A}} \ell(J_\alpha), \text{ with } A \subset \bigcup_{\alpha \in \mathcal{A}} J_\alpha, \tag{2}$$

where $\mathcal{A}$ is an index set, J_α is an open interval (a_α, b_α), $\ell(J_\alpha) = b_\alpha - a_\alpha$, and $m(A)$ is the as-yet-undefined measure of A. For example, with B as defined above, $B = \cup_{\alpha \in \mathcal{A}} J_\alpha$ where $\mathcal{A} = \{1, 2\}$, $J_1 = (0, 1/2)$, and $J_2 = (3/4, 1)$. To continue with his definition, Lebesgue defined the *outer measure* of A, $m^*(A)$, as the *infimum*, or *greatest lower bound*, of $\sum_{\alpha \in \mathcal{A}} \ell(J_\alpha)$ over all indexing sets for which $A \subset \cup_{\alpha \in \mathcal{A}} J_\alpha$. As an obvious lemma, we invite the reader to prove (Exercise 2) that for any two sets of numbers A and B

$$m^*(A) \leq m^*(B) \text{ whenever } A \subset B. \tag{3}$$

As an illustration using outer measure, we show that the outer measure of $[a, b]$ is $b - a$. To do this, let $J_n = (a - 1/n, b + 1/n)$, where n is a positive integer. By (2), $m^*([a, b]) \leq \ell(J_n) = b - a + 2/n$. The infimum of such numbers is $b - a$. Therefore $m^*([a, b]) \leq b - a$. By (3), $m^*([a, b]) \geq \ell((a, b)) = b - a$. Thus $m^*([a, b]) = b - a$.

As a more interesting example, we find the outer measure of the positive rational numbers: the set $\mathcal{Q} = \{m/n \mid m, n \in \mathcal{Z}\}$, where $\mathcal{Z}$ is the set of positive integers. As Lucretius pointed out on p. 52, any number can be halved; that is, $1 = 1/2 + 1/2$, $1/2 = 1/4 + 1/4$, and so on. Therefore

$$1 = 1/2 + 1/4 + 1/8 + \cdots = \sum_{n=1}^{\infty} \frac{1}{2^n}. \tag{4}$$

The members of $\mathcal{Q}$, where the members may recur, can be arranged as a sequence q_n as indicated in Figure 5. That is, the sequence begins as $1, 2, 1/2, 1/3, 1, 3, 4, 3/2$, and so on. Every element of $\mathcal{Q}$ appears in this sequence.

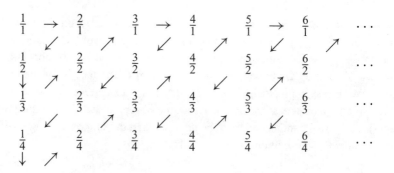

Figure 5. The rational numbers as a sequence

Now imagine being a landowner in Lineland, a place where the universe has collapsed to the real number line, and inviting all members of $\mathcal{Q}$ for a sleep-over. We imagine that the guests can be teleported to any location on the line. As a landowner, you have decided to bed your guests in an interval of length two units. Is there enough room for each guest to lie down without disturbing another guest? We define a Lineland sleeping mat as an open interval of finite length. We need a collection of mats so that guests have their own mats and no two mats overlap. In accordance with (4) we issue guest q_n a mat of length $1/2^n$ units. Each guest has a mat, and the sum of their lengths is one unit. That is, yes, the guests have plenty of room in the guest bedroom. Furthermore, since the length unit was arbitrary, what we have just shown is that the measure of $\mathcal{Q}$ is zero! That is, $m^*(\mathcal{Q}) = 0$.

The implications of $m^*(\mathcal{Q})$ being zero gives rise to some bizarre conclusions. Toward that end, the reader is invited to prove the following lemma (Exercise 2).

$$m^*(A) = m^*(A \setminus B), \qquad (5)$$

where A is any set of real numbers, B is a set of outer measure zero, and $A \setminus B = \{x \mid x \in A \text{ and } x \notin B\}$.

Now we use (5) where A is the unit interval, $I = (0, 1)$, and B is the set of rational numbers $\mathcal{Q}$. By (5), the set of irrational numbers between zero and one has measure 1. This result seems to imply that if one randomly chooses a number in the unit interval, it will be irrational! This is Archimedes's paradox, in that even though between any two rational numbers an irrational number can be found and between any two irrational numbers a rational number can be found, the number of irrational numbers that exist dwarves the number of rational numbers that exist. That is, there are different magnitudes of infinity. Surprisingly, when we focus on the very small with special care so as to avoid any difficulty or confusion, we inadvertently unveil a difficulty with the very large! Such a dilemma is an inescapable conclusion when using a sufficiently meaningful extension of the natural length of an interval.

However, Lebesgue's definition is yet incomplete. Wild sets of real numbers exist for which the sum of the outer measures of its (finite number of) parts exceeds the outer measure of the whole. Exercise 5 introduces the set B for which $m^*(B \cap I) = 1 = m^*(B' \cap I)$ where B' is the complement of B and I is the unit interval. To avoid such wild sets, Lebesgue said that a set A is *measurable* if for every set of real numbers C, $m^*(C) = m^*(A \cap C) + m^*(A' \cap C)$. Thus, if A is a measurable set, its measure $m(A)$ is its outer measure. As straightforward exercises, the reader may show that

$$\text{If } m^*(D) = 0 \text{ then } m(D) = 0 \qquad (6)$$

and that any interval is a measurable set (Exercise 2). Fortunately, many phenomena of the universe can be well-modeled, it seems, using measurable sets—otherwise the quest to understand the universe would probably be hopeless indeed. For most of this book, we content ourselves with measuring the tamest of measurable sets, the interval.

Small, Medium, and Large

Voltaire describes a universe of great variety. Micromégas is twenty-three miles tall; his Saturnian friend is one mile tall; humans are about five to six feet tall. Micromégas has visited worlds whose inhabitants are to him as he is to Voltaire. And considering bees and microscopic bacteria, Voltaire suggests there are worlds whose inhabitants are to a human as a human is to Micromégas. More so, the inhabitants of these worlds look alike; their outward appearance is a scaled version of man. Self-similarity all the way up and all the way down! To be sure, the universe contains little stars and big stars, little men and big men. But each special kind of thing appears to have a range beyond which it cannot go. When stars get too big, they explode as novas or collapse into black holes. When they get too small, they cool and become planets or rocks. When a man is too big, the muscle-bone structure cannot support the attendant increase in mass, as volume increases with the cube of height. When a man is too small, among many other difficulties, the brain cannot remember anything worth remembering. That is, it is impossible to map the knowledge contained in a typical human brain into the ganglia of an earthworm. There is not enough room. Nevertheless, within a certain range of being, the universe appears to be self-similar. And this limited self-similarity of the physical universe is one of the reasons why mathematics can be used to model the dynamics of the universe with some degree of confidence.

To illustrate, we explore the notion of a *fractal*.

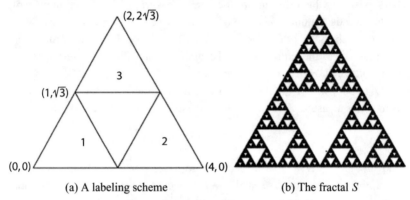

(a) A labeling scheme   (b) The fractal S

Figure 6. The Sierpinski triangle

Informally, a fractal is a planar set whose parts are similar to its whole. The Sierpinski triangle S of Figure 6(b) is a classic example. At its outermost level, S is contained in an equilateral triangle with vertices (0, 0), (4, 0), and (2, 2$\sqrt{3}$). As indicated in Figure 6(a), S appears to be the union of three smaller equilateral triangles each of side length 2 and each of which is a translate of a $\frac{1}{2}$-scaled copy of S. We use these observations to sharpen our definition of a fractal. We say that a function T is a *contraction* of the plane if there is a 2 × 2 matrix M and a vector b for which $T(x) = Mx + b$ with $|Mx| < |x|$ for all x in the plane except for the zero vector. Let $\mathcal{F}$ be a family of contractions. More formally, we define a *fractal* A corresponding to $\mathcal{F}$ as a set of points in the plane for which

$$A = \bigcup_{T \in \mathcal{F}} T(A),$$

a standard definition as appears in [**13**] and [**40**].

A corresponding family of contractions for S has as its members:

$$T_1(x) = x/2, \qquad T_2(x) = x/2 + (2, 0), \qquad T_3(x) = x/2 + (1, \sqrt{3}).$$

(The matrix M in each of these contractions is $\frac{1}{2}$ times the identity matrix.) According to the labeling scheme of Figure 6(a), T_1 maps S into triangle 1, T_2 maps S into triangle 2, and T_3 maps S into triangle 3. Although we omit a proof, S is indeed equal to $T_1(S) \cup T_2(S) \cup T_3(S)$. One way to graph the fractal corresponding to this family is to start with an arbitrary point X; for convenience, we take a point inside the large equilateral triangle of Figure 6(a), such as (0.3, 0.1). Randomly choose a contraction T (from the family), plot $T(X)$, and rename the point $T(X)$ as X. Then repeat the instructions of the last sentence a suitable number of times. As the repetition number increases, so does the resolution of the fractal. Exercise 6a gives some computer code to generate Figure 6(b).

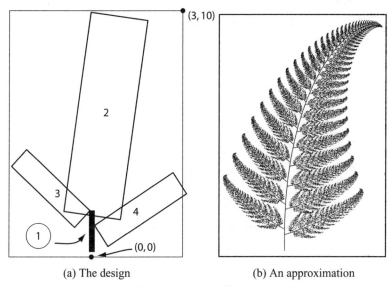

(a) The design (b) An approximation

Figure 7. A wilderness fern

Now we consider a more interesting, true-to-life example: a fern as shown in Figure 7(b). Each of its fronds looks like the fern itself. At first glance, its complexity may seem to be beyond characterization. However, four contractions are enough to encode its self-similar nature. In particular, we re-create the steps in designing the fern of Figure 7(b). The fronds on the right of the central vein look like mirror images of the ones on the left. We decompose the fern into four parts: the lower part of the stem, the lower right frond, the lower left frond, and the remainder of the fern. From the layout of the parts labeled 1, 2, 3, and 4 in Figure 7(a), we construct contractions T_1, T_2, T_3, and T_4. We adopt the following conventions.

- The fern will be contained within a rectangular grid G whose lower-left and upper-right corners are $(-3, 0)$ and $(3, 10)$.

- The root of the fern will be at the origin, O.
- The stem should be about 1/10 of the height of the fern.
- The fern will twist to the right as height increases.
- The fronds of the fern will branch off at an angle of about 60° from its central vein. The lower two fronds will have length of about 4 units. The fronds will branch out in pairs with the left-hand frond slightly above the right-hand frond.
- The fronds will be narrower than the fern; that is, the length-to-width ratio of the fern will be greater than for its fronds.

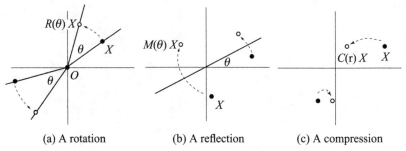

(a) A rotation   (b) A reflection   (c) A compression

Figure 8. Elementary transformations of the plane

To execute these design plans, we adopt some notation. Let $R(\theta)$ be the matrix corresponding to a counterclockwise rotation of the plane about O by θ radians, $M(\theta)$ be the matrix corresponding to a reflection of the plane about a line passing through O with slope $\tan\theta$ (M stands for mirror), and $C(r)$ be the matrix corresponding to a compression of the x-coordinates by r that leaves the y-coordinates intact. The formulas for these transformations are

$$R(\theta) = \begin{bmatrix} \cos\theta & -\sin\theta \\ \sin\theta & \cos\theta \end{bmatrix},$$

$$M(\theta) = \begin{bmatrix} \cos(2\theta) & \sin(2\theta) \\ \sin(2\theta) & -\cos(2\theta) \end{bmatrix}, C(r) = \begin{bmatrix} r & 0 \\ 0 & 1 \end{bmatrix}. \quad (7)$$

To form T_1, compress G by a factor of 0, and scale the result by a little more than 10% so as to obtain a line segment of about height 1. To form T_2, scale G by 0.9, rotate by about 3° (0.05 radians) clockwise, and translate up by (0, 1). Since the distance from the base line of G to the bottom of part 2 is 1 (10% of G's height) and since part 2 has height of about 90% that of G, then the top of part 2 should be somewhere near $y = 10$. But since part 2 has been rotated, the top of the fern will fall shy of $y = 10$. To form T_3, first compress G by 0.5 (to obtain a narrow petal), rotate 60° (to obtain the frond at the desired branching angle), scale by 0.35 (to obtain a frond-length of 35% that of the fern), and finally translate up. To form T_4, compress G by 0.5, reflect about the y-axis, rotate by 60° clockwise, scale by 0.39, and translate up.

To recap, we have

$$\begin{cases} T_1(x) = 0.12\,C(0)x, \\ T_2(x) = 0.9R(-0.05)x + (0,\ 1), \\ T_3(x) = 0.35R(\pi/3)C(0.5)x + (0,\ 0.8), \\ T_4(x) = 0.39R(-\pi/3)M(\pi/2)C(0.5)x + (0,\ 0.6). \end{cases} \qquad (8)$$

Iterating this family of contractions in the same manner as we did to generate the Sierpinski triangle gives the remarkably life-like fern of Figure 7(b). But a closer look exposes it as a counterfeit. The fractal fern looks the same all the way down, no matter how many times it is magnified, whereas, when the natural fern is magnified, we soon reach the cellular level, beyond which is the molecular level, the atomic level, and the subatomic level—to the very levels that Lucretius speculated must surely exist.

How does nature generate the intricate pattern of the fern, as well as all the other patterns in the universe? Somehow, the information embedded in (8) must also be embedded in the DNA strands of the fern. Somehow nature knows the limitations of its building blocks. Self-similarity, although remarkable, bottoms out and tops out in real life, so that only at most a half-dozen levels of self-similarity occur in the actual fern. Nevertheless, this fractal is a good example of the range of self-similarity we see in nature. In like fashion, the optimistic analyst expects that any observable phenomenon in the universe can be modeled by mathematics, albeit within a limited range of applicability.

The earth, moon, and sun

The mechanics of measuring seem to work well with terrestrial objects such as measuring heights of people and trees and depths of rock quarries. Mathematical models readily extend to differing scales. According to the ancients, the earth is a shadow of the perfection in the heavens. Can we measure objects beyond the earth? The section in this chapter on fractals was a caveat. Does the universe behave in the same way at grander scales with respect to measure? The only way to know is to try and see what can be seen, keeping in mind Micromégas's injunction, "More things are possible than man is able to imagine."

We review how the diameters of the earth, moon, and sun, and the distances between them can be measured. That is, we determine the diameter of the earth d_e, the distance between the earth and the moon r_m, the diameter of the moon d_m, the distance between the earth and the sun r_s, and the diameter of the sun d_s. In this section, we distort history and allow Eratosthenes (276–194 BC) and Archimedes (287–212 BC), who were friends in real life, to work as a team so as to measure these quantities. To make this fantasy even more informal, we nickname Eratosthenes and Archimedes as Beta and Alpha. In fact, Eratosthenes was known as Beta during his lifetime, as he—as well as everyone else—was second to Archimedes. Furthermore, the name Archimedes, when translated from Greek can mean "number one mind" [**94**, p. 36]. We allow Alpha and Beta some luxuries that they did not have in 240 BC when Eratosthenes indeed determined the diameter of the earth.

- We allow both Alpha and Beta to have synchronized, functional cell phones. That is, we allow them instant, remote communication and the ability to measure time.

- Alpha and Beta assume that the earth, moon, and sun are perfect spheres, and that the moon rotates about the earth in a circle whose center is earth's center having period 29.53 days with respect to the sun, and that the earth rotates about the sun in a circle whose center is the sun's center with period 365.25 days and spins on its axis with period 24 hours.

- Alpha and Beta can position themselves anywhere on earth and they implicitly know when and where solar and lunar eclipses can be seen from earth's surface.

- Finally, at each stage in their calculations, once they estimate a distance, Alpha and Beta exchange that estimate for a modern day value, so that the results of their subsequent calculations are more likely to harmonize with reality.

Alpha and Beta first find d_e. Beta learns that on summer solstice at noon, the sun illuminates the depths of a deep artesian well in what is now Aswan, Egypt. He also knows that, 500 miles north in Alexandria, towers cast shadows of 7.2° at noon. Assuming that the earth has radius R and that light

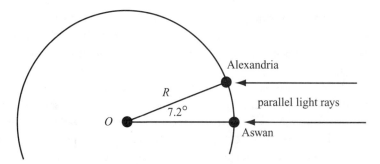

Figure 9. Shadows at noon near the summer solstice

arrives from the sun in parallel streams, he sketches Figure 9, and forms the ratio $500/7.2° = 2\pi R/360°$. That is, $R \approx 3978$ miles, or a diameter of 7956 miles—a very good guess, considering that the equatorial diameter of the earth is $d_e = 7927$ miles and the polar diameter of the earth is 7900 miles. Of these calculations, Pliny the Elder says

> [It was] achieved by such subtle reasoning that one is ashamed to be skeptical. [**106**, book ii, section cxii, p. 371]

Next, to find r_m, Alpha and Beta use a solar eclipse. Figure I.5 is a photo of a solar eclipse on Jupiter; any Jovian positioned within the shadow would be unable to see the sun. Fortuitously, Alpha and Beta determine that a complete solar eclipse will soon occur in east Africa along the equator. Journeying there, they position themselves near Lake Victoria and the Indian Ocean, and determine that the moon's shadow moves across the earth's surface at 1040 miles per hour. They know that according to the earth's rotation, a point on the equator travels at πd_e distance per day (or, if the reader insists on a geocentric universe, that the noon daylight moves across the face of the earth at πd_e distance per day). The moon's shadow on a non-rotating earth will ideally travel at $2\pi r_m/29.53$ miles/day, where r_m is the distance from earth's center to the moon's center. Alpha and Beta again assume that rays of sunlight are parallel. Thus we have

$$2\pi r_m/29.53 - \pi d_e = 1040 * 24 \text{ miles/day},$$

which gives the distance between the earth and moon as about 234,000 miles. The actual mean distance between the earth and the moon is $r_m = 238,857$ miles, the value Alpha and Beta henceforth use. However, in Ptolemy's *Almagest*, which served as the world's astronomy handbook from the second century until after the time of Copernicus, Ptolemy has the moon's distance from the earth vary in its orbit from 34 to 64 earth radii [51, p. 35]. Exercise 9 outlines the method used by Hipparchus to obtain an estimate of $r_m \approx$ 280,000 miles.

In his treatise *The Sand Reckoner*, Archimedes mentions some guesses for the relative sizes of the earth, moon, and sun. In particular, he said that the early Greek astronomers were in agreement that the moon was smaller than the earth. Eudoxus (c 410–355 BC) said that the sun's diameter was nine times that of the moon and Aristarchus of Samos said that it was between 18 and 20 times. Archimedes then argues that the sun's diameter is about 30 times that of the moon. Aristarchus in his treatise, *On the Sizes and Distances of the Sun and Moon*, concluded that the sun was about 19 times as far from the earth as the moon was, and that the sun's diameter was about seven times that of the earth's.

Given the luxuries with which we have equipped them, how could Alpha and Beta have improved upon these guesses? To find d_m, Alpha and Beta journey to the north pole and measure the time it takes for the moon to move its diameter against the background of the fixed stars. Then they use an idea of Aristarchus:

Let X be a heavenly body subtending an angle of ω degrees as viewed from the north pole (or from a stationary earth). Let r_X be the distance between X and the earth. Let t_X be the time it takes for X to move its diameter length d_X along its orbit against the stars. Let p be the period of X about the earth. Then $t_X/\omega = p/360°$. Thus the ratio of t_X to p is the same as d_X to the circumference of X's orbit, which upon simplification gives

$$\frac{r_X}{d_X} = \frac{180°}{\omega\pi}. \tag{9}$$

Since the moon subtends an angle of about $0.5°$ at earth's surface, (9) gives the moon's radius as 2290 miles. Its actual diameter is $d_m = 2160$ miles, the value Alpha and Beta henceforth use. Exercise 9 outlines how Hipparchus obtained the estimate $d_m \approx 2666$ miles.

Alpha and Beta's last goal is to find r_s and d_s. By a fortuitous coincidence, the sun also subtends an angle of about $0.5°$ in the sky. But Alpha and Beta use an improved approximation for this subtension angle. By (9), the stated values for r_m and d_m give the subtension angle as $0.518°$ for the moon, which we now take as the subtension angle for the sun. From (9), the ratio of the

The Micro and the Mega 63

sun's distance from the earth to the sun's diameter is $r_s/d_s \approx 111$, which we write as $d_s \approx r_s/111$.

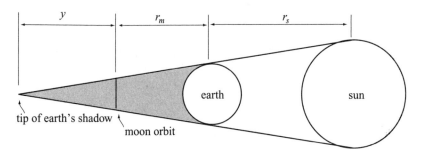

Figure 10. The earth's penumbra

Alpha and Beta meanwhile observe several lunar eclipses, and determine that the shadow cast by the earth onto the moon's orbit about the earth is 5800 miles. They reach this conclusion by knowing the times when the moon enters and exits earth's shadow. With this information they sketch Figure 10, which they perceive as a diagram of three similar triangles, with common vertex at the tip of earth's shadow and whose opposite sides are the shadowed 5800 miles of the moon's orbit (which they approximate as a straight line), the diameter of the earth, and the diameter of the sun. That is,

$$y/5800 = (y + r_m)/d_e = (y + r_m + r_s)/d_s, \qquad (10)$$

where y is the distance from the tip of earth's shadow to the moon's orbit. Solving this equation gives r_s as about 76 million miles, 17 million miles too short. What happened?

The main problem is that r_s as a function of r_e is very sensitive. That is, solving (10) for r_s in terms of d_e gives a singularity at $d_e \approx 7951$ miles, as shown in Figure 11.

Upon further thought, Alpha and Beta add a few more miles to d_e to account for earth's atmosphere. Although Pliny the Elder says that "the majority of [ancient] writers state that clouds rise to a height of 111 miles" [**106**, book ii, section xxi, p. 229], Alpha and Beta take an atmospheric depth of about 2 miles—so that d_e now has value 7931 miles. This time, they obtain 91 million miles for the distance between and earth and the sun, and a sun diameter of 820 thousand miles, very close to the actual values of $r_m \approx 93$ million miles, and $d_s \approx 865{,}370$ miles.

In summary, these guesses are very good even though we have ignored any number of factors that influence these measurements such as the moon-earth's

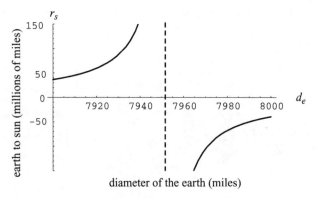

Figure 11. A delicate ratio problem

barycenter (their common center of mass) failing to be at earth's center, and sunlight rays failing to be parallel. Simple proportional reasoning does extremely well. Of course, the correction factors at each stage of the calculations were pivotal. Yet such are the dynamics of successive generations of inquiry.

Exercises

1. (a) Figure 12(a) is Dürer's woodcut showing proportions for the human body. With the use of a rule, how many heads tall is this human figure?

   (b) In the *Inferno*, canto XXXI, lines 61–64, Dante gives a second clue for Nimrod's height:

   > So that the bank which aproned him [Nimrod] from zone
   > To foot, still showed so much, three Friesians
   > Might vainly boast to lay a finger on
   > His hair.

   In this passage, Nimrod is embedded in a bank at his "zone" (hip region); Dante imagines three Dutchmen (three Friesians), who were known as a tall people, standing on each other's shoulders. The bottom most one's feet are level with Nimrod's zone; the top most one when reaching upward fails to reach Nimrod's hair. How tall is Nimrod using this clue? Figure 12(b) shows my Fall 2008 electricity and magnetism class staging this clue; the two young ladies at the base are in the place of the first Friesian.

   (c) In lines 64–67, Dante gives a third clue for Nimrod's height:

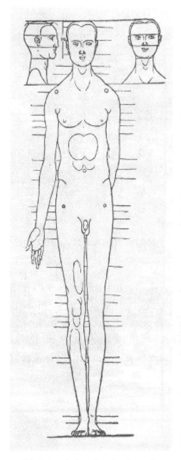

(a) Dürer's head to body ratio [**38**, cut 346]

(b) Stacking three Friesians

Figure 12. Dante's clues for Nimrod's height

> For from the place at which a man's
> Mantle is buckled, downward, you may call me
> Liar if [Nimrod] measured not full thirty spans.

If the mantle is buckled at the shoulder and a span is nine inches, obtain a third estimate of Nimrod's height. Interpret the phrase "thirty spans" as going from the shoulder to the hips.

(d) Use the result of parts (b) and (c) and (1) to determine the height of Dante's Satan. Compare your results with Galileo's estimate of 2000 armlengths.

2. (a) Prove statements (3), (5), and (6) about outer-measure.

   (b) Prove that any interval of real numbers is a measurable set.

3. Show that if the set of rational numbers Q is extended to the smallest field F containing π, then $m(F) = 0$.

4. **A set wholly holely.** Let $D_n = \bigcup_{k=1}^{2^n} \left(\frac{k}{2^n} - \frac{1}{4^n}, \frac{k}{2^n} \right)$,

   $$E_n = \bigcup_{k=1}^{2^n-1} \left(\frac{k}{2^n}, \frac{k}{2^n} + \frac{1}{4^n} \right),$$

   $$A_n = D_n \setminus \bigcup_{k=n+1}^{\infty} E_k, \quad \text{and} \quad A = \bigcup_{n=1}^{\infty} A_n.$$

   Prove that $0 < m(A \cap J) < m(J)$ for all non-empty open subintervals J of the unit interval $(0, 1)$.

5. **A non-measurable set.** Define an equivalence relation $\sim$ on the set of real numbers wherein $x \sim y$ if and only if $x - y$ is a rational number with odd denominator when expressed as a fraction in lowest terms. From each equivalence class choose an α, and let B be the set of all numbers $\alpha + 2p/q$ for all odd integers p and q and all α. Let B' be the complement of B.

   (a) Show that for any pair of odd integers p and q, $B' + p/q = B$.

   (b) Show that for any integer p and odd integer q, $B + 2p/q = B$ and $B' + 2p/q = B'$.

   (c) Show that $m^*(B \cap I) = 1 = m^*(B' \cap I)$, where I is the unit interval $(0, 1)$.

6. (a) The following *Mathematica* code was used to generate Figure 6 (b).

   ```
   F[x_, n_] := If[n ≤ 33, x/2,
       If[n ≤ 67, x/2 + {2, 0}, x/2 + {1, √3}] ];

   Block[{bag, i, X}, X = {0.3, 0.1}; bag = {};
       Do[X = F[X, Random[Integer, 100]];
           bag = Append[bag, X], {i, 1, 1000}];
       ListPlot[bag, AspectRatio → Automatic] ]
   ```

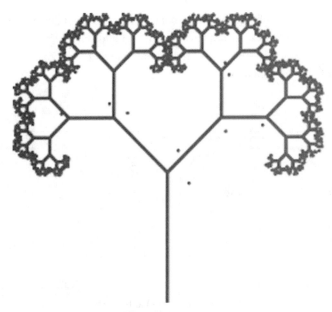

Figure 13. Binary tree

For this code, the initial value of X is (0.3, 0.1) and the repetition number is 1000 so as to plot 1000 points. Adapt this code to your computer algebra system and generate a Sierpinski right triangle (rather than an equilateral one).

(b) Design a family of three contractions that defines the binary tree in Figure 13.

7. (a) Re-create Figure 7(b), by changing the F function in the *Mathematica* code of Exercise 6a as indicated below, and execute the (unchanged) *Block* command. Remember to first define the functions C, R, M, and the T_i's in your CAS.

```
F[x_ , n_] := If[n ≤ 1, T₁[x],
    If[n ≤ 86, T₂[x], If[n ≤ 93, T₃[x], T₄[x] ] ] ];
```

(b) Modify the contractions used in generating Figure 7 so that the generated fern has its left-hand fronds forking 45° from its central vein, its right-hand fronds at 60°, its fronds compressed by a factor of one-fourth, and the fern itself twisted to the left.

8. (a) Construct a family of twelve contractions corresponding to a fractal that says "Yes," using the schematic of Figure 14(a).
   (b) Construct a family of thirteen contractions corresponding to the fractal depicted in Figure 14(b).

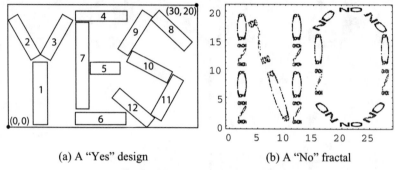

(a) A "Yes" design (b) A "No" fractal

Figure 14. Single-minded fractals

9. (a) Following Kline [74, p. 154–156], we sketch how Hipparchus estimated r_m. Hippparchus imagined two points on earth's equator, P and Q so that at time T, the moon M is directly overhead P and a line from the moon to Q is tangent to the earth, with center O, as shown in Figure 15(a). Take the central angle at O subtending P and Q as $\theta = 89°4'$, and take d_e as 7927 miles. Estimate r_m.
   (b) If Hipparchus estimated r_m as 280,000 miles, what was his value for θ?
   (c) Using the same idea as in part (a), Hipparchus estimated the radius of the moon. He imagined standing on the earth at a point E with the moon directly overhead. Take the angle ϕ of subtension at E

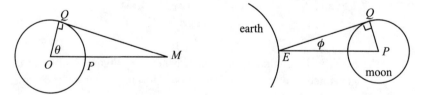

(a) Estimating r_m, earth-moon distance (b) Estimating d_m, moon diameter

Figure 15. Hipparchus triangles

between the moon's center P and the point of tangency Q of a line from P to the moon—as shown in Figure 15(b)—as $\phi = 15'$, (or $\phi = 0.25°$). Estimate d_m, the moon's diameter.

(d) Under ideal conditions the shadow cast by the moon onto the earth during a solar eclipse is a circle of diameter 167 miles. Use this fact (rather than the 5800 mile shadow width of a lunar eclipse) to modify the argument and the diagram of Figure 10 so as to recalibrate the distance between the earth and the sun.

(e) Use the method of Hipparchus as discussed in Exercise 9(a) so as to estimate the distance between the earth and the sun, r_s. What value would you take as θ so as to obtain an estimate of $r_s \approx 93$ million miles?

10. In a passage about the time of day being a local phenomenon, Pliny recounts the feat of Philonides, an athlete/soldier of Alexander the Great. Philonides runs 80 miles due west in nine hours (6 a.m. to 3 p.m., local time) and makes the return trip in fifteen hours (6 a.m. to 9 p.m., local time) even though the road west is more uphill than downhill [**106**, book ii, section lxxiii, pp. 314–315].

(a) Comment on the credibility of the story.

(b) Assuming that this event takes place at latitude 35° and that the course is on level ground and that Philonides runs at constant speed, estimate what the difference in times should be.

(c) At what latitude could the story be true? (Hint: We allow the runner to use snowshoes.)

Vignette III: The Bastille

The Bastille was a series of towers built in the fourteenth century as a Parisian armory. From the seventeenth century up until its demolition, it was used to imprison people for political, religious, social, or literary impropriety. On July 14, 1789, the Bastille was mobbed by crowds looking for weapons, marking the beginning of the French Revolution.

Voltaire spent eleven months in one of its cells in 1717 when he was twenty-two years old. Before he was encarcerated his name had been François-Marie Arouet. Forced to follow in his father's profession, he had been trained as a lawyer. However, he enjoyed writing poetry, had a quick wit, and loved the praise of men: a recipe for a turbulent life in eighteenth century Paris. Soon "everything that appeared in print against the government began to be laid at Voltaire's door" [**87**, p. 35], for much of what was written of a political/social nature at that time was done under pen-names. Voltaire had been censured on several occasions, exiled for months at a time to various country estates. He usually denied the charges. In fact, Voltaire often used denial of writing a work as a public relations ploy to further the distribution of his work. The more controversy encircling a work, the better he liked it. For example, of *Candide*, the best-known of his works, Voltaire said [**141**, back-cover],

They must have lost their minds if they think that I wrote this trash.

Despite previous warnings, Arouet eventually went too far, writing six lines mocking the regent for having an ongoing affair with his own daughter. The regent could ignore being heckled for mismanagement, poor judgment, and ruining France, but being lampooned for incest crossed the line, even if it was true. One morning, two gendarmes rustled the young man from his bed and escorted him to prison. There was no appeal, no trial—he could be held indefinitely at the pleasure of the government.

The cell walls were ten feet thick and had no windows. As he had been at work revising a play on Oedipus, he was allowed two books of Homer, one copy in Latin and another in Greek. Denied paper so as to discourage him from writing additional satirical couplets, he wrote in the space between the lines and in the margins of the two books. When he later retold this story to the English philosopher George Berkeley, "he claimed to have made paper by chewing his clothes" [**101**], but he loved to tell stories to the credulous. Some of *The Henriade*, an epic poem about Henry IV of France—the work for which he was best known during his lifetime—was written in the Bastille. Having plenty of time to think, he composed large sections in his mind, and memorized them, to be written down after he was released.

Meanwhile an early version of his play, *Œedipe*, that he had submitted to the Comédie-Française, the state theatre of Paris, was selected for production. With this timely development aided by some prodding of the government by his aristocratic friends, Arouet was released to another country-estate-exile-from-Paris. The play was a smash hit. The regent forgave Voltaire, issuing him a gold medal. As he accepted it, Arouet told the regent [**101**, p. 52],

> Sir, I would be most pleased if His Majesty were henceforth to provide for my board, but I beg your Royal Highness no longer to see to my lodging.

He was now a celebrity.

He also changed his name. His time in the Bastille became a rite-of-passage. The American Indian teen-age boy of yesteryear enters the wilderness alone to fast until having a vision of who he is. The young Maasai slays a lion so as to become a man. So too, with François-Marie Arouet. Forever after he was Arouet de Voltaire, Monsieur de Voltaire, or simply Voltaire.

There are several stories behind the name change choice. My favorite involves his childhood nickname of *le petit volontaire* which means *determined little thing*, an apt description of his joy in reciting verse to guests of his father [**101**, p. 17]. Removal of two letters from *volontaire*, like removal of training wheels on a bicycle, gives *Voltaire*.

With this name change, Voltaire resolved to be a man of letters, forgetting the law profession until returning to it very late in life. More plays tumbled from his pen. Controversy followed. Seven years later he was again escorted to the Bastille, this time for challenging a chevalier, a French knight, to a duel. After two weeks in prison, his request for a two-year exile to England was granted. For the remainder of his long life, Voltaire was ever looking over his shoulder and planning avenues of rapid exit just in case the authorities came for him again. However, rather than muting his criticism, the image of the Bastille, and all that it represented with respect to injustice, merely sharpened his quill.

Vignette III: The Bastille

In this next chapter, we look at the work of a figure from the same genre as Voltaire's *Micromégas*, a Mr. A. Square from Flatland who writes from his prison cell of the mechanics of falling in his world and beyond.

Figure 1. Storming of the Bastille, watercolor (1789) by Jean-Pierre Louis Laurent Houel, Musée Carnavalet, Paris, France

CHAPTER III

Fragments from Flatland

Flatland, a book published in 1884, is the purported memoirs of an amateur mathematician named A. Square, edited by a London school master, Edwin A. Abbott (1838–1926). A. Square lived in a two-dimensional universe. One evening, he was visited by a sphere who transported him to a three-dimensional realm. Afterwards, A. Square tried convincing his fellow Flatlanders that a third dimension really does exist. Failing to do so, and condemned by his society as a heretic, he wrote his memoirs in prison. His commentary is vintage Voltaireian-style satire on class structure, narrow-mindedness, and freedom of thought.

Most of *Flatland* is a description of Flatland customs along with A. Square's accounts of what he saw or imagined when he visited alternate dimensions. Besides these memoirs, he also speculated on Flatland mechanics: how things moved, and how that motion contrasts with motion in other dimensions. By no means did he finish his work. His notes are fragments, being pin-prick like indentations on coils of parchment filaments only now made public. The documents were recently found by isosceles triangle craftsmen refurbishing A. Square's prison cell as a national monument. In this chapter, we first sketch some Flatland background and then present a few of A. Square's computations and models: how to play ball in Flatland, orbits of heavenly bodies, and a comparison of planetary trajectories in other dimensions.

Some Flatland lore

A. Square was literally a square. His portrait from a three dimensional perspective appears in Figure 2; in Flatland, his portrait appears as a glimmering line segment. A later chronicler of Flatland determined that A. Square's first name was Albert [137]. We refer to him by that name. Albert's countrymen

Figure 2. Portrait of A. Square, martyr to the imagination

are polygons, whose width, in any direction, is no more than a foot. The working class are isosceles triangles. The criminal class are triangles with no two sides having equal length. The upper classes are those regular polygons that are indistinguishable from circles. The strength of one's intellect is directly proportional to the angle between one's equal sides. Since each woman is a line segment, a woman's intellect is unmeasurable. Some say nonexistent [1, p. 53]:

> [Women are] wholly devoid of brain-power, and have neither reflection, judgment nor forethought, and hardly any memory.

Others say profound [1, preface to the second edition, p. 25]:

> Straight Lines [women] are in many respects superior to the Circles, [the highest class].

The true state of affairs is probably a state secret; too much inquiry on this and other equally sensitive issues traditionally led to prison or exile. However, as a testimony to the mental prowess of line segments, Stewart [137], in a sequel called *Flatterland*, chronicles the adventure of V. Line, Albert's intellectually adept great-great granddaughter. Whereas women of all generations are line segments, the sons of an isosceles triangle are isosceles triangles of greater intellect by half of a degree, while the sons of a regular polygon of n sides, under normal circumstances, are regular polygons of $n+1$ sides, as depicted in Figure 3. Thus Flatland citizens can look forward in the normal course of events to a brighter and brighter future.

The gravity of Flatland is southward. When a man and a woman meet going east and west, the lady passes to the north lest the stronger gravity to the south cause her to slip southward. All rain falls south. Houses therefore have roofs on the northward side. A Flatlander can make a loop around the outside of any free-standing house. How then does a house, or anything else, stay in one

Fragments from Flatland

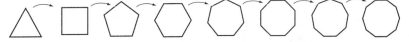

Figure 3. The generation of a decagon

place in Flatland? Why don't houses slip southward? Albert acknowledged this puzzle and similar issues, but he had no space in his memoirs to explain such things as "the means by which we give fixity to structures" [1, p. 97]. Nevertheless, it is natural to ask how far south a Flatlander can explore. Is there a line where gravity is most extreme, to the south of which gravity then points north? How does one play catch in Flatland? How do masses attract each other in n-space? How does a 2-D world rotate about its sun, if indeed 2-D suns exist? From his notes, it is clear that Albert was intrigued with these questions. From a 3-D perspective, he had seen into the inside of everything in his 2-D universe. Wanting more, he had asked the sphere to take him into a 4-D universe so that he could see the inside—"thy stomach, thy intestines"—of the sphere. But to repeated pleas, his Virgil-like guide turned a deaf ear to his Dante-like self, saying [1, preface to the second edition, p. 25],

> There is no such land [as a 4-D universe]. The very idea of it is utterly inconceivable.

Was the sphere correct? Indeed, if such a universe existed, how would a 4-D world rotate about its sun? Fortunately, the sphere had outlined Newton's 3-D approach to the problem for Albert. Albert's notes are a hodge-podge of insights, false starts, and intermittent musings about being jailed, some apologetic, others passionate, all jumbled together. What follows is an edited rendition of his manuscripts. We present three models he used as his understanding of gravity developed, and we close with his reflections about life in a 4-D universe.

A first Flatland model

For a Flatlander to play catch, a ball is thrown northward, until gravity slows it down, whereupon it slides south to the catcher. In his analysis, Albert ignores all issues involving resistance.

As a first model of the gravity dynamics of Flatland, Albert imagined Flatland as an infinitely long rectangle of finite width w. For ease of discourse with the sphere, Albert arbitrarily assigned Flatland's width as 100 *yards* and adopted *seconds* as the unit of time. Thus, latitude s is the line parallel to the southern edge of Flatland, s yards to the north. As a simple first model for southward gravitational acceleration $a(s)$, Albert considered $a(s) = -(100-s)$, so that at the northern edge of Flatland, a is 0, and at the southern edge, a is -100 yd/sec^2.

What is the path of a thrown ball? And what is the escape velocity for this model; that is, how fast must a ball be thrown so that it reaches latitude 100? Albert positioned himself at latitude h and threw the ball due north with initial velocity v_0.

Acceleration a times velocity v can be written as

$$a\frac{ds}{dt} = v\frac{dv}{dt}, \tag{1}$$

where s is distance and t is time. Integrating (1) as a dummy variable τ goes from 0 to t gives

$$\int_0^t a\frac{ds}{d\tau}\,d\tau = \int_0^t v\frac{dv}{d\tau}\,d\tau. \tag{2}$$

Since both s and v are monotonic functions of t, as the ball goes north and with the ball's initial velocity being v_0, (2) becomes

$$v^2 - v_0^2 = 2\int_h^s a\,dy,$$

where y is a dummy variable representing latitude y. Since the ball's velocity going north is positive,

$$v(s) = \sqrt{2\int_h^s a\,dy + v_0^2}. \tag{3}$$

To find a formula for the time T it takes for the ball to go from latitude h to latitude s, start with $ds/dt = v$, which gives $dt = (1/v)ds$, which means that

$$T = \int_h^s \frac{1}{v}\,dy. \tag{4}$$

With $a(s) = s - 100$, $v(s) = \sqrt{s^2 - 200s + 200h - h^2 + v_0^2}$ by (3). Let α be the non-negative number with

$$\alpha^2 = |(100-h)^2 - v_0^2|. \tag{5}$$

By (4) and (5),

$$T(s) = \begin{cases} \cosh^{-1}\left(\dfrac{100-h}{\alpha}\right) - \cosh^{-1}\left(\dfrac{100-s}{\alpha}\right), & \text{if } v_0 < 100 - h \text{ and} \\ & h \leq s \leq 100 - \alpha, \\ \ln(100-h) - \ln(100-s), & \text{if } v_0 = 100 - h \text{ and } h \leq s < 100, \\ \sinh^{-1}\left(\dfrac{100-h}{\alpha}\right) - \sinh^{-1}\left(\dfrac{100-s}{\alpha}\right), & \text{if } v_0 > 100 - h \\ & \text{and } h \leq s \leq 100. \end{cases} \quad (6)$$

Example 1. If $v_0 < 100 - h$, then $100 - s \geq \alpha$ by (5), which means that the largest s value for the ball is $100 - \alpha$, and it reaches this latitude in a time lapse of $\cosh^{-1}((100-h)/\alpha)$. For example, if $h = 10$ yards and $v_0 = 10$ yd/sec, then $\alpha = \sqrt{8000}$ by (5), which means that the ball goes north until latitude $100 - \sqrt{8000} \approx 10.55$ yards in 0.11 sec. Thereafter, the ball glides south and returns to latitude 10 in another 0.11 sec. However, if $h = 10$ and $v_0 = 90$, then the ball always goes north as time goes on, but never quite reaches latitude 100. If $h = 10$ and $v_0 = 100$, then $\alpha = \sqrt{1900}$ and the ball reaches latitude 100 in time $\sinh^{-1}(90/\alpha) \approx 1.47$ sec, after which time the ball has left Flatland.

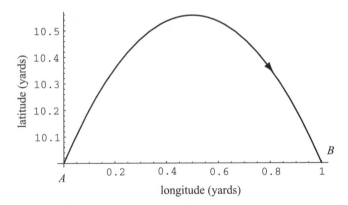

Figure 4. A. Square throws a ball to B. Pentagon

Example 2. Albert and his son Benjamin Pentagon play ball. Albert is at longitude 0 and latitude 10, with coordinates $A = (0, 10)$, and Benjamin at $B = (1, 10)$. Albert nudges the ball with initial velocity $\mathbf{v}_0 = (v_x, v_y)$.

To find just one trajectory (of many) for the ball to go from A to B we take $|v_y| = 10$, which means that $\alpha = \sqrt{8000} \approx 89.4$ by (5), which in turn means that the ball goes from latitude 10 yards to about latitude 10.6 yards in time $\cosh^{-1}(90/\alpha) - \cosh^{-1}(1) \approx 0.116$ seconds by (6). Thus the total flight time from A to B is 0.23 sec, which in turn means that $v_x \approx 1/0.23 \approx 4.35$ yds/sec. The resulting trajectory, as shown in Figure 4, the parametric graph of

$$\left(4.35t,\ 100 - \alpha\cosh\left(t - \cosh^{-1}\left(\frac{100-h}{\alpha}\right)\right)\right)$$

for $0 \leq t \leq .23$, whose second component is obtained by inverting the first rule of (6), looks very much like a parabola.

This first model has some shortcomings in that, according to Albert's grandson, A. Hexagon, whose adventures are chronicled in [20], as latitude increases, the atmosphere thins and gravity weakens, but never vanishes altogether. From his notes, Albert anticipated this phenomenon, and worked at length to find a better model.

A universal law of gravitation

Albert proposed an intuitive gravity function, one that he extended to multiple dimensions, so as to compare his universe with that of $\mathcal{R}^3$ and with $\mathcal{R}^n$ in general. To explain this model, let $B_n(r)$ be the set of all points in $\mathcal{R}^n$ that are no more than r units from the origin. In Flatland, $B_2(r)$ is a disk of radius r, the shape of Flatland priests. $B_3(r)$ is a solid sphere of radius r in $\mathcal{R}^3$. Let $S_n(r)$ be the hyper-surface area of $B_n(r)$, a measure of the set of all points r units from the origin. For example, $S_2(r) = 2\pi r$ and $S_3(r) = 4\pi r^2$.

Let us imagine, Albert continued, that gravity is like light. If a light source at point P flashes momentarily, light proceeds outward in a concentric wavefront about P as shown in Figure 5. He defined $I(r)$, the brightness of the light source as it would appear to an observer at distance r, to be the quotient of the total number of photons emitted in the flash of the light source and of $S_n(r)$, (since all of the photons emitted in the flash would be distance r from the source when the observer sees the light). That is, the intensity of the light in the wavefront is inversely proportional to $S_n(r)$. Furthermore, this relationship is valid when the light source at P shines with constant brightness rather than giving a single flash.

In the same way, Albert concluded, the gravitational acceleration $a(r)$ on a point mass at distance r from P exerted by the point mass at P must be

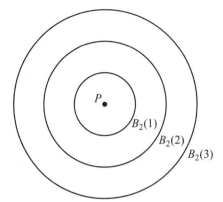

Figure 5. An expanding wave-front from a point source

inversely proportional to $S_n(r)$. For example, $a(r)$ is inversely proportional to r in Flatland, inversely proportional to r^2 in $\mathcal{R}^3$, and inversely proportional to r^3 in $\mathcal{R}^4$. Since the magnitude of the mass m of the point mass P should also influence the value of $a(r)$ and since the simplest model is a linear one, Albert argued that $a(r)$ also varies directly with m. Finally, since he wished to compare the effect of gravity on motion in all dimensions, Albert whimsically imagined that

$$a(r) = -\frac{Gm}{S_n(r)}, \qquad (7)$$

where, for the yard unit of length, G has the same value in every dimension even though the form of its units, yd^n/(kg·sec^2), changes with dimension n. Under such a fancy, G would truly be a meta-universal constant. The negative sign indicates that the acceleration is toward P.

Next, Albert generated a formula for the hyper-volume $V_n(r)$ of $B_n(r)$ so as to find $S_n(r)$. Since $V_2(r) = \pi r^2$ and $V_3(r) = \frac{4}{3}\pi r^3$, Albert suspected that $V_n(r)$ varied with r^n. But what was the constant of proportionality? Rather than follow his derivation we appeal to Appendix, Item 8:

$$V_n(r) = \frac{\pi^{n/2} r^n}{\Gamma(1+n/2)} \qquad (8)$$

where Γ is the gamma function as presented in Appendix, Item 6. In particular, $\Gamma(\frac{1}{2}) = \sqrt{\pi}$ and $\Gamma(1+n/2) = (n/2)\Gamma(n/2)$ for all positive integers n.

Table 1. Hyper volumes and surface areas of the unit ball in $\mathcal{R}^n$

n	0	1	2	3	4	5	6	7	8
$V_n(1)$	1	2	π	$\frac{4}{3}\pi$	$\frac{1}{2}\pi^2$	* $\frac{8}{15}\pi^2$	$\frac{1}{6}\pi^3$	$\frac{16}{105}\pi^3$	$\frac{1}{24}\pi^4$
$S_n(1)$	0	2	2π	4π	$2\pi^2$	$\frac{8}{3}\pi^2$	π^3	* $\frac{16}{15}\pi^3$	$\frac{1}{3}\pi^4$

Now for each n and every $r \geq 0$, $V_n(r) = \int_0^r S_n(\rho)\,d\rho$. By the fundamental theorem of calculus,

$$S_n(r) = \frac{n\pi^{n/2} r^{n-1}}{\Gamma(1+n/2)}. \tag{9}$$

Table 1 lists the first few values of $V_n(1)$ and $S_n(1)$. The * alongside the $V_5(1)$ entry indicates that the greatest volume of the unit sphere occurs in dimension 5, whereas the * alongside the $S_7(1)$ entry indicates that the greatest surface area of the unit sphere occurs in dimension 7. Since Albert visited Lineland, at least in his dreams, he was intrigued with the value $S_n(1)$ being 2. "How could the hyper-area of a line segment in $\mathcal{R}$ be other than 0?" he wondered at first. But each line segment has two endpoints. "That must be what $S_n(1)$ is counting," he concluded. Similarly in Pointland, the hyper-volume of the unit ball in $\mathcal{R}^0$ simply counts the only point that exists, and so its volume is 1. The results of Table 1 appear to be counter-intuitive. In particular, if G is a meta-universal constant, then by (7) gravity is least in dimension seven at distance one yard between fixed masses. With surprising results like this, Albert set aside his work for a season, wondering whether he had the ability to understand the very thing he was trying to understand.

Nevertheless, once the manuscript resumed at a later date, Albert melded (7) with his first model in an effort to improve his intuition. In particular, in Flatland with $n = 2$, by (7) and (9), Albert took the gravitational acceleration induced by a point mass Q on another point mass P of mass m and distance r from Q to be

$$-\frac{Gm}{r} \tag{10}$$

for some constant G.

Fragments from Flatland

A second geophysical model for Flatland

For a more realistic geophysical model of Flatland, Albert imagined that almost all of the mass of Flatland is along a line segment called the *base line* in a plane, as shown in Figure 6. He assumed that the rest of Flatland, including its inhabitants, houses, atmosphere, and rain, has negligible mass. He positioned the base-line along the x-axis with its midpoint at the origin. We determine the gravitational acceleration on a point-mass $P(s)$ at $(0, s)$ where $s > 0$. Partition the interval $[0, b]$ by $0 = x_0 < x_1 < x_2 < \cdots < x_n = b$ so that $\Delta x = (x_i - x_{i-1}) = 1/n$ for all positive integers i where n is some positive integer. We assume that the base line has unit density (unit mass per unit length).

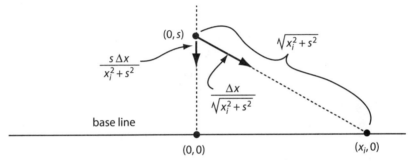

Figure 6. The base line of mass in Flatland

Let us assume that all the mass along the base line in the interval $(x_{i-1}, x_i]$ is coalesced at the point $(x_i, 0)$. The distance between $(x_i, 0)$ and $(0, s)$ is $\sqrt{x_i^2 + s^2}$. By (10), Albert reasoned that the acceleration due to gravity on $P(s)$ is $-G\Delta x / \sqrt{x_i^2 + s^2}$ and is directed along the ray from $(0, s)$ through $(x_i, 0)$. So the downward component of this acceleration is

$$-\frac{G\Delta x}{\sqrt{x_i^2 + s^2}} \frac{s}{\sqrt{x_i^2 + s^2}} = -\frac{Gs\Delta x}{x_i^2 + s^2}. \tag{11}$$

We simplify matters by assuming that the yard has been defined so that G is 1. We also ignore all difficulties involving the notion of mass, which may very well be a phenomenon varying from one dimension to another. However, when the Sphere first interviewed Albert, he mentioned that thickness in Flatland was infinitesimal, not zero. Perhaps this phenomenon may serve to explain the Flatland phenomenon of the fixity of structures [1, p. 140].

The sum over all the point masses of (11) gives the total downward acceleration on $P(s)$ exerted by the base line, for any n, as

$$-2\sum_{i=1}^{n}\frac{s\Delta x}{x_i^2+s^2}.$$

Letting $n \to \infty$, we find that the total acceleration on $P(s)$ is the integral

$$a(s) = -2\int_0^b \frac{s}{x^2+s^2}\,dx, \qquad (12)$$

which means that

$$a(s) = -2\tan^{-1}\frac{b}{s}. \qquad (13)$$

If the base line of Flatland were unbounded, then $a(s) = -\infty$ for all s, which would mean that no Flatlander could move at all. Thus, if the model mirrored reality, Albert concluded that Flatland must be bounded.

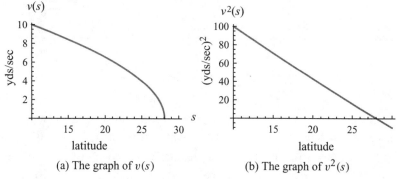

(a) The graph of $v(s)$ (b) The graph of $v^2(s)$

Figure 7. Finding an approximation for $v(s)$

Example 3. We throw a ball straight up in this gravity model where $h = 10$ yards, $v_0 = 10$ yds/sec, and $b = 100$ yards. From (3) and (13) and with the help of a CAS, the latitude at which $v = 0$ is $s_m \approx 28.07462$ yards. Since (4) in this case appears to be an intractable integral, we proceed by using approximations. The graph of (3) appears in Figure 7(a). Its shape resembles a parabola whose central axis is horizontal. With this idea in mind, we graph $v^2(s)$ as shown in Figure 7(b), which resembles a line, as we might expect. However the match is not especially good. So we try a second-order approximation, and fit a parabola through the data $\{(10, 100), (20, v^2(20)), (s_m, v^2(s_m))\}$ obtaining

$$160.732 - 6.26578s + 0.0192559s^2.$$

Let

$$V(s) = \sqrt{160.732 - 6.26578s + 0.0192559s^2}$$

be the function we use to approximate $v(s)$. Graphs of the two are almost indistinguishable. In (4), replace v with V, obtaining a function $T(s)$ that approximates the time T at which the ball is at latitude s. Figure 8 shows the parametric plot of $(T(s), s)$, the trajectory of the ball in time. The dotted curve is that of a parabola corresponding to the trajectory of a ball thrown straight up with initial velocity $v_0 = 10$ yds/sec in a constant gravity field with acceleration $a(10) \approx -2.94226$ yds/sec^2. Exercise 2 asks for the trajectory given that Flatland's base line mass is a homogeneous rectangle rather than a line segment.

A third geophysical model for Flatland

After Albert finished his calculations for model two, he was troubled by the implications of a finite east-west expanse of his world. "Flatland should have no end," he wrote. The Sphere had shown him a view of Flatland from above. He had seen no end. But just because he hadn't seen boundaries wasn't enough evidence that they failed to exist. Travelers in his day had gone far to the east and west. None had reported an end to Flatland. Since Albert was in jail for heresy, he had no qualms about imagining a more unorthodox model, namely that Flatland was perhaps part of a very large disk. Thus, the inhabitable part of Flatland could possibly be an annulus as in Figure 9. Lines of latitude, Albert mused, would then be circumcircles about the planet's center. They would appear to be straight because the planet's radius was so large. In later years, A. Hexagon confirmed this hypothesis by relating the tale of two travelers who actually met, one going west, the other east, always following a path of constant gravity, as determined by a Flatland gravimeter [20].

In order to perform his next set of calculations, Albert needed to know the cumulative gravitational acceleration on a particle induced by all point masses within a Flatland planet. Using the same reasoning as he used to

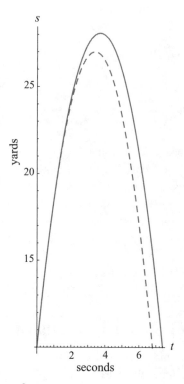

Figure 8. A. Square tossing a ball to himself

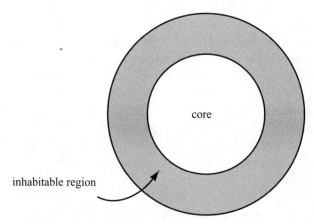

Figure 9. The inhabitable, annular region on a Flatland planet

deduce (11) and assuming a homogeneously dense planet of mass density δ and radius R and with center at the origin O, he set up the integral

$$a(s) = 2 \int_0^R \int_{-\sqrt{R^2-x^2}}^{\sqrt{R^2-x^2}} \frac{G\delta(y-s)}{x^2 + (y-s)^2} \, dy \, dx$$

which gives the gravitational acceleration on a point mass at $(0, s)$, where $s > R$.

As shown in Exercise 3a, the value of this integral varies inversely with respect to s. That is, when determining accelerations induced by the type of Flatland planet described, one may treat the entire planet as a point with the same mass as the planet and centered at the planet's center. Furthermore, he extended this result to dimensions three and four, finding that

$$a_3(s) = -\frac{k_3}{s^2} \text{ and } a_4(s) = -\frac{k_4}{s^3},$$

respectively, for some constants k_3 and k_4. He also showed that from Flatland, there was no escape; any ball hurled upwards will eventually come down, Exercise 3b.

More than that, he derived the equations of motion for a planet circling a sun in $\mathcal{R}^n$, as outlined below, supposing of course that suns exist in dimensions other than 3.

Position the polar plane with the sun's center at the origin and let a planet's position be $P = (r, \theta)$, where both r and θ are functions of time t. Let $\mathbf{s}$ be the rectangular coordinate equivalent to (r, θ). The two natural orthogonal unit vectors are $\mathbf{u}_r = \cos(\theta)\mathbf{i} + \sin(\theta)\mathbf{j}$ and $\mathbf{u}_\theta = -\sin(\theta)\mathbf{i} + \cos(\theta)\mathbf{j}$, as depicted in Figure 10.

Since $\mathbf{s} = r\mathbf{u}_r$, $\frac{d\mathbf{u}_r}{d\theta} = \mathbf{u}_\theta$, and $\frac{d\mathbf{u}_\theta}{d\theta} = -\mathbf{u}_r$, earth's velocity is

$$\mathbf{v} = \frac{d\mathbf{s}}{dt} = \frac{d(r\mathbf{u}_r)}{dt} = r\mathbf{u}_\theta \frac{d\theta}{dt} + \mathbf{u}_r \frac{dr}{dt}.$$

Similarly, earth's acceleration is

$$\mathbf{a} = \left(r\frac{d^2\theta}{dt^2} + 2\frac{dr}{dt}\frac{d\theta}{dt} \right) \mathbf{u}_\theta + \left(\frac{d^2 r}{dt^2} - r\left(\frac{d\theta}{dt}\right)^2 \right) \mathbf{u}_r. \qquad (14)$$

Assume that P's only acceleration is that induced by the gravitational field of the sun. Let $f(r, n) = -k_n/r^{n-1}$ be the gravitational acceleration on P

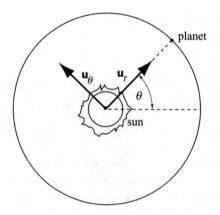

Figure 10. The natural directions

when it is r units from the sun in $\mathcal{R}^n$ where k_n is some constant. Acceleration is solely in the radial direction, so

$$0 = r\frac{d^2\theta}{dt^2} + 2\frac{dr}{dt}\frac{d\theta}{dt} = \frac{1}{r}\frac{d\left(r^2\frac{d\theta}{dt}\right)}{dt} \quad \text{and} \quad f(r, n) = \frac{d^2r}{dt^2} - r\left(\frac{d\theta}{dt}\right)^2. \tag{15}$$

The first of the equations in (15) is the law of the conservation of angular momentum. Since $1/r$ is never 0,

$$r^2\frac{d\theta}{dt} = h, \tag{16}$$

where h is a constant. In terms of h, the second equation in (15) is

$$\frac{d^2r}{dt^2} - \frac{h^2}{r^3} = f(r, n). \tag{17}$$

To solve (17) we need some initial conditions. We imagine at some point in its trajectory a ball will reach an equilibrium with respect to r, if only for a moment, which we call time $t = 0$ and angle $\theta = 0$ so that dr/dt and $dr/d\theta$ are both zero, whereas $d\theta/dt = \omega_0 > 0$, where ω_0 is a constant. To summarize,

$$r = \rho \text{ when } t = 0 \text{ and } \theta = 0, \quad \text{and} \quad \left.\frac{dr}{dt}\right|_{t=0} = 0 = \left.\frac{dr}{d\theta}\right|_{\theta=0},$$

$$\text{and } \left.\frac{d\theta}{dt}\right|_{t=0} = \omega_0. \tag{18}$$

Fragments from Flatland

A clever way to rewrite (17) for r in terms of θ is to let $z = 1/r$. Then, using (16),

$$\frac{dr}{dt} = \frac{d\left(\frac{1}{z}\right)}{d\theta}\frac{d\theta}{dt} = -\frac{1}{z^2}\frac{dz}{d\theta}\frac{d\theta}{dt} = -r^2\frac{d\theta}{dt}\frac{dz}{d\theta} = -h\frac{dz}{d\theta}.$$

Similarly,

$$\frac{d^2r}{dt^2} = -h^2 z^2 \frac{d^2z}{d\theta^2}.$$

Thus (17) is equivalent to

$$\frac{d^2z}{d\theta^2} + z = -\frac{f(1/z, n)}{h^2 z^2}. \tag{19}$$

In Flatland, $f(r, 2) = -k_2/r$, so (19) becomes

$$\frac{d^2z}{d\theta^2} + z = \frac{k_2}{h^2 z}, \tag{20}$$

a nonlinear differential equation.

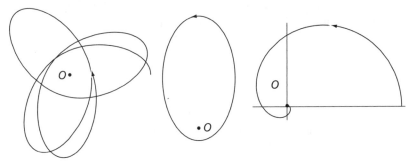

(a) A precessing ellipse in 2-D   (b) An ellipse in 3-D   (c) Spiralling into the sun in 4-D

Figure 11. Planetary orbits in alternate universes

Using chalk on his cell floor, Albert drew the shapes of these orbits for dimensions 2, 3, and 4 as given by (17) and (19). Of course, part of his drawings was erased each time he crossed a curve. Workmen swear that etchings of his drawings are embedded in his cell floor. In a three-dimensional universe, Figure 13(b) is an ellipse with the sun at a focus, as the Sphere had told him, and as we detail in the next chapter. In a two-dimensional universe, Figure 13(a), a numerically generated solution for (r, θ) using (20), appears to be a *precessing* ellipse, the curve traced by a point moving along an elliptical path while the path itself is rotating about another point. A succession of exercises (Exercise 5 to 9) outlines Albert's attempt to show that these orbits are closely approximated by such a curve where the ellipse rotates about a (non-focal, non-central) point along its major axis. In a four-dimensional universe (in which the equations of motion are easiest to solve, and which is Exercise 4(b)), any planet spirals into its sun, as illustrated in Figure 13(c). Although some circular orbits are solutions to (19), such trajectories are unstable whenever n exceeds three, as shown in Exercise 10.

Therefore Albert concluded that the sphere had been correct about life in a 4-D universe:

Life in four space is truly inconceivable. Planetary orbits are simply too short-lived in four dimensions for life to flourish. Unless the physics of four dimensions is altogether more strange than in three, in which case...

The remainder of his speculation is lost, for in the midst of this thought Albert's manuscript ends. Aprocryphal stories abound about Albert being whisked into various n-spaces by assorted meta-creatures, but beyond $n = 3$ we find no hint of such adventures in his extant writings.

Exercises

1. For the model of Flatland as depicted in Figure 6, draw the gravitherm through the point $(0, 10)$. That is, for each x with $-b < x < b$, find that y so that the magnitude of the acceleration of gravity at (x, y) is the same as at $(0, 10)$.

2. One of the difficulties with the second model for Flatland's gravity is that if a Flatlander slides to the base line, $a(0)$ is infinite. He will be forever stuck on the base line. One way to improve the model is to give the base a depth of w yards, as shown in Figure 12. As in the second model, each point (x, y) in the base induces a downward acceleration of $(s - y)dx\, dy/(x^2 + (s - y)^2)$, where $dx\, dy$ is the mass of unit density in a rectangular region with infinitesimal length dx and infinitesmal depth dy. So $a(s)$ is the sum of all of these accelerations over the

Fragments from Flatland

homogeneously dense base rectangle of length $2b$, which is

$$a(s) = -2 \int_0^b \int_{-w}^0 \frac{s-y}{x^2 + (s-y)^2} \, dy \, dx.$$

Show that b must be finite in order for Flatlanders to be able to move. Furthermore, for the b value for which $a(10) = -2\tan^{-1} 10$ yards/sec^2 with $w = 2$, plot a trajectory of a ball thrown in this gravity with $s_0 = 10$, $v_0 = 10$, and compare it with the graph of Figure 8.

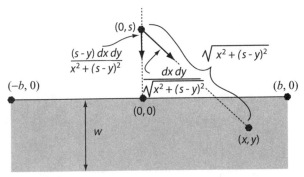

Figure 12. The base rectangle of mass in Flatland

3. (a) Let P be a homogeneous planet in Flatland of radius R and mass m. Show by way of evaluating an integral that $a(r) = -Gm/r$, when $r \geq R$, where $a(r)$ is the gravitational acceleration induced by P on a point mass at radial distance r from P's center.

   (b) Show that in model 3, any ball thrown to the north will ultimately return to the south.

4. Find an analytic solution to (19) in

   (a) Three dimensions—that is, take $f(r, 3) = a_3(r) = -k/r^2$.

   (b) Four dimensions—that is, take $f(r, 4) = a_4(r) = -k/r^3$.

5. The equations of motion for a Flatland planet P orbiting a fixed sun O where the gravitational acceleration on P is inversely proportional to its distance from O given $k = 5$, the constant of proportionality; $h = 1$, the constant of angular momentum; $r(0) = 1$, the initial radial distance of P from O; $\theta(0) = 0$, the initial polar angle; $\frac{d\theta}{dt}|_{t=0} = 0 = \frac{dr}{d\theta}|_{\theta=0}$;

$\frac{d\theta}{dt}|_{t=0} = 1$, and $z = 1/r$, become

$$\frac{d^2z}{d\theta^2} + z = \frac{5}{z}.$$

Generate a numerical solution for z and so obtain the polar plot of $r(\theta)$, as in Figure 13. Find the least positive constant λ for which $r(\lambda\theta)$ has period 2π. Furthermore, show that this oval is not an ellipse of the form $r = p/(1 - q\cos\theta)$, for any choice of constants p and q. Trajectories for Flatland planets are also sketched in Dewdney [34, pp. 132, 233].

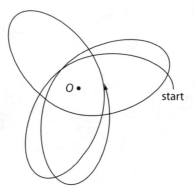

Figure 13. Presessing oval orbits

6. Find the semi-major and semi-minor axial lengths of the oval in Exercise 5. Superimpose the graphs of the oval and the ellipse with these semi-axial lengths. Are they the same curves? Repeat the exercise and test your intuition, changing the proportionality constant to $k = 10$ rather than 5.

7. The ellipse whose parametrization is $(a\cos\phi + p, b\sin\phi)$ where a, b and p are positive with $p < a$ has its center at $(p, 0)$ and is aligned with both the x-axis and the y-axis, as shown in Figure 14. Show that a polar parametrization $Q(\theta)$ of this ellipse in terms of the polar angle θ, the angle between the positive x-axis, the origin $(0, 0)$, and points on the ellipse, is

$$Q(\theta) = \frac{b^2 p \cos\theta + \sqrt{a^2 b^4 \cos^2\theta + a^2 b^2 (a^2 - p^2)\sin^2\theta}}{b^2 \cos^2\theta + a^2 \sin^2\theta}.$$

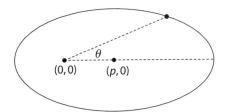

Figure 14. An uncommon elliptical parametrization

8. For a polar curve $f(\theta)$, its ρ-*precession* is the polar curve $f(\rho\,\theta)$. Plot the ρ-precession of the ellipse of Exercise 6 using the parametrization of Exercise 7 and $\rho = 1/\lambda$ from Exercise 5. Superimpose the polar plot of this ρ-precession, $f(\theta/\lambda)$, onto the orbit of planet P of Exercise 5, as shown in Figure 13.

9. Let $Z(\theta) = 1/f(\theta/\lambda)$, where $f(\theta/\lambda)$ is defined in Exercise 8. Plot the graph of $y = ZZ'' + Z^2$, and compare it with the graph of $z = 5$. Can you find an elementary function better than this $Z(\theta)$ that approximates the solution to the differential equation of Exercise 5?

10. (a) Let n be a positive integer. Show that $z = a$ is a solution to (19) for some positive constant a.

    (b) If $n \geq 4$, show that $z = a$ corresponds to an unstable trajectory for a planet.

Vignette IV: A want-to-be mathematician

For a season in his life, Voltaire tried being a mathematician.

Exiled from France, and arriving in London in 1726, Voltaire focused on mastering English. Within three months he was somewhat conversant. After eighteen months, he was confident enough to publish essays in English, although he confessed to weaknesses in pronunciation and comprehension in conversation [**152**, p. 76]. Wined and dined by the literary community, he was welcomed as a hero and an accomplished artist.

Shortly after Voltaire's arrival in England, Newton died. Nobles vied to be pall bearers. Eight dukes carried his coffin. He was buried in Westminster Abbey with great ceremony. Such honor given to a common man overwhelmed Voltaire. A similar event could never happen in Paris. Who was this man for all England to revere? He resolved to find out.

He interviewed Newton's niece who told him that in 1666 when Newton was at his mother's estate while the black plague raged in London, he saw an apple fall and thereby imagined gravity and its attendant dynamics, all of which led to the *Principia* of 1689.

Recognizing a good story, Voltaire used it in an essay in English on epic poetry that he published in London early in December 1727. In context, Voltaire characterizes the epic poetry of various European countries. Last on his list is England, from which Voltaire features Milton's *Paradise Lost*. He argues that Milton's inspiration for retelling the mythic story of man's fall upon eating a Garden of Eden apple grew from his travels to Florence, Italy as a youth where he attended a play, the theme for which was the fall of man. Voltaire then gave two analogies, both involving the inspiration of legendary mathematical masters [**152**, pp. 130–131]:

> In like manner, Pythagoras owed the invention of music to the noise of the hammer of a blacksmith. And thus in our days Sir Isaac Newton walking in his gardens had the first thought of his system of gravitation, upon seeing an apple falling from a tree.

Apparently Voltaire was the first to tell the story in print. Thirty-three years later, in Letter LII of his *Letters to a German Princess*, Euler embellished the story, saying that the apple actually landed on Newton's head.

The more Voltaire learned about Newton, the more intrigued he became, so much so that he resolved to write a popular introduction to Newton's discoveries. While in England, he wrote *The Philosophical Letters* about life in England, which some have characterized as the first major work of the Enlightenment. Four of these letters were about Newton. By then, his two years of exile were over, and Voltaire was allowed to return to France.

At home again he began preparations for a deeper understanding of Newton, and recruited various tutors to help. One of the first of these was Charles-Marie de La Condamine, who persuaded Voltaire along with others to form a syndicate so as to play the French lottery. La Condamine had found a flaw in the system. Low-cost tickets had the same probability of winning as high-cost tickets. The trick was to buy as many low-cost tickets as possible. Sure enough, the syndicate won systematically for six months until the government realized its error. Voltaire's share in the winnings was a half-million livres. From then on, Voltaire was a wealthy man, as was La Condamine. In fact, when La Condamine was in Quito on an expedition a few years later to measure three degrees of arc and official funds ran low during a protracted stay in the field, his credit was ever good, and he was able to carry the team. Meanwhile Voltaire hired Maupertuis and Clairaut to help him learn more Newtonian mathematics.

By happenstance, he met Gabrielle Émilie Le Tonnelier de Breteuil Du Châtelet (1706–1749) in 1733. We shall call her Emily. She was an unusual woman: a marquise educated in mathematics and physics. Emily in turn was intrigued with this poet-turned-natural-philosopher. For the next fifteen years until her death, they were nearly inseparable. While Voltaire worked on his introduction to Newton's work, Emily resolved to translate the *Principia* from Latin into French. Voltaire's tutors became her tutors as well, and they found her to be an apt student. At her somewhat run-down country estate, Voltaire underwrote significant repairs and added an entire new wing to the house for himself. He purchased the best of scientific equipment and a library of scientific books. Together, they conducted experiments so as to understand Newton and the universe. They read. They discussed. They debated. In Emily's coined word, they *newtonized*. Each year the French Academy offered a prize essay question. The 1737 prize was on the nature of fire and how it spread. Voltaire and Emily ran more experiments. Each submitted an entry.

Vignette IV: A want-to-be mathematician

Figure 1. From Newton to Emily to Voltaire, the frontispiece to Voltaire's 1738 *Elements of Newton's Philosophy* [**143**, plate facing p. 195]

The result?

Euler won first place. Emily and Voltaire each earned honorable mention and their papers appeared in the Memoires of the French Academy. In fact, Emily's entry was the first paper by a woman to be published by the French Academy. Some say that she was the first woman scientist.

In 1738, Voltaire's *Elements of Newton's Philosophy* was published. The frontispiece in Figure 1 shows Newton in heaven filtering the light of knowledge down to Emily who in turn reflected it with a mirror to Voltaire writing at his desk. The secretary of the French society, de Fontenelle, whom Voltaire regarded as a friend, gave a disheartening review of the book. Voltaire, he said, you really need another three years of study before writing about Newton [148, p. 24]. Voltaire then asked Clairaut for a second opinion, who concurred, saying that his gifts lay with poetry and not equations, for "even with the most stubborn labour [you will] not be able to attain anything beyond mediocrity" [87, p. 103], [90, pp. 108–109].

From then on, Voltaire focused on writing plays, romances, histories, letters, and essays. However, he continued to support Emily's work on Newton's *Principia*. She finished the manuscript in 1749, gravely ill, and died on the night of posting it to the publisher. Her translation and commentary on it remains the definitive French translation of the *Principia* to this day.

As can be seen, Voltaire had been a want-to-be mathematician-savant for about eight years. He wrote the first draft of *Micromégas* in 1739, during that period. He knew the members of the expeditions in both the arctic and the equatorial expeditions who tested Newton's theory of a flattened earth, and he followed their progress with interest. And his role in mathematics though tangential helped the non-English mathematical community to read and accept Newtonian mechanics. As the Voltaire Foundation concludes in *The Complete Works of Voltaire* [143, p. 3],

> Voltaire's *Eléments de la philosophie de Newton* was one of a small number of published works which contributed significantly to the acceptance and adoption of Newtonian theory in France. [His book] is a rare exception to the general rule [about popularizations being ephemeral], being the popularization of the work of genius by a man of genius.

Interestingly, in the Micromégas story Voltaire often uses proportional reasoning, and his calculations sometimes appear to be faulty. As one example, Voltaire says that the giant's nose was one-third of his head, his head was one-seventh of his body, and his body was 120,000 feet. He concluded that the nose was 6333 feet, which is about 619 feet too much. Indeed, Wade

Vignette IV: A want-to-be mathematician

[**148**] says, "It is a curious fact that there is not one single proportion in the whole story accurately given."

Were these intentional mistakes? Was he mocking the results of long series of precise measurements and convoluted logical arguments? Voltaire is a satirist of satirists. Or, were they honest mistakes, and evidence that justified Clairaut's opinion of Voltaire's mathematical gifts? Of course, Voltaire was a poet, and a charm of poetry is that it is ambiguous, and can be interpreted on many levels. Voltaire loved good stories, beauty, and truth. As a poet, he glimpsed all of those elements in Newton's discoveries; and he used his gifts to convey what he saw to the public at large. As he says in his fifteenth philosophical letter from England, "I will now acquaint you (without prolixity if possible) with the few things I have been able to comprehend of [Newton's] sublime ideas."

In that same spirit, we briefly derive Kepler's laws—Newton's great achievement—and illustrate its predictive power with a few case studies.

CHAPTER IV

Newton's Polar Ellipse

> Ye comets, dreaded like the bolts of Jove,
> In vast ellipses regularly move!
> Cease with your motions mortals to affright;
> Remount, descend near the great orb of light;
> Newton has marked the limits of thy race,
> March on, illumine night, we know thy place.
> —*On the Newtonian Philosophy* [**144**, vol. xxxvi, p. 301]

These lines of poetry are from Voltaire's preface to his book of 1738, *Elements of Newton's Philosophy*. As A. Square reminded us in the last chapter, Isaac Newton solved the equations of motion for a planet, comet, or rock orbiting a sun in a three dimensional universe. In fact, a comet in some ways precipitated the *Principia*. In 1682, a comet appeared in the skies. Edmond Halley proposed that it was a recurrent visitor, coming by the earth once every seventy-five years. With this on his mind, he asked Newton about trajectories of comets, supposing that gravitational force followed an inverse-square law. Newton replied that he had already calculated them: the trajectories were the ellipses of Johannes Kepler with the sun at a focus. Encouraged by Halley and ignoring the apparent ridiculousness of the idea of masses exerting force across empty space, Newton reworked the details and published his great work, the *Principia*, in 1689.

In this chapter, we outline Newton's calculations and consider two case studies: determining spring-time for a planet of given orbital eccentricity and polar orientation, and checking the route of the comet on which Micromégas toured the solar system.

Kepler's laws of planetary motion

From Chapter III, the equations describing the position of any three dimensional planet X with respect to a fixed sun at the origin O are

$$\text{a. } \frac{d^2r}{dt^2} - \frac{h^2}{r^3} = -\frac{k}{r^2} \quad \text{and} \quad \text{b. } \frac{d^2z}{d\theta^2} + z = \frac{k}{h^2}, \tag{1}$$

where r is the radial distance of X from O, $z = 1/r$, θ is the counterclockwise polar angle to X, t is time,

$$h = r^2 \frac{d\theta}{dt} \tag{2}$$

is the constant of angular momentum, and $-k/r^2$ is the inverse square law for gravity, with k being a positive constant.

To solve (1) we need initial conditions. Throughout its orbit of period T, X's distance from O varies from a minimum value, the perihelion, ρ, to a maximum value, the aphelion. At time $t = 0$, we take $\theta = 0$, $r(0) = \rho$, which also means that $\frac{dr}{dt}|_{t=0} = 0 = \frac{dr}{d\theta}|_{\theta=0}$. The rate at which θ changes with respect to time is non-constant. It is a maximum at perihelion and a minimum at aphelion. Since θ changes by 2π radians over one rotation about the sun, we let $\frac{d\theta}{dt}|_{t=0} = \omega_0$, where $\omega_0 \geq 2\pi$ radians/T. To summarize,

$$r = \rho \text{ when } t = 0 \text{ and } \theta = 0, \quad \text{and} \quad \frac{dr}{dt}\bigg|_{t=0} = 0 = \frac{dr}{d\theta}\bigg|_{\theta=0},$$

$$\text{and } \frac{d\theta}{dt}\bigg|_{t=0} = \omega_0. \tag{3}$$

As shown in Example 2, Item 5 of the appendix, the solution to $d^2z/d\theta^2 + z = k/h^2$ is $z = A\cos(\theta + \phi) + k/h^2$, so the solution to (1a) is

$$\frac{1}{r} = A\cos(\theta + \phi) + \frac{k}{h^2}. \tag{4}$$

Differentiating (4) with respect to θ gives $1/r^2 \frac{dr}{d\theta} = A\sin(\theta + \phi)$, which by (3) means that $A\sin\phi = 0$, so we can take the phase angle ϕ to be 0. Using (3) and (4) gives $A = 1/\rho - k/h^2$. Thus the orbit of planet X is

$$r = \frac{h^2/k}{1 + \left(\frac{h^2}{k\rho} - 1\right)\cos\theta}. \tag{5}$$

Provided $0 < h^2/(k\rho) < 2$, this is the polar equation of an ellipse, Kepler's first law of motion.

The area $A(t)$ of this ellipse swept out in time from 0 to t is, from (2),

$$A(t) = \int_{\theta(0)}^{\theta(t)} \frac{1}{2} r^2 \, d\theta = \int_0^t \frac{1}{2} r^2 \frac{d\theta}{d\tau} \, d\tau = \int_0^t \frac{1}{2} h \, d\tau = \frac{ht}{2}. \quad (6)$$

This is Kepler's second law of motion.

The standard form of an ellipse in polar coordinates is

$$r = \frac{b^2/a}{1 + e \cos \theta}, \quad (7)$$

where a is the length of the semi-major axis, b is the length of the semi-minor axis, and $e = c/a$ is the eccentricity, where $c = \sqrt{a^2 - b^2}$. Its area is πab.

From (6), the area of the ellipse is $hT/2$, where T is the orbital period of the body. Comparing (5) and (7), we see that $b^2/a = h^2/k$ so $b/h = \sqrt{a/k}$. Thus

$$T = \frac{2\pi}{h} ab = 2\pi a \sqrt{\frac{a}{k}}.$$

This is Kepler's third law of motion,

$$T^2 = \frac{4\pi^2}{k} a^3. \quad (8)$$

Case study: finding spring on planet X

How long is winter? That is, how many days lie between the winter solstice and the spring equinox in the northern hemisphere? From the calendar, the winter solstice is December 21 and the spring equinox is March 20, so the answer is 89 or 90 days (because February has 29 days during leap years). However, since there are 365 days in a year, why isn't the answer 91 or 92 days?

Moreover, suppose that the earth's orbit had a different eccentricity or that the winter solstice is far from the time when the earth is at perihelion. How do these factors affect the answer? For a given orbital eccentricity, what are the maximum and minimum possible time lapses between winter and spring on a planet?

To help answer these questions, we review what is meant by a solstice and an equinox. With the sun's center at the origin O, planet X orbits O in

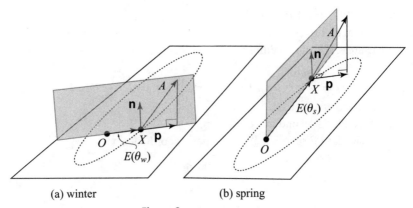

Figure 2. Winter and spring

an ellipse traced out in the *ecliptic plane*, which we take as the xy-plane, as illustrated in Figure 2. Let $E(\theta)$ be the vector from O to X's center where θ is the polar angle to X. Let **n** be the unit vector in the positive z direction. In order to have seasons, X's axis of rotation must lie in a direction other than **n** because seasons are caused by the inclination of the planet to its ecliptic plane. Let Q be a point in the northern hemisphere. A *winter solstice* occurs when daytime is as short as possible at Q. A *summer solstice* occurs when daytime is as long as possible at Q. An *equinox* occurs when the length of daytime is the same as nighttime at Q. The spring equinox follows the winter solstice, and the fall equinox follows the summer solstice.

Let A be the vector from the X's south pole to its north pole. Let **p** be the projection vector of A onto the ecliptic plane. Figure 2(a) shows the geometry of a winter solstice occuring at angle θ_w. When **p** and $E(\theta)$ point in the same direction, a winter solstice occurs. When **p** and $E(\theta)$ point in opposite directions, a summer solstice occurs. Furthermore, Figure 2(b) shows the geometry of a spring equinox occurring at angle θ_s. In this case, **p** is perpendicular to $E(\theta_s)$. Therefore, $\theta_s = \theta_w + \pi/2$.

To see graphically why the length of daylight changes in the northern hemisphere, Figure 3(a) shows the view of earth from the sun at winter solstice, with A tilted 23.5° away from **n**. To accentuate the view of latitude 45°, the region between latitudes 30° and 45° has been cut away in the northern hemisphere. Figure 3(b) is a simplified version of 3(a), with a dotted ellipse representing the circle of latitude 45°. At winter solstice, the arc of this circle illuminated by the sun has chordal distance d, a length that

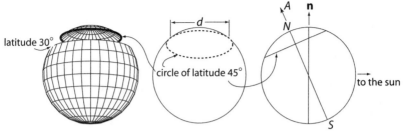

(a) cut-away (30°N – 45°N)   (b) view from the sun   (c) view from the side

Figure 3. A winter solstice

is less than the diameter of the circle. The rest of the circle, more than half of its circumference, is on the dark side of the earth. So, at this latitude, the length of daytime is less than nighttime. Similarly, in a summer solstice, the arc of the dotted circle on the dark side of the earth has chordal distance d. Figure 3(c) shows the view of the earth from a direction perpendicular to the plane determined by **n** and $E(\theta_w)$.

Figure 4 shows several elliptical orbits and possible positions for the winter solstice and spring equinox. Figure 4(a) depicts earth's orbit, with eccentricity $e \approx 0.017$ and the approximate location of the solstice and equinox with respect to earth's perihelion and aphelion. (The eccentricity of an ellipse is the ratio of the distance between its two foci and its major axial length. Since earth's orbit is almost circular, the ellipse's two foci are close together, making earth's eccentricity almost zero.)

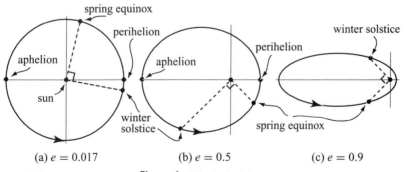

(a) $e = 0.017$ (b) $e = 0.5$ (c) $e = 0.9$

Figure 4. Elliptical orbits

Planet X's elliptical orbit has polar equation

$$r = \frac{h^2/k}{1 + e\cos\theta}, \qquad (9)$$

from (5), where $e = h^2/(k\rho) - 1$ is the orbital eccentricity.

Combining (2) and (9) gives

$$\frac{dt}{d\theta} = \frac{r^2}{h} = \frac{h^3/k^2}{(1+e\cos\theta)^2}. \qquad (10)$$

With $t = 0$ when $\theta = 0$, integrating gives

$$t(\theta) = \frac{h^3}{k^2(1-e^2)^{\frac{3}{2}}}\left(2\tan^{-1}\left(\sqrt{\frac{1-e}{1+e}}\tan\frac{\theta}{2}\right) - \frac{e\sqrt{1-e^2}\sin\theta}{1+e\cos\theta}\right), \qquad (11)$$

where $-\pi < \theta < \pi$ and $|e| < 1$, as outlined in Exercise 7. Exercise 8 shows that (11) can be written as

$$t(\theta) = \frac{h^3}{k^2(1-e^2)^{\frac{3}{2}}}\left(\theta - 2\tan^{-1}\left(\frac{e\sin\theta}{1+\sqrt{1-e^2}+e\cos\theta}\right) - \frac{e\sqrt{1-e^2}\sin\theta}{1+e\cos\theta}\right), \qquad (12)$$

for all θ.

Newton wrestled with the form of equation (12), stating it in Book I, Section 6, Proposition 31 of the *Principia* [**95**, pp. 513–517]:

> If a body moves in a given elliptical trajectory, to find its position at an assigned time.

After his geometrical construction in proposition 31, Newton concludes that "the description of this curve [(12)] is difficult," and he goes on to approximate position with respect to time, introducing what is now known as *Newton's method* for finding solutions to the general equation $f(x) = 0$ where f is smooth, starting with an initial approximate guess x_0. Furthermore, Colwell [**26**], chronicles the various approaches used since Kepler to approximate angle versus time as exemplified in (12). Standard methods for solving this problem involve calculating three angles called the *mean*, *eccentric*, and *true anomalies*.

Now we model the motion of a planet whose period is 365 earth days. To change (12) into a function that gives the day of the year for a given angle,

Newton's Polar Ellipse

let $T(\theta) = 365\, t(\theta)/t(2\pi)$ so that

$$T(\theta) = \frac{365}{2\pi}\left(\theta - 2\tan^{-1}\left(\frac{e\sin\theta}{1+\sqrt{1-e^2}+e\cos\theta}\right) - \frac{e\sqrt{1-e^2}\sin\theta}{1+e\cos\theta}\right). \quad (13)$$

That is, dividing $t(\theta)$ by $t(2\pi)$ normalizes the total time for one revolution of X about the sun, and multiplying by 365 gives the relation between θ and time in days. In effect, (13) sidesteps the need to find the values of h and k, giving us the day of the year at which X is at polar angle θ.

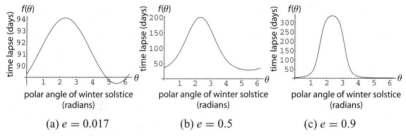

(a) $e = 0.017$ (b) $e = 0.5$ (c) $e = 0.9$

Figure 5. Time lapse from winter to spring, $f(\theta) = T(\theta + \pi/2) - T(\theta)$

A winter solstice occurs when **p** (the projection of planet X's axis onto the ecliptic plane) and $E(\theta)$ have the same direction. Thus if θ is the polar angle at which the winter solstice occurs, then the time lapse from winter solstice to spring equinox is $f(\theta) = T(\theta + \pi/2) - T(\theta)$. For a planet with orbital eccentricty 0.5 and period 365 days with winter solstice at 1 radian, the time lapse until spring equinox is $f(1) \approx 77$ days.

Plotting $f(\theta)$ for $0 \leq \theta \leq 2\pi$ shows the range of possible time lapses between winter and spring on planet X for an elliptical orbit with given eccentricity and period 365 days. On earth, with $e = 0.017$, Figure 5(a) shows that the possible time lapse from winter to spring varies from 88.5 days to 94.1 days. For $e = 0.5$, Figure 5(b) shows that the time lapse varies from 28.3 days to 199.6 days, and for $e = 0.9$, Figure 5(c) shows that the time lapse varies from 2.3 days to 336.1 days. The maximum time lapse occurs when the winter equinox is at $\theta = 3\pi/4$, since a planet moves slower in its orbit when it is farther from the sun.

Will the maximum ever occur for earth?

Yes, it will!

Why?

Table 1. Planetary data

	earth	Mars	Jupiter	Saturn
radius (a.u.)	1	1.524	5.203	9.540
period (years)	1	1.88	11.86	29.46
angle θ (radians)	−1.758	−1.335	−0.643	−0.435

The answer as we will see in Chapter VI, involves a flattened earth and the gravitational attraction by the moon and sun on the bulge at earth's equator.

Case study: Micromégas's tour of the solar system

For Micromégas's route through the solar system, Voltaire describes the comet's path from Saturn to Jupiter as 150 million leagues, from Jupiter to Mars as 100 million leagues (a league is about 3 miles), and from Mars to earth as "a very long time." How good were Voltaire's guesses?

To model Micromégas's route, we take a particular comet: Halley's Comet, which has a period of about 75 years, with aphelion 35 a.u. and perihelion 6/10 a.u., where an a.u. is an astronomical unit, the distance of the earth from the sun, about 93 million miles. Halley's comet has a *retrograde* motion about the sun (clockwise with respect to a view from Polaris, earth's current north star). For simplicity, we assume that the orbits of all the planets and the comet are in the same plane, even though the orbit of Halley's Comet is inclined about 18° to the plane of earth's orbit. We also assume that the planets have circular orbits about the sun with radii and periods as given in Table 1. (With these units, by Kepler's third law (8), the period of a body in its orbit is the 1.5 power of the semi-major axis of the body's orbit.) We furthermore ignore any planetary gravitational influence on the comet's trajectory.

We take the polar orbit of Halley's comet as

$$r = \frac{k}{1 - e\cos\theta}, \qquad (14)$$

where k is a constant, $e = c/a$ is the eccentricity of the ellipse, where c is the semi-focal length of the ellipse and a is the semi-major axis. The semi-minor axis, b, is given by $a^2 = b^2 + c^2$. Since $2a \approx 35.6$ and $a - c \approx 0.6$, then $a \approx 17.8$ a.u., $b \approx 4.58$ a.u., $c \approx 17.2$, $e = 0.996$, and $k \approx 1.180$ a.u.. The orbits of the planets and Halley's Comet are illustrated in Figure 6. When $\theta = 0$ at point B in the figure, Halley's Comet is beyond Neptune's orbit. With this information, we find the angles θ for which the comet will

Newton's Polar Ellipse

be at Saturn, Jupiter, Mars, and earth, as indicated in Table 1. The arclength along the orbit of (14) between angle α and β is

$$\int_\alpha^\beta \sqrt{r^2 + (r')^2}\, d\theta. \qquad (15)$$

By (15), the distances $\widehat{SJ}$ between Saturn and Jupiter, $\widehat{JM}$ between Jupiter and Mars, and $\widehat{ME}$ between Mars and earth are

$$\widehat{SJ} \approx 4.58 \text{ a.u.}, \quad \widehat{JM} \approx 4.17 \text{ a.u.}, \quad \text{and} \quad \widehat{ME} \approx 0.74 \text{ a.u.}.$$

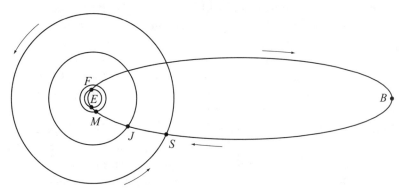

Figure 6. Orbits of earth, Mars, Jupiter, Saturn, and a comet

In miles, Voltaire says that $\widehat{SJ}$ and $\widehat{JM}$ should be 450 million miles and 300 million miles, respectively, whereas the above computations give $\widehat{SJ} \approx 426$ million and $\widehat{JM} \approx 388$ million miles. The ratio of Voltaire's two guesses is 1.5 whereas $\widehat{SJ}/\widehat{JM} \approx 1.10$. That is, Voltaire's guess for the Jupiter-Mars leg is too short. Relative to his guess of $\widehat{SJ}$ as 450 million miles, his guess for $\widehat{JM}$ should have been about 409 million miles.

Now let's consider Micromégas's time between these planets. How long does it take to go from Mars to earth when astride Halley's Comet? By Kepler's laws of planetary motion, the area swept out by a ray from the origin to the comet varies directly with time. The area A of an ellipse with semi-major axis a and semi-minor axis b is πab, which for Halley's comet is $A \approx 256.3$ a.u.^{2}. The area swept out along a polar curve between angles α and β is

$$\int_\alpha^\beta \frac{1}{2} r^2\, d\theta,$$

where r is given by (14). The time for the comet to travel from α to β is therefore

$$\frac{T}{2A} \int_\alpha^\beta r^2\, d\theta,$$

where T is the comet's period. Thus the time from Saturn to Jupiter is 1.6 years, from Jupiter to Mars is 0.9 years, and from Mars to earth is 0.1 years or about 34.5 days. That is, Voltaire's description of the travel time from Mars to earth as "very long," especially when compared to the transit times from Saturn to Jupiter and from Jupiter to Mars, is inflated. Of course, Micromégas could rendezvous with the earth at F rather than E in Figure 6, when the comet has already whipped around the sun. In this case the time from Mars to earth is nearly four months.

Did any comet pass near earth in 1737? Yes: Comet Kegler of 1737 was first observed by Ignatius Kegler (1680–1746), a Jesuit missionary in Beijing, China. The comet was later identified by Harvard-Smithsonian astronomer Brian Marsden (b. 1937) as being the same as Comet Swift-Tuttle-$109P$, named after amateur astronomer Lewis Swift and Harvard astronomer Horace Tuttle who observed it in 1862. Not long after, the Italian astronomer Giovanni Schiaparelli (1835–1910) determined that Comet Swift-Tuttle, as is indicated by its alpha-numeric code $109P$, is part of what makes up the Perseid showers, a meteor swarm through which earth passes each August 11-12. Indeed, the Perseid showers are debris that Comet Swift-Tuttle loses whenever it is near the sun. In the early 1980's, with the appearance of the catastrophe novel *Lucifer's Hammer* by Larry Niven and Jerry Pournelle in 1977 describing a modern-day earth-comet impact and with the realization that Kegler 1737, alias Swift-Tuttle, of projected period 120 years, is part of the Perseid showers, concern grew in the popular media about an impending collision. Where was this comet? Its arrival was imminent. It was finally spotted in 1992 and missed earth by about 110 million miles. We now know it is six miles in diameter so Micromégas would have little room to sit on it (Halley's Comet, whose long diameter is 10 miles, is not much roomier). At present, Comet Kegler has period 130 years, eccentricity 0.964, inclination 113.4°, which means it has a retrograde orbit inclined to earth's at 66.6°. According to Marsden [**134**], Comet Kegler's next return in 2126 will miss earth by two weeks and 15 million miles, while in 3044 AD, unless its orbit alters, Comet Micromégas-Kegler-Swift-Tuttle will come within a million miles of earth.

Newton's Polar Ellipse

Figure 7. Close-up of a comet, courtesy of NASA Photo by the Cassini spacecraft flyby of Saturn in 2004

Figure 7 is a close-up photo of Phoebe, a moon of Saturn. Since it has a retrograde orbit, it is thought that Phoebe once was a comet. Phoebe has diameter of about 200 km, large enough to have accommodated Micromégas and his retinue on his tour of the solar system.

Finally, one of the reasons that Voltaire wrote Micromégas—featuring a comet as the preferred mode of transport in the solar system—was the growing anticipation about Edmond Halley's prediction of 1682 about his comet. Talk at Frederick's court during the tenure of Voltaire as poet-in-residence would surely have included speculation of the predicted return. However, Halley's date came and went. Finally, the comet was spotted on Christmas day in 1758; it reached its perihelion in March, 618 days late. Saturn and Jupiter were to blame. Their gravitational influence, as further discussed in Exercise 4, changed the trajectory of the great comet, as anticipated by Alexis Clairaut.

Exercises

1. For an orbital period of 365 days, find the orbital eccentricity for which the time lapse between winter and spring is 150 days, given that the winter solstice occurs at angle $\theta = 0$.

2. Find the maximum possible time lapse between winter and spring on Jupiter.

3. Comet Kirch, discovered in 1680, is a sun-grazing comet. Around November of that year, it could be seen during the day. Its perihelion is 0.00622 a.u. and it has a period of about 9356 years. In 1681, Jacob Bernoulli (1655–1705) in his first mathematical paper predicted that Comet Kirch would return in May 17, 1719, and destroy the earth.

   (a) How far above the surface of the sun did this comet approach?

   (b) What is the comet's aphelion?

4. Edmond Halley predicted that the comet of 1682 would return in 1757. Meanwhile, Clairaut determined that in passing by Jupiter and Saturn the comet would slow down. He revised the arrival date prediction to 1758, missing the exact perihelion date by about a month. To wrestle with this 4-body gravity problem, Clairaut used some of the same ideas he used in his 3-body (the sun, earth, and moon) solution in the letter to Euler mentioned in Vignette I. Solve this much simpler problem: Imagine a clone of Comet Halley, Comet Galley: perihelion 0.586 a.u., aphelion 35.1 a.u., and period 75.3 years, with trajectory in the ecliptic plane.

   (a) Suppose that Comet Galley's last aphelion occurred on January 1, 2000. Without being influenced by the gravity of any planet, at what date (call it Q) will it cross Jupiter's orbit? (Let Z be the ideal point at which Comet Galley intersects Jupiter's orbit.)

   (b) Now suppose that at time Q of part (a) Jupiter is 10 million miles west of Z along its orbit. Estimate Jupiter's influence on Comet Galley's orbit. That is, recalculate its aphelion, perihelion, and period after it goes beyond Jupiter's orbit. Assume that Comet Galley's influence on Jupiter is negligible.

5. In 1456, Comet Halley passed near earth. Its tail arced in the form of a sabre over about 60° of the night sky. It is estimated that its tail near perihelion was about 55 million km long. Assuming half this tail length at Halley's closest approach to the earth, determine how close Comet Halley approached earth.

6. In 837 AD, Comet Halley's tail covered 90° of the night sky. Its closest approach to earth was 3.2 million miles. On this basis, estimate the length of the tail and compare with the estimate given in Exercise 5.

Newton's Polar Ellipse

7. **Cracking an integral.** Follow the outline below to derive (11). Use the substitution $z = \tan(\theta/2)$ to simplify

$$\int \frac{1}{(1+e\cos\theta)^2}\,d\theta, \tag{16}$$

with $-1 < e < 1$.

(a) Show that

$$\theta = 2\tan^{-1} z, \quad d\theta = \frac{2}{1+z^2}\,dz, \quad \sin\theta = \frac{2z}{1+z^2},$$

$$\text{and } \cos\theta = \frac{1-z^2}{1+z^2}.$$

(b) Use (a) to show that (16) is equivalent to

$$\frac{2}{(1+e)^2}\int \frac{1+z^2}{(1+\alpha^2 z^2)^2}\,dz,$$

where α is a positive number with $\alpha^2 = (1-e)/(1+e)$.

(c) Show that

$$\alpha^2 + 1 = \frac{2}{1+e}, \quad \alpha^2 - 1 = -\frac{2e}{1+e}, \quad \text{and } \frac{1+\alpha^2 z^2}{1+z^2} = \frac{1}{1+e}(1+e\cos\theta).$$

(d) With $\tan\omega = \alpha z$, show that

$$dz = \frac{1}{\alpha}\sec^2\omega\,d\omega \quad \text{and} \quad \sin 2\omega = 2\sin\omega\cos\omega = \frac{2\alpha z}{1+\alpha^2 z^2}.$$

(e) Use (b) and (d) to show that (16) is equivalent to

$$\frac{2}{\alpha(1+e)^2}\int \cos^2\omega + \frac{1}{\alpha^2}\sin^2\omega\,d\omega.$$

(f) Use the double angle formulas for $\sin^2\omega$ and $\cos^2\omega$ and (c) to show that the integral in (e) becomes

$$\frac{2}{\alpha^3(1+e)^3}\int 1 - e\cos 2\omega\,d\omega.$$

(g) Use the definition of α in (b) and perform the integration in (f) to get

$$\frac{1}{(1-e^2)^{\frac{3}{2}}}(2\omega - e\sin 2\omega) + C,$$

where C is a constant of integration.

(h) Use (d) and (g) to show that (16) is equivalent to

$$\frac{1}{(1-e^2)^{\frac{3}{2}}}\left(2\tan^{-1}(\alpha z) - \alpha e \frac{2z/(1+z^2)}{(1+\alpha^2 z^2)/(1+z^2)}\right) + C,$$

which in terms of the original variable θ, using parts (a) and (c), is

$$\frac{1}{(1-e^2)^{\frac{3}{2}}}\left(2\tan^{-1}\left(\sqrt{\frac{1-e}{1+e}}\tan\frac{\theta}{2}\right) - e\sqrt{1-e^2}\frac{\sin\theta}{1+e\cos\theta}\right) + C.$$

Since $t = 0$ when $\theta = 0$ in (11), then $C = 0$, thus giving the desired result.

8. **Extending an identity.** To show that (11) can be extended to (12):

   (a) Show that the expression $\tan^{-1}(c\tan\beta)$ can be rewritten as

   $$\tan^{-1}(c\tan\beta) = \beta + \tan^{-1}\left(\frac{(c-1)\sin(2\beta)}{(1+c) + (1-c)\cos(2\beta)}\right). \quad (17)$$

   (b) Use part (a) with $\beta = \theta/2$ and $c = \sqrt{(1-e)/(1+e)}$ to show that

   $$\tan^{-1}\left(\sqrt{\frac{1-e}{1+e}}\tan\frac{\theta}{2}\right) = \frac{\theta}{2} - \tan^{-1}\left(\frac{\sin\theta}{(1+\sqrt{1-e^2})/e + \cos\theta}\right).$$

   (c) Now use part (b) to show that (11) can be extended to (12).

9. As variations on Micromégas's tour problem, find the distances and time lapses between Saturn, Jupiter, Mars, and earth along trajectories of

   (a) Halley's Comet with perihelion: 3/10 a.u. and aphelion: 40 a.u.,
   (b) Comet Kegler (rotated into the equatorial plane).

10. In Voltaire's story, once Micromégas reaches Jupiter he remains there for a year before continuing on to skirt Mars and to land on earth. Imagine that upon arriving at Jupiter, he finds that the sun, earth, Mars, and Jupiter are in conjunction (all in a line). Suppose also that he can conjure a Halley's Comet clone at a moment's notice so asto catch a

ride toward the sun. Define a rendezvous with Mars or earth as coming within ten thousand miles of the planet's center.

(a) How long should Micromégas spend on Jupiter before setting off to rendezvous with both Mars and earth?

(b) Answer part (a) if the comet's orbit is counterclockwise (as viewed from Polaris) rather than clockwise.

(c) Compare the answers to parts (a) and (b). Are they the same?

Vignette V: A Bourgeois Poet in the Temple of Taste

From early on, Voltaire wished to be a poet. His father, like many fathers, was skeptical, wondering how he would be able to support himself. The stereotype of the starving artist has a long tradition even as Voltaire later commented,

> [I had] seen so many men of letters who were poor and held in contempt that long ago [I] decided not to increase their number. [**101**, p. 87]

Although his father insisted that he attend law school, Voltaire continued to write. At age 22, his play *Œdipe*—a retelling of the Oedipus story—was produced by the Comédie-Française, the proceeds from which were 3000 livres, twelve times the annual wage of a common laborer at this time, 250 livres. Voltaire's father was suitably impressed, and gave Voltaire his reluctant blessing. Although his father died the next year, his will prevented any share of the estate from being transferred to Voltaire until his thirty-fifth birthday. If Voltaire was to survive in the literary world, he would do so on his own.

Voltaire's financial low occurred during a two-year exile to England, (1726–1728). His credit evaporated. Fortunately, various kind souls opened their homes to him. Nevertheless, some say that that one reason Voltaire subsequently left England so abruptly was that had he remained any longer in London, in the words of the English poet Thomas Gray, "he would have been hanged for forging banknotes" [**101**, p. 84].

His luck improved upon returning to France in 1728. As mentioned in Vignette IV, he joined with La Condamine and others in playing a flawed state lottery, ultimately realizing winnings totaling about a half million livres. He invested in importing and exporting grain, and in supplying the French army with material. As the years went on, he resorted to granting loans as annuities; while he was alive, the borrower agreed to pay him 10% of the principal per year; if and when Voltaire died, the interest rate would fall to a fraction of the original rate. Since Voltaire had a reputation of being

Figure 1. Voltaire on a ten franc note of 1966

continually ill—in fact, he was a hypochondriac—many strapped-for-cash noblemen took advantage of these attractive terms, and wound up ever more disheartened as the years passed with Voltaire yet breathing. In the last twenty-some years of his life, Voltaire established a quasi-utopian society at Ferney on the eastern border of France in the Alps, fostering a watch-making community; there, like the hero at the end of *Candide*, he tended his own garden. Figure 1 shows Voltaire's image on the French ten franc note, which was printed from 1962 to 1979; his father would have been very proud.

Meanwhile the successful businessman continued his word magic. In his 2005 biography of Voltaire, Pearson [**101**, p. 32] estimates that Voltaire wrote about fifteen million words. Play followed play with epic poems and histories interspersed. He had many fans. Not the least of these was Frederick the Great, who, in one of his letters to Voltaire, appraises his style [**144**, vol. xxxvi, p. 96]:

> I sometimes am vexed at the puerilities, the trivial remarks, and the dry style of certain books. You spare your readers that trouble. Pray, my dear friend, tell me how you pass your time at Cirey [the home of Emily and Voltaire].

Voltaire answered in verse [**144**, vol. xxxvi, *The Answer*, pp. 98–99]:

> You ask me, and I'll tell in rhyme,
> How we at Cirey pass our time:
> What need I to you this relate?
> Our master, you we imitate:

> Such a comparison may show
> That some philosophy I know,
> That I've read Newton and Kirkherus,
> Authors both learned, profound, and serious.
>
> Let others in their lyric lays
> Say the same thing a thousand ways,
> The world with ancient fables tire,
> I new and striking truths admire.
>
> This rhapsody, great prince, excuse,
> 'Tis but the folly of my muse,
> Judgment was from my breast expelled,
> For fair Emilia I beheld.

In this poem, the name Kirkherus might refer to the Jesuit scholar Athanasius Kircher (1602–1680), whom Voltaire describes as "one of the greatest mathematicians and most learned men of his age" [**142**, p. 150]. As one of the biographers of this curator of papal museums and this author of over thirty diverse, encyclopedic books says, "the single most memorable fact about Kircher was that no one could spell his name correctly" [**45**, p. 10]. Apparently, Voltaire added to the list of spelling variations.

Not everyone who read Voltaire's work applauded, of course.

Voltaire's wit often prompted reprisals, raising controversy which more often than not prompted great sales. For in the clandestine publishing and distribution arrangements for books and pamphlets, Voltaire was a master. Even as a poet, Voltaire was a businessman.

For example, when Voltaire said of Jean-Baptiste Rousseau's "Ode to Posterity" that "it was unlikely to reach its destination," Rousseau lambasted Voltaire's currently running plays, which prompted Voltaire to write and publish *The Temple of Taste*, a mixture of poetry and narrative in which Voltaire raises the question of what constitutes good literature. That is, what works from "writers of every rank, age, and condition, scratching at the door and begging of Criticism to permit them to enter" are allowed entrance into the Temple of Taste? As may be suspected, Rousseau is at first denied entrance, but then enters under the proviso that his work be condensed and rewritten. Voltaire's *The Temple of Taste* is somewhat like Dante's *The Divine Comedy* in that Dante peoples heaven and hell with his heroes and enemies. The Temple of Taste is also like the mythic idea in the mathematical community of *The Book:* an ideal book that contains the best of mathematical proofs. A legendary itinerant mathematician of the last century, Paul Erdős, spoke often and fondly about such a book. In his memory, two of his friends produced a first draft, first hard-copy installment of this ideal book, calling it *Proofs from the Book* [4], which we commend to the reader.

Voltaire concludes *The Temple of Taste* with a helpful observation for himself and anyone else who converses in the marketplace of ideas [**144**, vol. xxxvi, *The Temple of Taste*, p. 40–69]:

> I then found that the God of Taste is very hard to be pleased, but that he is never pleased by halves. I perceived that the works which he criticizes the most are those which he likes best.

CHAPTER V

A Mandarin Orange or a Lemon?

Aristotle had imagined that the earth was motionless, and argued that its shape was a sphere: since everything is attracted to the earth's center and since this attraction is the same in all directions, the earth must be radially symmetric, namely a sphere [**10**, p. 251]. For two millennia, people believed that Aristotle had gotten it right and his proof belonged in *The Book*, or better yet, Archimedes' improved version of the proof from *On Floating Bodies*.

In *The Temple of Taste* when Voltaire satirizes scholastic thought, he depicts the sages of yesteryear as "poring over Greek authors" far from the Temple of Taste. Passing by, to them he hailed,

> "To Taste's famed Temple do you bend?"
>
> "No, sir, we no such thing intend.
> What others have with care expressed,
> With accuracy we digest,
> On others' thoughts we spend our ink,
> But we for our part never think."

Against these traditions, Copernicus, Galileo, and Kepler argued for a heliocentric solar system and a rotating earth. In such a system, some mechanism must exist to keep the planets in their orbits. Would earth's rotation affect its shape? Had Aristotle's brilliant proof been a mere special case of a more complicated question? From the 1504 title page of *De scientia motus orbis* by Masha'allah Ibn-Athari (c. 740–815), a reference book for medieval astrologers, Figure 2 shows an old woodcut of an astronomer seemingly wrestling with the question: what is the shape of our globe?

René Descartes proposed a system of vortices to keep the planets moving in their elliptical orbits. Voltaire, in his fifteenth philosophical letter from England, explains Descartes' idea this way:

121

Figure 2. Measuring heaven and earth (1504), woodcut attributed to Dürer [**38**, cut 171]

This is the cause of gravity according to the Cartesian system: All space is filled with a vast whirlpool of subtle matter in which the planets are carried around the sun; another vortex floats in the great one, and turns daily round the planets. When all this is done, it is pretended that gravity depends on this diurnal motion.

In such a system, the heavenly bodies should be shaped like lemons, squeezed into elongated spheroids by the whirlwinds of vortices.

When Newton proposed that gravity arose from attractions between each bit of matter and every other bit of matter in the universe, he saw that it also determined the shape of a rotating earth. He said that earth's rotation would lead to a flattening at the poles, to a shape like a mandarin orange, and estimated earth's degree of flattening in proposition 19, book iii of the *Principia*. He claimed that the ratio of the diameter of earth's pole to a diameter at the equator was 229 to 230 [**95**, p. 824].

Figure 3 is a cartoon characterizing this debate, with the lemon camp labeled Cassini rather than Descartes since the leading French proponents of Descartes' vortices in Newton's day were Giovanni and Jacques Cassini, both accomplished astronomers.

A Mandarin Orange or a Lemon?

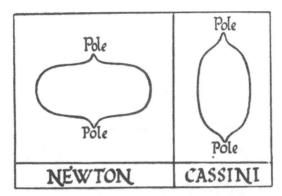

Figure 3. "An old-time caricature of the controversy" [24, p. 3]

Who was right?

Almost every member of the French Academy of Sciences favored Descartes' model even as late as 1736, nearly fifty years after the *Principia* appeared. Perhaps they were being nationalistic, opting to follow a French philosopher rather than an English one, as Voltaire suggested in his *Elements* [142, p. 103],

> Is it because they are born in France, that they are ashamed of receiving Truth at the Hands of an Englishman? Such a thought would ill become a Philosopher.

But the members of the French Academy of Sciences intuitively reasoned thusly: from measurements taken around the globe, weight is less at the equator; therefore at the equator there must be less earth between the surface and the center than at the poles. Besides, measurements taken by Giovanni and his son Jacques Cassini (when Jacques was 23) along a line of longitude through France seemed to support the French conclusion. However, there was evidence to the contrary. The telescope showed that Jupiter was flattened at her poles; in 1691 Giovanni Cassini estimated its flattening ratio as seven percent [85, p. 58]. (See Exercise 7.) Some comets, such as Halley's Comet, follow a retrograde (clockwise) orbit, clearly at odds with swirling counterclockwise Cartesian vortices about the sun. To settle the matter once and for all, the Academy proposed an expedition to the equator to measure the arclength along a line of longitude near Quito in South America.

Some were less than enthusiastic. Here is Johann Bernoulli's opinion in a letter to Maupertuis:

> Tell me, do the observers have a predilection for one or the other of the two sentiments? Because if they favor the flattened earth, they will

find it flattened; if on the contrary, they are imbued with the idea of the elongated earth, their observations will not fail to confirm its elongation; the difference between the compressed spheroid and the elongated is so slight, that it is easy to be mistaken if one wants to be mistaken in favor of one or the other opinion. [**140**, pp. 94–95]

Figure 4. Maupertuis, honored by Finland

Nevertheless, Louis XV authorized the trip. Louis Godin, Pierre Bouguer, Charles-Marie de La Condamine, and a supporting crew of technicians set off for the Caribbean. Within a year they were in what is now Ecuador and taking measurements. Meanwhile in Paris, Maupertuis (whose image is on the Finnish stamp of Figure 4) lobbied for a second expedition to go north and take measurements at the arctic circle. A definitive test to determine whether the earth was shaped like an orange or a lemon needed data from both regions. The Academy agreed, as did the King. Maupertuis recruited Alexis Clairaut, Réginald Outhier, Anders Celsius, and others to join him. Voltaire nearly volunteered, saying that he "would like to be the poet of the expedition, but it is too cold there" [**54**, p. 651]. By the summer of 1736, the expedition set sail in a fishing boat called the *Prudent*. They stopped in Stockholm on the way north to apply for authorization from Swedish King Frederick I (1676–1751) to survey a degree of his territory. With Sweden being mostly Lutheran, the French delegation sought a special provision so that Outhier, a priest as well as a savant, could lead Catholic mass. The King granted them dispensation to do so but stipulated that anyone except Frenchmen at mass would be fined 1500 dahlers [**68**, p. 105].

If Newton was correct, the measure of one degree of arclength along the earth near the north pole should be greater than near the equator. Figure 5

A Mandarin Orange or a Lemon?

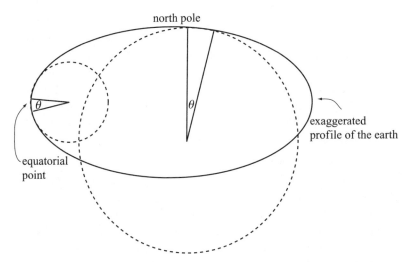

Figure 5. Arc length near a pole and the equator for the same central angle

illustrates this phenomenon. Let θ be a small angle in radians. To approximate the radius of a circle or of a circular-like curve at a point P, construct two normals, perhaps with a quadrant much like the one on the stamp of Figure 4, along the curve's arc near P which intersect in an angle of measure θ, and consider the ratio of subtended arc to θ, so getting an estimate of the radius of the circle best approximating the curve at P. Thus the radius of the (dotted) circle approximating the curve of the earth near the pole is more than the radius of the (dotted) circle approximating the curve of the earth at the equator, which in turn means that the measure of θ radians of arc near the pole exceeds that of the measure of θ radians of arc near the equator. See Exercise 9 for a more formal presentation of the notion of the curvature of a curve.

In Lapland at 66° latitude, the polar team identified two points one degree apart along a line of longitude: a church steeple in the town of Tornio on the Gulf of Bothnia, and the mountain top of Kittisvaara, about 100 km north. Between the two points they formed a network of eight triangles. The map of Figure 6 shows the southernmost triangle of this network, with vertex in Tornio. Two angles in each triangle of the network had to be measured; as a consistency check, the third angles were measured as well. The altitude at each vertex was determined by barometer. One edge of the network needed to be measured so as to determine the lengths of all the sides, from which they could then deduce the distance between the church steeple and the mountain top. Battling mosquitoes in summer and enduring cold in winter, Maupertuis's team completed their measurements within one year. The triangular leg they chose to measure came from the southernmost triangle, and ran mostly along a river bed, as can be seen in Figure 6(a). When the river froze over, the team measured it without overmuch cutting of forest paths and of building risers to accommodate going up and down hills and valleys. Returning to Paris a year after he had left, Maupertuis published his report in 1738, from which we sample his commentary on the arctic cold:

> You may imagine what it is to walk in two feet of snow, carrying heavy measuring sticks [thirty-foot-long spruce poles], which must be continually set down in the snow and retrieved. All this in a cold so great that when we tried to drink *eau de vie*, the only drink that could be kept liquid, the tongue and lips froze instantly against the cup and could only be torn away bleeding. A cold so great that it froze the fingers of some of us [**140**, p. 123].

The southern expedition was less lucky. Attaining their objective of measuring three degrees took eight years. Whereas Maupertius was the clear leader of the arctic expedition, the southern expedition was at times of three different minds. In fact, after the equatorial team completed their measurements, each member returned to France separately. La Condamine journeyed down the Amazon River for eight months as the first leg of his trip home, collecting plant specimens and mapping the territory; Bouguer recrossed Panama, from where he took passage to Paris; and Godin, delayed by personal debts in Lima, Peru, returned to Paris sixteen years after he had left [**151**, pp. 164, 184, 287].

We defer the expeditions' results to the end of the chapter. First we show how Newton's calculus and gravity model predict a flattened earth.

Newton's estimate

To reach Newton's conclusion, we imagine that earth consists of an incompressible fluid, such as ideal water. A non-spinning earth will be a homogeneous spherical ball with radial gravity function $f(r) = -gr/R$ where r is the distance from the earth's center, and R is the radius of the ball. We use this function in our model for the shape of the earth as an approximation to the oblate spheroid's actual gravity function.

For a spinning earth, the acceleration on any surface point mass is the sum of gravitational acceleration and centripetal acceleration. Let R be the radius of the earth at the pole, g be the acceleration due to gravity at the north pole,

A Mandarin Orange or a Lemon?

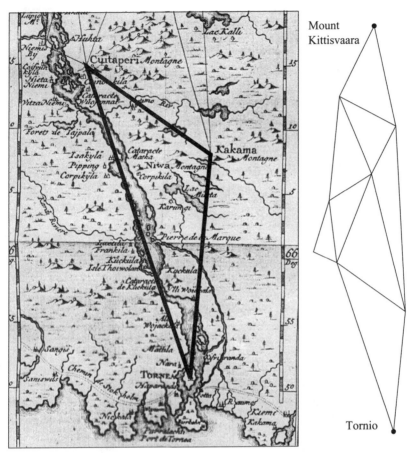

(a) The southernmost triangle, map adapted from Outhier's journal [99]

(b) The entire network

Figure 6. The triangular network for the polar expedition

and T be the period of the earth. Let r be the radius of the earth at the equator. The effective acceleration on a surface point mass at the equator is

$$\frac{gr}{R} - r\left(\frac{2\pi}{T}\right)^2. \qquad (1)$$

As a second approximation, with the shape of the watery earth in equilibrium we will assume that the effective acceleration on any surface point mass is g—for we naïvely expect that water should flow so that the effective gravity at the surface is a constant. (In actuality, the value Q of the effective acceleration at the equator for our earth is slightly less than g as discussed in Exercise 1. But imagine that Q is unknown for our homogeneous earth model. We take Q as the best guess available, namely, g.) With $R = 6400$ km, $g = 9.8$ m/sec^2, and earth period $T = 1$ day, equating g to (1) and solving for r gives $r - R \approx 22.2$ km. What is remarkable is that this differs from the actual physical value by about 0.8 km. Now we determine how much of this value is a flattening at the poles and how much is a bulging at the equator.

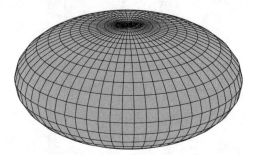

Figure 7. Earth's shape corresponding to a period of 2 hours

To do this, we generalize (1) to be valid at any angle ϕ, equate it to g, and solve for $r(\phi)$ as a function in terms of ϕ. At angle ϕ, where $\phi = 0$ corresponds to the equator and $\phi = \pi/2$ corresponds to the north pole, (1) becomes

$$g = \left\| \frac{g r(\phi)}{R}(\cos\phi,\ \sin\phi) - r(\phi)\left(\frac{4\pi^2}{T^2}\right)(\cos\phi,\ 0) \right\|,$$

when equated with g, since gravity is directed towards earth's center whereas centripetal acceleration is direction away from the vertical axis. Solving for r gives

$$r(\phi) = \frac{g}{\sqrt{\left(\frac{g}{R} - \frac{4\pi^2}{T^2}\right)^2 \cos^2\phi + \frac{g^2}{R^2}\sin^2\phi}}. \qquad (2)$$

With $R = 6400$ km, $g = 9.8$ m/sec^2, and $T = 1$ day, the polar graph of (2) reduced to this page-size would be indistinguishable from a circle. However, if the earth were to rotate once every two hours, the acceleration dynamics in this model would result in the squashed oval as shown in Figure 7. Exercise 2 explores some gravity function issues involved in pushing this model to such periods.

We assume that the polar radius is R, the equatorial radius is r_0, and the volume of the earth is V. We then find the radius ρ of the sphere with volume V. Thus the earth should be flattened by $\rho - R$ at the poles and bulging at the equator by $r - \rho$. The next section gives a convenient way of determining volumes of solids whose surfaces arise from the rotation of a polar curve about the vertical axis.

Volumes of revolution, a polar version

Let $S(r, \theta)$ be a sector of radial length r and central angle θ at O, the center of the circle from which the sector has been cut. We say that the *central axis* of a sector is along the ray from O that bisects it into congruent sectors. We find where the centroid of $S(r, \theta)$ lies as θ gets small. The sector of radial length r and central angle θ when θ is small can be approximated both by the triangle having the same three vertices as the sector and by the larger similar triangle whose side opposite θ is tangent to the sector's arc, as shown in Figure 8.

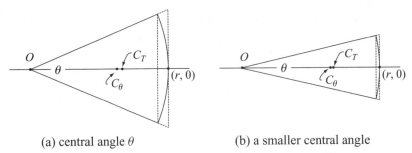

(a) central angle θ (b) a smaller central angle

Figure 8. The centroid C of a sector as its central angle collapses

Position the sector so that its central axis lies along the positive x-axis with O at the origin. The larger approximating triangle has its centroid C_T at $(2r/3, 0)$, as follows from geometry (see Exercise 3). Let $C_{(r,\theta)}$ be the centroid of sector $S(r, \theta)$ and $\hat{C}_{(r,\theta)}$ be the centroid of the smaller approximating triangle. It is clear that when $\theta \to 0$, then $\hat{C}_{(r,\theta)} \to C_T$. Since the sector's centroid lies between $\hat{C}_{(r,\theta)}$ and C_T for any θ, $0 < \theta < \pi$, then $C_{(r,\theta)} \to (2r/3, 0)$ when $\theta \to 0$. That is, the centroid of a sector of radial length r with infinitesimal central angle is at a point two-thirds of the way out along its central axis.

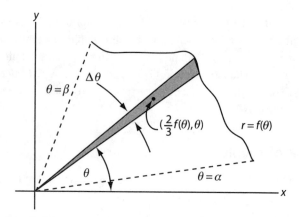

Figure 9. A sector of radial length $f(\theta)$ and central angle $\Delta\theta$

Consider the polar curve $r = f(\theta)$ on the interval $\alpha \leq \theta \leq \beta$, with $0 < \beta - \alpha < 2\pi$, and $r \geq 0$ for all θ. Let Q be the region bounded by the rays $\theta = \alpha$ and $\theta = \beta$ and by the path as given by f. Let P be the solid obtained by rotating Q about the y-axis. By Pappus' theorem in Item 4 of the appendix, the volume of the solid obtained by rotating the infinitesimal sector of radial length $f(\theta)$ and central angle $d\theta$, as illustrated in Figure 9, is the product of its area and of 2π times the distance of its centroid from the y-axis, which is $2\pi(\frac{2}{3}f(\theta)\cos\theta)(\frac{1}{2}f^2(\theta))\,d\theta = \frac{2}{3}\pi f^3(\theta)\cos\theta\,d\theta$. Therefore the volume of the solid of revolution P is

$$\int_\alpha^\beta \frac{2}{3}\pi f^3(\theta) \cos\theta\, d\theta. \qquad (3)$$

Example 1. We verify (3) for finding the volume of a sphere of radius R. Its polar equation is $f(\theta) = R$. With $\alpha = -\pi/2$ and $\beta = \pi/2$, (3) is

$$\int_{-\pi/2}^{\pi/2} \frac{2\pi R^3}{3} \cos\theta\, d\theta = \frac{4}{3}\pi R^3,$$

as it should be.

Example 2. We verify (3) for finding the volume of the solid of revolution obtained by spinning the right-most petal Q of the polar flower $r = \cos(2\theta)$, as displayed in Figure 10(a), about the y-axis. The petal Q is traced out by

A Mandarin Orange or a Lemon?

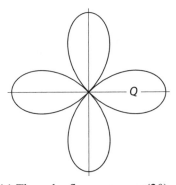

(a) The polar flower $r = \cos(2\theta)$ (b) The solid of revolution of petal Q

Figure 10. Rotating the polar flower

f as θ varies over the interval $[-\pi/4,\ \pi/4]$. By (3), the volume of the solid shown in Figure 10.b is

$$\int_{-\pi/4}^{\pi/4} \frac{2}{3}\pi \cos^3(2\theta) \cos\theta\, d\theta = \frac{32\sqrt{2}\,\pi}{105}.$$

We contrast this with the standard technique using the parametrization $(x(\theta), y(\theta)) = (\cos(2\theta)\cos\theta,\ \cos(2\theta)\sin\theta)$. The volume of the solid of revolution of Q about the y-axis is the integral

$$\int_0^1 2\pi x(2y)\, dy = \int_0^{\pi/4} 2\pi \Big(\cos(2\theta)\cos\theta\Big)\Big(\cos(2\theta)\sin\theta\Big)$$
$$\Big(2\sin(2\theta)\cos\theta + \cos(2\theta)\sin\theta\Big)\, d\theta,$$

since $dy/d\theta = -(2\sin(2\theta)\cos\theta + \cos(2\theta)\sin\theta)$. Its value is also $32\sqrt{2}\pi/105$. As this example illustrates, (3) is an efficient way to compute volumes of revolution if the bounding curve of the rotated region has a convenient polar parametrization.

The flattening and bulging components

The volume V of the earth is given by

$$V = 2\int_0^{\pi/2} \frac{2\pi}{3} r^3(\phi) \cos\phi\, d\phi,$$

where $r(\phi)$ is given by (2). As in Exercise 5,

$$\int \frac{\cos\phi}{(a\cos^2\phi + b\sin^2\phi)^{3/2}}\,d\phi = \frac{\sin\phi}{a\sqrt{a\cos^2\phi + b\sin^2\phi}} + C \quad (4)$$

where C is a general constant and $0 < a \le b$, which means that

$$\int_0^{\pi/2} \frac{\cos\phi}{(a\cos^2\phi + b\sin^2\phi)^{3/2}}\,d\phi = \frac{1}{a\sqrt{b}}.$$

With $a = (g/R - 4\pi^2/T^2)^2$ and $b = g^2/R^2$,

$$V = \frac{4\pi g^3}{3}\left(\frac{1}{a\sqrt{b}}\right) = \frac{4\pi}{3}\left(\frac{g^2 R}{\left(\frac{g}{R} - \frac{4\pi^2}{T^2}\right)^2}\right) = \frac{4\pi}{3}\rho^3,$$

where ρ is the radius of the sphere with volume V. Thus

$$\rho = \frac{g^{2/3} R^{1/3}}{\left(\frac{g}{R} - \frac{4\pi^2}{T^2}\right)^{2/3}}. \quad (5)$$

From satellite measurements, it has been determined that $R = 6356.8$ km, [**81**, p. 46]. Thus, from (2), the equatorial radius r of the earth is $r = r(0) \approx 6378.7$ km, (whereas satellite measurements give 6378.2 km). By (5), $\rho \approx 6371.4$ km, which means that, according to this model, the earth is flattened by $\rho - R \approx 14.8$ km and bulges at the equator by $r - \rho \approx 7.3$ km, remarkably close to the actual values of 14.2 km and 7.1 km.

Estimating the arclength along a meridian

Now we use (2) to estimate the length corresponding to one degree of arc along the earth's surface at the arctic circle and at the equator. First we must find the latitude $L(\phi)$ associated with angle ϕ. The slope of the tangent line to the polar curve of $r(\phi)$ from (2) is

$$\frac{dy}{dx} = \frac{dy/d\phi}{dx/d\phi} = \frac{r'(\phi)\sin\phi + r(\phi)\cos\phi}{r'(\phi)\cos\phi - r(\phi)\sin\phi},$$

which means that the slope of a normal to the earth's surface corresponding to angle ϕ is $-dx/dy$, which in turn means that the latitude $L(\phi)$ corresponding to angle ϕ is

$$L(\phi) = -\tan^{-1}\left(\frac{r'(\phi)\cos\phi - r(\phi)\sin\phi}{r'(\phi)\sin\phi + r(\phi)\cos\phi}\right). \quad (6)$$

Since the flattening of the earth is small compared with the radius of the earth, ϕ and $L(\phi)$ will be close to one another. For example, with $\phi = 1$ radian, the corresponding latitude is 1.00312 radians. Figure 11 shows the graph of $L(\phi)$ versus the identity function, with radian measurements converted to degree measurements, near the arctic circle at latitude 66.5° N. Let $\alpha_1 = 66\pi/180 \approx 1.15192$, $\alpha_2 = 67\pi/180 \approx 1.16937$, and $\alpha_3 = 0.5\pi/180 \approx 0.00872665$, the latitudes corresponding to 66°, 67° and 0.5° N, respectively. With the use of a computer algebra system, these latitudes correspond to angles $\phi_1 \approx 1.14934$, $\phi_2 \approx 1.16688$, and $\phi_3 \approx 0.0086688$, respectively, that is, $L(\phi_i) = \alpha_i$, with $1 \le i \le 3$.

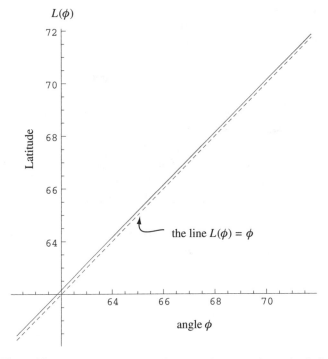

Figure 11. The latitude corresponding to angle ϕ near the arctic circle

With respect to our model, the distance M (for Maupertuis) along one degree of arc at earth's surface from latitude α_1 to α_2—at the arctic circle—is

$$M = \int_{\phi_1}^{\phi_2} \sqrt{r^2(\phi) + (r'(\phi))^2}\, d\phi \approx 111.592 \text{ km},$$

whereas the distance B (for Bouguer) along one degree of arc at the equator from latitude $-\alpha_3$ to α_3 is

$$B = 2 \int_0^{\phi_3} \sqrt{r^2(\phi) + (r'(\phi))^2} \, d\phi \approx 110.567 \text{ km}.$$

The results of Maupertuis and Bouguer

Maupertuis's polar team concluded that one degree of longitude in Lapland measures 111.86 km, whereas Bouguer's equatorial team concluded that one degree of longitude near Quito measures $\hat{B} = 110.64$ km [**68**, p. 227]. Measuring expeditions conducted years later determined that Maupertius had erred by about 0.4 km too much, thus we modify the field results of Maupertuis's team to $\hat{M} = 111.46$ km. Therefore our simple model yields results that are off by $M - \hat{M} \approx 0.13$ km and $B - \hat{B} \approx -0.07$ km.

However, the results of the two expeditions failed to change everyone's opinions. For example, even though Bouguer had collected the data, analyzed it, and reached the conclusion that yes, Newton's guess was correct, he remained a committed Cartesian to his grave, primarily because he failed to see how any force could travel across a vacuum.

About fifty years later, during the French Revolution, two astronomers, Pierre-François-André Méchain and Jean-Baptiste-Joseph Delambre, measured the meridian arc from Dunkirk to Barcelona with a triangulation scheme very much like the geodesic scheme used by the polar and equatorial teams of the 1730's, a scheme first pioneered by Willebrord Snell, "the Dutch Eratosthenes" in 1617 [**6**]. The results helped establish the length of the meter, giving the world a universal standard.

Exercises

1. In the model of the earth in this chapter, Emily, of mass 50 kilograms, steps on a weight scale at the pole and at the equator. Are the readings the same? Explain your answer.

2. Plot the graph of $r(\phi)$, (2), for $0 < T < 24$ hours. For what value of T does the model break down? Assume that the gravity function at the earth's surface disregarding earth's oblate shape, is the constant function 9.8 m/sec^2. With $T = 2$ hours, generate the profile of the earth in this new model and compare it with that of Figure 7. Assume that the earth is homogeneously dense, and that its profile is that of an ellipse with polar radius of 4000 km and equatorial radius of 7000. Find a

function giving the surface gravity of this planet in terms of angle ϕ. Then find the period for which these two radii dimensions are stable. Generate the shape of the earth in this gravity model at this period, and compare the shape with that of the ellipse. For a more general approach to finding the shape of planets, consult a geophysics text such as Stacey [**133**, pp. 20–27]. For a survey of the approaches up through Clairaut, see Greenberg [**57**]. For a modern approach, see Chandrasekhar [**24**].

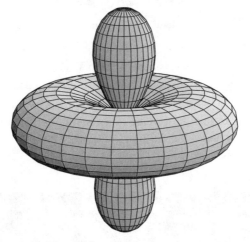

Figure 12. A polar space station

3. A median of a triangle is a line segment from a vertex to the midpoint of the triangle's opposite side.

   (a) Show that the medians of a triangle meet at a point.

   (b) Show that the point of intersection is the centroid of a flat homogeneous triangle.

   (c) Show that a centroid of a triangle partitions its medians into a 2 to 1 ratio, where the distance from the vertex to the centroid is twice the distance from the centroid to the other endpoint of the median.

   (d) Find an expression for the x-coordinate of the centroid of a flat, homogeneous sector, oriented as in Figure 8, by using a straightforward integral. Show that the limit of this expression when $\theta \to 0$ is $2r/3$.

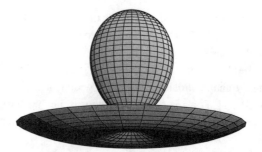

Figure 13. Hard-boiled egg, upright on a plate

4. (a) Use (3) to find the volume of the solid of revolution obtained by spinning the entire polar flower $r = \cos(2\theta)$ about the y-axis, as shown in Figure 12.

   (b) Use (3) to find the volume of the solid of revolution obtained by spinning the cardioid $r = 1 + 2\sin\theta$, $-\pi/2 \leq \theta \leq 0$, about the y-axis, as illustrated in Figure 13.

Figure 14. Jupiter, courtesy of NASA

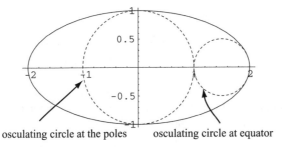

Figure 15. Osculating circles

5. Derive (4) using basic integration techniques.

6. (a) Let a and b be positive numbers. Show that $r(\theta) = \dfrac{1}{\sqrt{\dfrac{\cos^2\theta}{a^2} + \dfrac{\sin^2\theta}{b^2}}}$ is a polar parametrization of an ellipse.

   (b) Find the centroid of the region $\dfrac{\cos^2\theta}{a^2} + \dfrac{\sin^2\theta}{b^2} \leq 1$ in the first quadrant.

   (c) Use the results of parts (a) and (b) above to find an alternate way of determining the flattening factor of the earth as given on p. 132.

7. With a compass and straightedge, find the center C of the disk of Jupiter's image in Figure 14. Then construct a circle with center C that best coincides with the circumference. Does Jupiter appear to be flattened? Using the model developed in this chapter, predict the amount of flattening of Jupiter, and contrast your result with the actual value. Also compare your results with Newton's values as found in the *Principia*, book 3, proposition 19 [95, p. 826].

8. (a) For the polar curve $r(\phi)$ as given in (2) corresponding to a rotation rate $T = 12$ hours, find the arc length along one degree of longitude at the equator and at the north pole.

   (b) Imagine that Planet Valhalla was once a motionless sphere of liquid water with radius 2500 km. But then Thor took a great swallow of its seas, and spat it out, creating a dynamic that set the planet spinning with period 15 hours. What is the polar radius and the equatorial radius using the model given in this chapter? Assume that 1 ml of water has mass 1 gram.

9. The curvature κ of a parametrized curve $\alpha(t) = (x(t), y(t))$ is

$$\kappa(t) = \frac{x''(t)y'(t) - x''(t)y'(t)}{\sqrt{(x'(t))^2 + (y'(t))^2}}.$$

The circle of radius $1/\kappa(t)$ whose center lies on the normal to the curve at t (and for which the center and the curve near t are both on the same side of the tangent line at t) is called the *osculating circle* to the curve, the circle most like the curve at that point. For example, if $\alpha(t) = (2\cos t, \sin t)$, then $\kappa(0) = 2$ and $\kappa(\pi/2) = 1$. The osculating circles at the equator and the north pole are shown as dotted curves in Figure 15.

Find $\kappa(\phi)$ for the polar curve $r(\phi)$ as given by (7). Determine the radii of the osculating circles at the equator and poles. Compare them to $r(0)$ and $r(\pi/2)$ for the rotation rate $T = 24$ hours.

10. (a) Find an elementary derivation of Kepler's third law of motion for a satellite orbiting an earth in a circle of radius r with period T given that the acceleration induced on the satellite by the earth's mass is $f(r) = -gR^2/r^2$, where R is the radius of the earth.

(b) Use Kepler's third law to find the distance of the moon from the earth.

(c) Use Kepler's third law to find the distance of a geostationary satellite from the earth, that is, a satellite whose period is one day.

Vignette VI: The Zodiac

The Greek word *zodiac* means circle of little animals. The animals refer to twelve constellations of stars as seen from earth. Each constellation, when its stars are imaginatively connected with line segments, forms the framework of a mythic symbol: goat, water-jug, fish, ram, bull, the twins, crab, lion, woman, balance-scales, scorpion, and centaur. For example, Figure 1(a) is a group of stars along with a dotted curve showing where the ecliptic plane cuts the celestial sphere (the stars marked α, γ, and η are the brightest of the group); (b) shows the line-segment connections; and (c) is Albrecht Dürer's image from 1512 of two fish connected at their tails by a ribbon, adapted from his woodcut print of the zodiac [**38**, p. 295]. A reason why this group of stars is associated with fish is that in ancient Egypt, at the time of year when Pisces (the fishes) was dominant, "fish were fattest and easiest to catch" [**108**, pp. 269–270].

The Chinese have twelve animals in their zodiac, too: rat, ox, tiger, rabbit, dragon, snake, horse, sheep, monkey, phoenix, dog, and boar. Ancient astronomers as early as 3000 BC, used them to track the year by mapping the apparent position of the sun at noon. Saying that the sun was in the house of the ram meant that the sun, as viewed at noon from the earth, was somewhere in the constellation Aries (the ram). Of course, the only time that the stars can be seen at noon is during a complete solar eclipse—a rare and special occurrence.

From a long tradition of associating perfection with the heavens and because of the human penchant of looking for signs, fortunes were told according to the arrangement of the planets against the zodiac. Indeed, astronomers, astrologers, and mathematicians were often one and the same. Both Kepler and Galileo regularly cast horoscopes to earn extra income and to fulfill their professional obligations [**138**, p. 135]. As a young man in 1597, Kepler cast his own horoscope, which reads like an extended Chinese fortune cookie. Besides speaking about the planets being in their various houses, Kepler concluded that

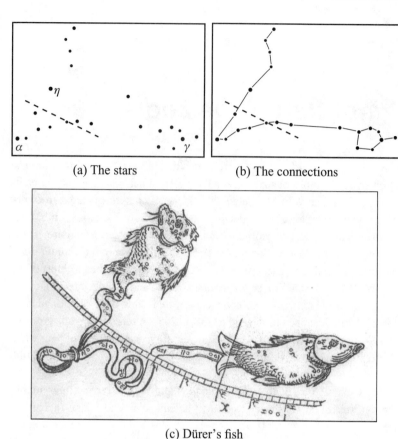

(a) The stars (b) The connections

(c) Dürer's fish

Figure 1. The zodiac constellation Pisces

I have been born with the destiny to spend most of my time working on the difficult things others shirk. [**27**, p. 31]

In a horoscope Galileo cast for his friend Giovanfrancesco Sagredo, whose personality and name are immortalized as one of the three debaters in Galileo's great dialogue about the motion of the earth [**116**, p. 9], he concluded

Venus illuminated by Jupiter and Mercury makes [Sagredo] kind, happy, merry, beneficent, pacific, sociable, pleasure-loving, a lover of God, [yet] impatient. [**18**, p. 121]

While men like Kepler and Galileo probably attached little credibility to a connection between the orientation of points of light in the sky and the affairs

Vignette VI: The Zodiac

of men, others then and now make decisions based on them. Astrologers usually couch their readings ambiguously so that a horoscope can be construed as accurate. To forecast specific events is reckless, because the lie is readily exposed if the event fails to occur.

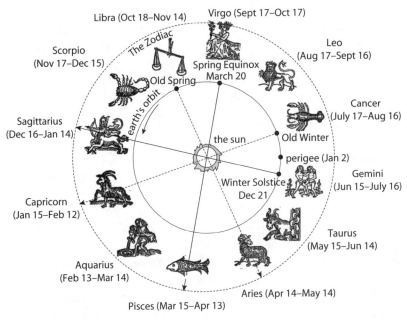

Figure 2. The sun's apparent position through the year embellished with sixteenth-century woodcuts (dates correspond to present-day alignments)

As Voltaire championed reason throughout his life, he routinely pointed out such delusions. In the following passage, he refers to an astrological prediction by astronomer-clergyman Johnannes Stöffler (1452–1531), some of whose work formed the basis for the Gregorian calendar.

One of the most famous mathematicians of Europe, named Stöffler, foretold a universal deluge for the year 1524. The deluge was to happen in the month of February, and nothing can be more plausible, for Saturn, Jupiter, and Mars were then in conjunction in the sign of the Fishes. Every nation that heard of the prediction was in consternation. A doctor of Toulouse named Auriol, had an ark built for himself, his family, and friends. At last the month of February arrived, and not a drop of rain fell, never was a month more dry, never were the astrologers more embarrassed [**144**, vol. vi, *Astrology*, p. 96].

Voltaire also points out, as Hipparchus, the father of astronomy, had observed in about 130 BC, that the astrological charts are long out of date [**144**, vol. vi, *Astrology*, p. 94]:

> The great misfortune of astrologers is that the heavens have changed since the rules of the art were laid down. The sun, which at the equinox was in the Ram in the time of the Argonauts, is now in [Pisces]; and astrologers, most unfortunately for their art, now attribute to one house of the sun that which visibly belongs to another.

That is, as we show in the next chapter, the apparent position of the sun against the zodiac precesses with time. In the days of Hipparchus, the summer solstice occurred when the sun appeared (from the earth) to be in Cancer (the crab). At the summer solstice the sun appears to be directly overhead at noon. So latitude 23.5° N is as far north as the sun ever appears to be directly overhead. Hence this latitude is called the Tropic of Cancer. However today when the summer solstice occurs, the sun appears to be in Gemini (the twins), which means that latitude 23.5° N should be renamed the Tropic of Gemini. Similarly, for the winter solstice, the Tropic of Capricorn (the goat), 23.5°S, should be renamed the Tropic of Sagittarius (the centaur).

Figure 2 illustrates the geometry of the precession. The inner circle represents the earth's orbit as seen from above earth's northern hemisphere. Earth's perihelion occurs near January 2. The outer circle depicts the cycle of the zodiac throughout the present-day year. The twelve constellations of the zodiac each have dominance in the sky for about 30° or 31 days. The dates for each constellation in the figure correspond to the approximate days when the sun, as viewed from the earth, appears to be in that constellation. For example, at the winter solstice, the sun as viewed from the earth appears to be in Sagittarius, indicated by an arrow from the earth's position on December 21 to the outer ellipse. Similarly, the summer solstice occurs on June 21, while the sun appears to be in Gemini. The dates at which the seasons occur drift earlier in the year as time continues. This phenomenon is caused by earth's equatorial bulge, as Voltaire outlines below in poetic fashion and as we explain more fully in this chapter.

> Earth, change your form; let Newton's law of matter
> Depress your poles, and heighten the equator.
> Avoid, you pole, that fixed to sight appears,

The frozen chariot of the northearn bears;
Embrace in each of thy immense careers,
Almost three hundred centuries of years.
On the Newtonian Philosophy
[**144**, vol. xxxvi, p. 301]

CHAPTER **VI**

Hipparchus's Twist

When Hipparchus collected all that was known about the heavens in his time, he noticed that the seasons shift through time with a period of 25,800 years, which Voltaire referred to in the poem above as "almost three hundred centuries of years." Since the seasons are determined by the orientation of earth's axis, as shown in Chapter IV, the direction of earth's axis changes with time. The earth is like a toy top spinning on the floor and just as the tip of the top's axis precesses in a circle parallel to the floor, so the north pole of the earth's axis precesses in a circle parallel to the ecliptic plane, as is shown in Figure 3.

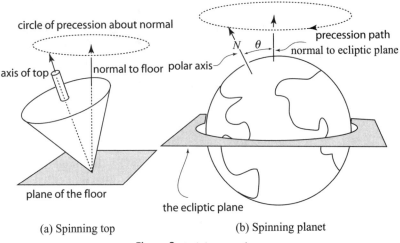

(a) Spinning top   (b) Spinning planet

Figure 3. Axial precession

Newton was able to derive this period from his theory of gravitation; it was good evidence that his idea about any two masses attracting one another

145

was correct. In this chapter we use the model of the precessing top to explain the twenty-six thousand year precession and to determine such dates as the spring equinox during the siege of Troy.

The notion of torque

The cross product of two vectors A and B is a vector perpendicular to them both with magnitude

$$|A \times B| = |A||B| \sin \gamma, \qquad (1)$$

where γ is the angle between the two vectors. Its direction is given by the right-hand rule of Figure 4(a): if the right-hand thumb represents vector A and the right-hand forefinger represents B the right-hand middle finger points in the direction of $A \times B$.

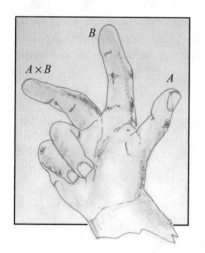 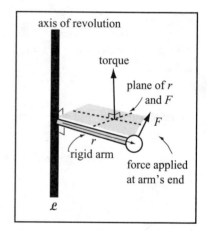

(a) The right-hand rule (b) Torque

Figure 4. The cross product

Let $\mathcal{L}$ be an axis with a rigid arm attached to it as in Figure 4(b). Apply a force F perpendicular to the axis to the end of the arm. The torque $\mathcal{T}$ about $\mathcal{L}$ induced by applying force F to the end of the arm is

$$\mathcal{T} = r \times F,$$

and points in the direction of $\mathcal{L}$.

A familiar equation of linear motion from physics is $F = ma$, where F is the force on a mass m experiencing an acceleration a. In rotational motion the analogs of force, mass, and acceleration are torque $\mathcal{T}$, rotational inertia I, and the instantaneous rate of change of the angular velocity ω. For an object with angular velocity ω, let $\mathcal{A}$ be the axis through the center of mass of the object in the direction of $\dot{\omega} = d\omega/dt$, let $\mathcal{T}$ be the torque about $\mathcal{A}$, and let I be the rotational inertia about $\mathcal{A}$. Then, the analog to $F = ma$ is the rotational motion formula

$$\mathcal{T} = I\dot{\omega}. \qquad (2)$$

We use (2) to show that the gravitational pull by the moon and the sun on the bulge of earth's equator results in earth's pole precessing.

Estimating the torque on earth's hinge

In Chapter V, we saw that the polar radius of the earth is $R = 6356.8$ km and its equatorial radius is $\rho = 6378.2$ km, as shown in Figure 5(a). Allow earth's mass beyond the radius R to coalesce into a homogeneous equatorial belt about the earth, as shown in Figure 5(b). Assuming that the earth is homogeneous gives the mass of the belt as $\alpha \approx 0.00685$ that of the earth, as shown in Exercise 1. When Newton attacked the problem of finding the precession period of earth's poles, he computed the ratio of the volume of "this ring gird[ing] the earth along the equator" to the volume of the earth as 459 to 52,441 or about 0.00876 [**95**, pp. 885–886]. Some of the reason for the discrepancy is that Newton used the ratio of R to ρ as 229 to 230.

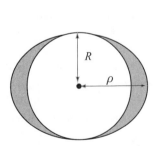
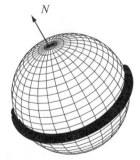

(a) Inscribed sphere within the earth   (b) Excess mass → the equator

Figure 5. A homogeneous belt around a spherical earth

Earth's axis is inclined at $\theta = 23.5°$ away from a normal to the ecliptic plane. For ease of analysis in this thought experiment, we position earth's center at the origin O and allow the sun to circle the earth in the ecliptic plane. The equatorial plane and ecliptic plane meet in a line through O. Let the xy-plane be the equatorial plane and let $z = y \tan \theta$ be the equation of the ecliptic plane. To find the rate at which the pole turns as induced by the sun's gravity, we find the torque about the ideal axis between the ecliptic and the equatorial planes which we call the hinge.

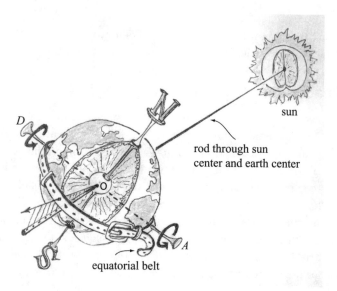

Figure 6. Hinge closing within a belted earth

Figure 6 is a sketch of the belted earth. The line $\overline{AD}$ is the hinge. With respect to the northern hemisphere, the earth is at the summer solstice in the figure. The sun's gravity exerts a northward force on the illuminated (daytime) part of the belt and a southward force on the shadowed (nighttime) part of the belt. Thus the sun's gravity is trying to close the hinge between the two planes. Just as a satellite in circular orbit about the earth ever falls to the earth yet gets no closer, so too, rather than closing the hinge, the torque forces the hinge to slide along the ecliptic plane. If we think of the pole and the hinge as being welded together, as the hinge slides on the ecliptic plane, the pole, being rigidly attached to the hinge, will rotate, as shown in Figure 7. What we determine is the rate at which the hinge and pole framework rotates. We will compute the torque about the hinge induced by the sun.

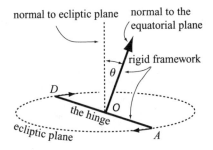

Figure 7. Pole and hinge framework sliding on the ecliptic plane

Assume that earth's belt is a circle with center O passing through $A = (R, 0, 0)$ and $B = (0, R, 0)$, as in Figure 8. Let ϕ be the angle between the sun at S and its projection (along $\overrightarrow{OE}$) onto the equatorial plane. Let δ be the angle between the positive x-axis and $\overrightarrow{OE}$. Let P be the point

$$P = R(\cos \epsilon, \sin \epsilon, 0),$$

$0 \leq \epsilon \leq 2\pi$, on the earth's belt. Let ψ be $\angle SOP$, and let Ω be $\angle PSO$. Thus

$$S = Q(\cos \delta \cos \phi, \sin \delta \cos \phi, \sin \phi),$$

where $0 \leq \delta \leq 2\pi$, with Q being the distance between the earth and the sun. Exercise 2 shows that ϕ is a function of δ,

$$\sin \delta \tan \theta = \tan \phi. \tag{3}$$

Since $\overrightarrow{OS}/Q$ and $\overrightarrow{OP}/R$ are unit vectors,

$$\cos \psi = (\cos \delta \cos \phi, \sin \delta \cos \phi, \sin \phi) \cdot (\cos \epsilon, \sin \epsilon, 0)$$
$$= \cos \phi (\cos \delta \cos \epsilon + \sin \delta \sin \epsilon). \tag{4}$$

Let H be the projection of P onto $\overrightarrow{OS}$ and let J be the projection of H onto $\overrightarrow{OE}$. Let μ be the angle between $\overrightarrow{PH}$ and the normal to the equatorial plane at P, which is also $\angle PHJ$.

Let F be the gravitational force of the sun on P. F's direction is $\overrightarrow{PS}$. We decompose F into two normal components: a component along $\overrightarrow{OS}$ and a

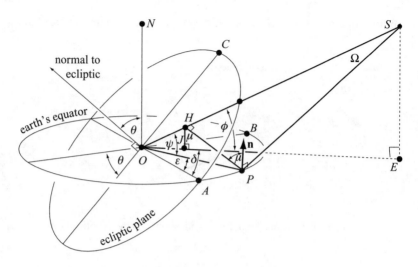

Figure 8. The ecliptic and equatorial planes.

component along $\overrightarrow{PH}$. The component along $\overrightarrow{PH}$ is the part of F pulling the equatorial plane into the ecliptic plane. We want that part of this force normal to the equatorial plane, which we call $\mathbf{n}$. Since $|\overline{PH}| = R \sin \psi$, $|\overline{OH}| = R \cos \psi$, and $|\overline{HJ}| = |\overline{OH}| \sin \phi$, then

$$\cos \mu = |\overline{HJ}|/|\overline{PH}| = \cot \psi \sin \phi. \tag{5}$$

Since Q is much larger than R, we assume that $|\overline{PS}| = |\overline{OS}| = Q$ for all P on the earth's belt. By the law of sines with respect to $\triangle SOP$, $\sin \Omega / R = \sin \psi / Q$, which means that $\sin \Omega = R \sin \psi / Q$. Hence the magnitude of the component of F normal to $\overrightarrow{OS}$ is $|F| \sin \Omega = R |F| \sin \psi / Q$. Thus, $\mathbf{n} = (R|F|/Q) \sin \psi \cos \mu \, \mathbf{k}$, where $\mathbf{k} = (0, 0, 1)$.

The torque about OA obtained by applying $\mathbf{n}$ to P is $R(0, \sin \epsilon, 0) \times \mathbf{n}$. Since the vectors are perpendicular, the total torque of the sun pulling on P so as to rotate the equatorial plane into the ecliptic plane is

$$R \sin \epsilon \, |\mathbf{n}| \, \mathbf{i} = \frac{R^2 \sin \epsilon \sin \psi \cos \mu}{Q} |F| \mathbf{i}, \tag{6}$$

where $\mathbf{i} = (1, 0, 0)$. By (4) and (5), (6) becomes

$$\frac{R^2}{Q} \sin \phi \cos \phi (\sin \epsilon \cos \epsilon \cos \delta + \sin^2 \epsilon \sin \delta) |F| \mathbf{i}. \tag{7}$$

Since Q is much larger than R, we assume that $|F|$ is constant for any P along earth's equatorial belt. We take the point-mass of P as $\alpha M_e \Delta\epsilon/(2\pi)$, where $\Delta\epsilon$ is the radian angle spanned by point P on earth's equatorial belt. Thus

$$|F| = \frac{\alpha G M_e M_s}{Q^2} \frac{\Delta\epsilon}{2\pi}, \qquad (8)$$

where G is the universal gravitational constant, M_e is the mass of the earth, and M_s is the mass of the sun. The definite integral of $a\sin\epsilon\cos\epsilon + b\sin^2\epsilon$ as ϵ varies from 0 to 2π, where a and b are arbitrary constants, is πb. Thus, since each of the torques (with arm $(0, R\sin\epsilon, 0)$ and vertical force $\mathbf{n}$ for any ϵ with $0 \le \epsilon \le 2\pi$) points in direction $\mathbf{i}$, then the combined sum, by integration with respect to ϵ for $0 \le \epsilon \le 2\pi$, of all the torques given by (7) and (8) along earth's belt is

$$\frac{\alpha G M_e M_s R^2}{2Q^3} \sin\phi \cos\phi \sin\delta\, \mathbf{i},$$

which by (3) becomes

$$\frac{\alpha G M_e M_s R^2}{2Q^3} \sin^2\phi \cot\theta\, \mathbf{i}. \qquad (9)$$

The direction of the torque throughout the year is in direction $\mathbf{i}$; at the spring and fall equinoxes, it falls to zero. Since the period of the precession of earth's pole is thousands of times greater than a year, we shall assume that (9) is a constant throughout the year for its torquing influence on the pole.

What constant shall we take? The average value is reasonable. By Exercise 4, the average value of $\sin^2\phi$ as δ varies from 0 to 2π is $1 - \cos\theta$, which means that the average value of $\mathcal{T}$ by (8) and (9) throughout the year is

$$\mathcal{T} = \frac{\alpha R^2 G M_e M_s}{2Q^3} \cot\theta (1 - \cos\theta)\mathbf{i}. \qquad (10)$$

Putting it Together

Now we are almost ready to find the rate at which earth's pole precesses.

We know two of the three ingredients of (2). The torque $\mathcal{T}$ is given by (10). Since the mass of earth's belt is negligible with respect to the mass of the earth, we may take the inertia I of the earth about any axis through O as

$$\frac{2M_e R^2}{5}, \qquad (11)$$

as shown in Exercise 5. To find the final ingredient $\dot{\omega}$ we reason intuitively.

We adopt the perspective of Figure 7, which shows a welded-together pole and hinge, with the hinge sliding along the ecliptic plane, which we take as the xy-plane. At time $t = 0$, we specify that the hinge lies along the x-axis and the north pole points into the first quadrant of the yz-plane. In this chapter we are assuming that the pole precesses in circular motion with constant speed as shown in Figure 3(b). Our goal is to find the angular frequency of the precession. In the last section, we assumed that the sun rotates clockwise about a stationary earth; so if the sun has any effect on the pole, it must drag the pole clockwise. In this section it will be convenient to change the perspective, and think of the earth rotating about its axis. Let $\omega_s = 2\pi$ radians per day, the rotation rate of earth about its axis with respect to the sun. Let $\hat{\omega}(t)$ be the angular velocity of the earth about its pole at time t. Thus $\hat{\omega}(0)$ points in the direction of the north pole and has magnitude ω_s. Let β_s be the angular velocity of the pole's precession (as accounted for by the sun). Since the pole is inclined at an angle of θ away from the positive z-axis,

$$\hat{\omega}(t) = \omega_s \Big(\sin(\beta_s t) \sin(\theta),\ \cos(\beta_s t) \sin(\theta),\ \cos\theta \Big),$$

where ω_0 and β_s are both positive.

Since

$$\dot{\hat{\omega}}(t) = \beta_s \omega_s \sin\theta \Big(\cos(\beta_s t),\ -\sin(\beta_s t),\ 0 \Big),$$

then

$$\dot{\hat{\omega}}(0) = \beta_s \omega_s \sin\theta\ \mathbf{i},$$

which by (11) means that

$$I\dot{\hat{\omega}}(0) = \frac{2\beta_s \omega_s M_e R^2 \sin\theta}{5}\ \mathbf{i}. \tag{12}$$

If we take $\mathcal{A}$ as the axis through O in direction $\mathbf{i}$, then $\mathcal{A}$ is the hinge axis, $\dot{\hat{\omega}}(0)$ is $\dot{\omega}$, and the torque about $\mathcal{A}$ is the torque $\mathcal{T}$ of (10). By (2), (10), and (12), we have

$$\frac{\alpha R^2 G M_e M_s}{2 Q^3} \cot\theta (1 - \cos\theta) = \frac{2\beta_s \omega_s M_e R^2 \sin\theta}{5}. \tag{13}$$

Solving for β_s gives

$$\beta_s = \frac{5\alpha G M_s \cot\theta(1-\cos\theta)}{4\omega_s Q_s^3 \sin\theta} \approx 2.24 \times 10^{-12} \text{ radians/sec},$$

where $Q_s \approx 1.496 \times 10^{11}$ meters, $M_s \approx 1.99 \times 10^{30}$ kg, and $G \approx 6.67 \times 10^{-11}$ Nm^2/kg^2.

The moon has a similar influence. We assume that the moon of mass M_m rotates about the earth in a circle in the ecliptic plane with radius Q_m, and that the period of the moon with respect to the fixed stars is 27.3 days, which means that ω_m, the rotation rate of the earth about its axis with respect to the moon, is 2π radians per about 1.04 days. Thus

$$\beta_m = \frac{5\alpha G M_s \cot\theta(1-\cos\theta)}{4\omega_m Q_m^3 \sin\theta} \approx 5.07 \times 10^{-12} \text{ radians/sec},$$

where $Q_m \approx 3.84 \times 10^8$ meters and $M_m \approx 7.35 \times 10^{22}$ kg.

The sun and the moon are like individual rockets giving an increase in the rate at which the pole precesses. Thus to find their combined influence on the pole precession, we add β_s and β_m. That is, our intuitive approximation of how quickly the pole precesses is

$$\beta_s + \beta_m \approx 7.31 \times 10^{-12} \text{ radians/sec},$$

or about one rotation in 27,300 years, which is a 5.8% relative error from the actual value of 25,800 years. Newton's conclusion for earth's theoretical polar precession rate was $50''0'''12^{iv}$ per year, or about one-sixtieth of a degree per year, giving a period of 25,920 years [**95**, p. 887].

To refine these intuitive answers, a better model should deal with the earth's bulge without coalescing it into an equatorial belt and by recognizing that earth is not homogeneous. A better model will also avoid any componentwise sifting of the arm-vector and the force-vector on each point-mass of the bulge. A semi-formal derivation for the precession period is given in [**133**, pp. 20–30] involving a series expansion of Laplace's equation of a gravitational potential model; a formal one is given in [**56**] starting with Euler's equations of moments of rotational inertia; both of these approaches involve more mathematics and physics than is presented in this book. Some of the exercises below suggest ways of improving the estimate.

An April-fool's riddle

In this section we characterize the seasonal drift through the years. In particular, we determine the year in which spring occurred on April 1. To simplify matters, we assume that the earth's period about the sun is 365 days and that dates are given with respect to the Gregorian calendar. We also ignore the fact that when today's calendar was initially adopted in 1582, October 5 through October 14, 1582 were dropped from the calendar so as to resynchronize it with earth's seasons.

We will find a function that outputs the date of the winter solstice for any given year. As one data point, let's assume that the winter solstice occurs on December 21 in the year 2000. Since perihelion occurred on January 3, then the winter solstice occurred on day $t = -13$. (We take January 3 as day 0.) From Chapter IV, recall that $T(\theta)$ gives the day of the year corresponding to when the earth is at a given polar angle along its orbit. Solving $T(\theta) = -13$ gives $\theta \approx -0.231$ radians. Since the rate ϵ at which the pole precesses is $\epsilon = -2\pi/25800 \approx -0.000244$ radians/year, then $-0.231/\epsilon \approx 950$ years ago earth's winter solstice occurred on January 3. That is, in 1050 AD, the winter solstice approximately coincided with earth's perihelion. Since ϵ is constant, the polar angle at which the winter solstice occurs in year τ is given by $\Omega(\tau) = \epsilon(\tau - 1050)$, where τ is the Gregorian year. Therefore, $T(\Omega(\tau))$ gives the day of the year when the winter solstice occurred or will occur in year τ.

For example, in the year 4000 AD, the polar angle at which the winter solstice occurs is $\Omega(4000) \approx -0.718$ radians, which means that the winter solstice will occur on day $T(-0.718) \approx -40$, which is November 26, while the spring equinox will occur on day $T(-.718 + \pi/2) \approx 49$, which is February 21. Under the current calendar, the winter solstice will amble backwards through the year, passing through each day, in 25800 years. For example, in the year 14,873 AD, the winter solstice will occur on June 21, the current summer solstice!

To find the year in which spring occurred on April 1, solve (with a computer algebra system) $T(\Omega(\tau) + \pi/2) = 88$, since April 1 is day 88 from perihelion. The answer is $\tau \approx 1140$ AD.

To solve for the occurrence date of the maximum possible time lapse between winter and spring of 94.1 days, we solve $\Omega(\tau) = 3\pi/4$, and get -8625, which is 8625 BC, (or 25,800 years after that date).

To generate further examples of determining conjunctions of dates with the equinoxes and solstices for the planets in the solar system, consult [41] where the dates for solstices and equinoxes are given to the nearest minute, Greenwich time. Because the actual year length is 365.2 days, the date of perihelion moves around from about January 1 to January 4. Leap year more or less resets the calendar so that perihelion stays fixed. Finally, there are other planetary phenomena which precess, and so influence the seasons. The perihelion of the orbit precesses and the ecliptic plane precesses. But these two motions are almost negligible in comparison to the effect of the precession of the planet's axis, at least for earth.

Exercises

1. Use (V.2) so as to determine the volume of earth outside an inscribed sphere.

2. Assume that the sun orbits the earth in a circle lying in the plane $z = y \tan \theta$, where $\theta = 23.5°$, with the earth at the origin, so that the xy-plane is the plane containing earth's equator. Let ϕ be the angle between the equatorial plane and the sun. Let δ be the polar angle measuring the angle between the positive x-axis and the projection of the sun onto the equatorial plane. This geometry is shown in Figure 9. Show that $\sin \delta \tan \theta = \tan \phi$.

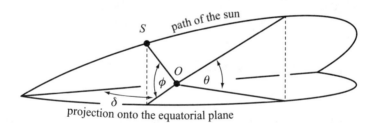

Figure 9. The ecliptic and equatorial planes.

3. Prove the identity
$$\int_0^{2\pi} \frac{\sin^2 \alpha}{a^2 \sin^2 \alpha + 1} \, d\alpha = \frac{2\pi(1 + a^2 - \sqrt{1 + a^2})}{a^2(1 + a^2)},$$
where a is a non-zero real number.

4. Using Exercises 2 and 3, show that the average value of $\sin^2 \phi$ for $0 \leq \delta \leq 2\pi$ is $1 - \cos \theta$.

5. (a) Show that the rotational inertia of a solid homogeneous sphere of radius R and mass M about its central axis is $2MR^2/5$.
   (b) Show that the rotational inertia of a homogeneous circular ring of radius R and mass M about its central axis is MR^2.

6. (a) Imagine that the moon was twice its mass. Everything else being the same, recalculate the period for the precession of earth's pole.
   (b) Imagine that the moon had a retrograde motion. Everything else being the same, recalculate the period for the precession of earth's pole.

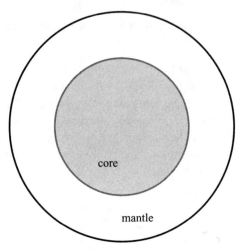

Figure 10. A two-tier earth

7. Planet X has an eccentricity of $e = 0.5$ and a rotation rate of 365 days about the sun. Its perihelion occurs on January 1, and the precession period of its pole is 100 years (clockwise). At $\tau = 0$ years, the winter solstice occurs on December 1. Find the date of winter and spring twenty-five years later.

8. Assume that the earth of radius R has a mantle and core with the density of the core twice the density of the mantle and that the core-mantle boundary occurs at radial distance $R/2$, as pictured in Figure 10. For this earth, find the precession rate of its north pole using the model given in this chapter. The mass of the earth is $M_e \approx 5.98 \times 10^{24}$ kg.

9. Try Exercise 8 in reverse. That is, suppose that the precession of the earth's pole is 25,800 years, and that the earth has a mantle and core

with the density of the core twice that of the mantle. At what radial distance is the core-mantle?

10. Jupiter's pole is inclined at 3.1° away from a normal to its ecliptic plane. It also has at least a dozen satellites. Estimate the period of Jupiter's polar precession.

Vignette VII: Love Triangles

> Love that makes its own misfortune
> Of such fire that enflames its heart
> In the frozen wastes of Bothnia [Lapland]
> Certain French gave in to it:
> A failing altogether ordinary.
>
> Frenchmen, show yourselves today
> As generous as ere you're faithless [**68**, p. 159].

The above lines from Voltaire were his public words of advice to his friend Maupertuis who was facing a scandal. Maupertuis had just returned from Lapland along with the rest of his team and soon thereafter two Lapland maidens, to whom he had sung love songs and recited poetry during the long winter in front of a cozy fireplace, arrived in Paris in expectation of promises made to them. Their father threatened a lawsuit. While Maupertuis sought to present at the Academy his team's findings that Newton was correct, that the earth was flattened at the poles, the public tittered about the star-crossed lover who masqueraded as an astronomer and mathematician in the service of the king. What could he do?

Here's how he explained the matter to his sister in a letter written from Voltaire's country retreat in Cirey.

> One artist [on our team] made one [of the two women] sacred promises of marriage and told so many lies about his wealth, that this poor girl came to find him; and her sister, counting on sharing this imaginary fortune, accompanied her. But [the artist] was unfaithful and married someone else. Everyone was touched by her trouble. All the company of the polar circle have contributed [to a fund for their support] [**140**, p. 145].

Maupertuis, as "a generous Frenchman," arranged for the two women to stay with the family of his friend Clairaut, whereupon one of the sisters was persuaded to enter a convent while the other quickly found a husband.

What about Voltaire? What was his experience with affairs of the heart to give such advice?

159

At age nineteen while a junior diplomat to Holland, Voltaire fell in love with a French Huguenot refugee. They vowed unending love one to another. However when the French ambassador learned of their plans, and interpreted them as unprofessional, he promptly sent Voltaire home to Paris. Three years later, and long after she had married another, Voltaire was incarcerated in the Bastille; the inventory of his personal effects totaled a pair of opera glasses, a notepad, a modest amount of money, and a much-read, old love letter to him from that sweetheart [**101**, p. 45]. He had a hard time letting go.

His second stint in the Bastille was over an opera singer. Voltaire and a chevalier vied for her attention. Foolish words were exchanged. The French knight commissioned a few thugs who duly thrashed the impertinent bourgeois commoner. Voltaire swore retribution, and began fencing lessons in earnest. At such news, the authorities incarcerated Voltaire, officially for his own protection, whereupon he successfully lobbied to substitute exile to England for continued lodging in the Bastille.

A few years later, upon returning from exile and in the midst of studying Newton's ideas, he re-met Émilie du Châtelet. Voltaire had known her since she had been eight years old, as she was the daughter of a close friend. Now he was intrigued with this young (age twenty-seven), beautiful woman who understood mathematics and philosophy. She in turn was intrigued with this internationally-known-playwright-turned-natural-philosopher. However she was a marquise, the wife of a marquis, and a mother of several children. And Voltaire was a commoner.

How could such a relationship continue?

Voltaire offers some insight from his *Philosophical Dictionary* [**144**, vol. v, *Adultery*, p. 79]:

> French mothers teach their daughters to look for liberty in marriage. Scarcely have they lived a year with their husbands when they become impatient to ascertain the force of their attractions. A young wife neither sits, nor eats, nor walks, nor goes to the play, but in the company of women who have each their regular intrigue. If she has not a lover like the rest, she is ashamed to show herself.

Whether or not Voltaire's assessment about women in high French society was correct, with respect to Emily it was accurate. Coincidentally, her former lover, a duke, was both a close friend of Voltaire from school days and the commanding officer of her husband. With a few words from the duke, and an understanding that Voltaire meant to bankroll the remodeling of the crumbling Châtelet country estate, and since the marquis had a string of

Vignette VII: Love Triangles

his own mistresses, the marquis consented to the arrangement. For the next fifteen years Voltaire and Emily were nearly inseparable. And many times Voltaire, Emily, and the marquis dined amicably at the same table.

However, due to Voltaire's hypochondriac disposition, which rivaled that of Descartes, and due to her ultra-strong libido, Emily maintained strings of intrigues, including flings with the mathematicians Maupertuis and Clairaut. Zinsser [**157**, p. 284] cautions that some of this romance may have been "embellished" by successive historians in the past two centuries, so much so that the number of her actual "liasons multiplied" significantly. Yet as a result of her last intrigue, she died of complications in child-birth at age forty-nine. Voltaire, the marquis, and her current beau were with her when she died. Her experience in life is best summed up by a phrase from one of her letters [**140**, p. 101]:

It is easier to do algebra than to be in love.

Voltaire's last attachment was with his own niece, the daughter of his half-sister, (as he himself was the love-child of one of his mother's intrigues). By this time Voltaire was about 60 and his new mistress was over 40, twice widowed. They resided in exile from Paris at Ferney on the French-Swiss border, a convenient locale in that he could escape to one country or the other should the powers that be attempt to imprison him again for various items that he might write. Together they kept house for twenty-five years, to which place many wandering artists and statesmen visited, so much so that Voltaire called himself "the inn-keeper of Europe."

As this vignette has been about romance, this next chapter is about a romantic family of curves.

CHAPTER VII

Dürer's Hypocycloid

Albrecht Dürer, the great Renaissance German artist, is credited with being the first to introduce the *hypocycloid* curve along with the more general family of *trochoid* curves, as presented in his 1525 four-volume geometry treatise, *The Art of Measurement with Compass and Straightedge*, one of the first printed mathematical texts to appear in German. In this chapter, we characterize the hypocycloid geometrically. We then characterize it algebraically as a system of parametric equations, and dynamically as a differential equation. Finally we show that the hypocycloid is the solution to a minor variation of one of the most famous of mathematical riddles. But first we ask a natural question:

Did Dürer use the trochoid in his woodcuts?

Dürer argues at length that "geometry is that without which no one can either be or become a master artist" [**39**, p. A1v]. If he truly believed what he said, we have a measure of hope of finding abstract curves in his artwork. For example, Figure 1 shows small sections of two works, "The Adoration of the Lamb" from 1498 and "The Circumcision" from 1505, [**38**, cuts 118, 184]. Is the lamb resting on a hypocycloidal arch within a circular halo and do the vines (adorning a synagogue wall) follow a *limaçon*? We superimpose computer generated trochoid curves upon his work so as to answer the question.

The trochoid family

We first agree upon what we mean by a trochoid, and adopt a standard, normalized parametrization in terms of t so that each curve is inscribed within the unit circle and so that at $t = 0$, the corresponding point on the curve is at $(1, 0)$. Informally, a *trochoid* is the curve traced by a tack stuck in a wheel as the wheel, without slipping, rolls along either the inside or the outside of a circle. When the tack is on the rim of the wheel rolling along the

163

(a) A lamb on a hypocycloid? (b) A camouflaged limaçon?

Figure 1. Trochoidal-like curves in Dürer's woodcuts

outside of a circle, the tack traces an *epicycloid* as shown in Figure 2(c) where the wheel's radius is one-tenth that of the circle. When the tack is offset from the rim, the generated curve is an *epitrochoid*, as shown in Figure 2(d) where the tack is outside the rim of the wheel. When the tack is on the wheel rolling along the inside of a circle, the tack traces a *hypocycloid*, Figure 2(a). When the tack is offset from the rim, the generated curve is called a *hypotrochoid*, Figure 2(b).

We consider the hypotrochoid first. Let T be a tack r units out along a ray from a wheel's hub through a point on its rim, where the wheel D is a disk of radius d rolling along the inside of a circle C of radius c, where $c > d > 0$ and $r > 0$. Position the origin O at C's center in the xy-plane. Interpret a parameter θ as the central angle at O between the positive x-axis and D's center, $D(\theta) = (c - d)(\cos\theta, \sin\theta)$. At $\theta = 0$, let $D(0) = (c - d, 0)$, with T at the point $(c - d + r, 0)$. Let $\phi(\theta)$ be the angle at the hub between direction $\mathbf{i}$ and the tack, where $\mathbf{i}$ is the unit vector $(1, 0)$, as illustrated in Figure 3 which shows the hypotrochoid for the case $3d = c$ with $r = 1.25d$. At angle θ, D has rolled along the perimeter of C by an arc length of $c\theta$, which means that D has rotated clockwise around its center by $c\theta/d$ radians, which is the angle between $c(\cos\theta, \sin\theta)$, $D(\theta)$, and T. Therefore $\phi(\theta) = -(c\theta/d - \theta) = -(c - d)\theta/d$. The position of the tack in the xy-plane is thus

$$(c - d)\begin{bmatrix} \cos\theta \\ \sin\theta \end{bmatrix} + r \begin{bmatrix} \cos\phi \\ \sin\phi \end{bmatrix}, \qquad (1)$$

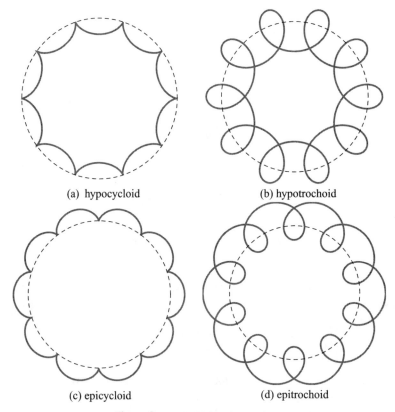

Figure 2. Trochoid family members

where $\phi = (d - c)\theta/d$. This is the standard parameterization of the hypocycloid.

Now let $\lambda = (c-d)/(c-d+r)$, $\mu = r/(c-d+r) = 1 - \lambda$, $p = d$, $q = c - d$, and let t be a new parameter with $\theta = pt$. An equivalent form to (1) is

$$(c - d + r)\left(\lambda \begin{bmatrix} \cos pt \\ \sin pt \end{bmatrix} + \mu \begin{bmatrix} \cos(-qt) \\ \sin(-qt) \end{bmatrix}\right). \qquad (2)$$

That is, given valid values for c, d, and r, (1) can be transformed into (2), and given λ, μ, p, and q with $0 < \lambda < 1$, $\mu = 1 - \lambda$, $p > 0$, $q > 0$, then $d = p$, $c = q + p$, $r = \mu(c-d)/\lambda$, and $\theta = pt$, which means that (2) can

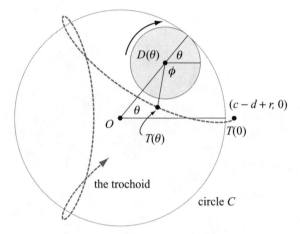

Figure 3. Generating a hypotrochoid

be transformed into (1). Hence, we say that a *normalized* parametrization of the hypotrochoid is

$$h(p, q, \lambda, \mu, t) = \lambda \begin{bmatrix} \cos pt \\ \sin pt \end{bmatrix} + \mu \begin{bmatrix} \cos(-qt) \\ \sin(-qt) \end{bmatrix}, \quad (3)$$

where $0 < \lambda < 1$, $\mu = 1 - \lambda$, and p and q have the same sign. The beauty of (3) over (1) is that its graph is inscribed in the unit circle for all valid values of λ, μ, p, and q. Thus any hypotrochoid is a scaled version of (3). In particular, when $\lambda = q/(p+q)$ (which corresponds to $r = d$), then

$$h(p, q, q/(p+q), p/(p+q), t) \quad (4)$$

is a hypocycloid.

Similarly, the standard parametrization of an epitrochoid is

$$(c+d) \begin{bmatrix} \cos \theta \\ \sin \theta \end{bmatrix} + r \begin{bmatrix} \cos \phi \\ \sin \phi \end{bmatrix}, \quad (5)$$

where $\phi = (c+d)\theta/d$, with $c > 0$, $d > 0$, and $r > 0$. With $\lambda = (c+d)/(c+d+r)$, $\mu = r/(c+d+r) = 1 - \lambda$, $p = d$, $q = c+d$, and $\theta = dt$, an equivalent form to (5) is

$$(c+d+r)\left(\lambda \begin{bmatrix} \cos pt \\ \sin pt \end{bmatrix} + \mu \begin{bmatrix} \cos qt \\ \sin qt \end{bmatrix}\right). \quad (6)$$

Dürer's Hypocycloid

Comparing (2) and (6), we can collapse the hypotrochoids and epitrochoids into a single normalized form,

$$\mathcal{T}(p, q, \lambda, \mu, t) = \lambda \begin{bmatrix} \cos pt \\ \sin pt \end{bmatrix} + \mu \begin{bmatrix} \cos qt \\ \sin qt \end{bmatrix}, \tag{7}$$

where $0 < \lambda < 1$, $\mu = 1 - \lambda$, and p and q are nonzero. If p and q have the same sign then $\mathcal{T}(p, q, \lambda, \mu, t)$ is an epitrochoid, and if p and q have opposite signs then $\mathcal{T}(p, q, \lambda, \mu, t)$ is a hypotrochoid. In particular, by (4), for any real number b with $0 \le b < 1$,

$$\mathcal{T}\left(1-b, -(1+b), \frac{1+b}{2}, \frac{1-b}{2}, t\right) \tag{8}$$

parametrizes a hypocycloid. Furthermore, a useful non-normalized parametrization of the hypocycloid is

$$\mathcal{Y}(t) = \begin{bmatrix} \cos t \\ \sin t \end{bmatrix} + \frac{1}{\lambda} \begin{bmatrix} \cos(-\lambda t) \\ \sin(-\lambda t) \end{bmatrix} \tag{9}$$

where $\lambda > 0$; to see this, reparametrize (4) so that $\tau = pt$ and $\lambda = q/p$, then divide by $q/(p+q)$, and finally replace τ with t.

The polar curve $r = 1 + a\cos\theta$, where a is a positive number, is called a *limaçon*. It belongs to the trochoid family in that it is a translate of the curve given by

$$\begin{bmatrix} \cos t \\ \sin t \end{bmatrix} + \frac{a}{2} \begin{bmatrix} \cos 2t \\ \sin 2t \end{bmatrix}, \tag{10}$$

a non-normalized form of (7). To see that this is so, in (10), replace $\cos(2t)$ with $2\cos^2(t) - 1$ and $\sin(2t)$ with $2\sin(t)\cos(t)$, which simplifies to $(1 + a\cos t)(\cos t, \sin t) - (a/2, 0)$, a representation for the polar curve $r = 1 + a\cos\theta$ shifted left by distance $a/2$.

Figure 4 depicts some of the trochoids characterized by (7) showing (a) a *limaçon*, (b) a polar rose, (c) a *cardioid* as well as an epicycloid, (d) a hypocycloid, (e) a typical epitrochoid, and (f) a typical hypotrochoid.

We call (7) a *bungee cord characterization* of the trochoid family because when two people run on a circular track holding the ends of a bungee cord between them, one running around the track at angular frequency p and the other at angular frequency q, then a tack λ of the way along the length of the cord between the runners traces a trochoid in time, as is explored in [121].

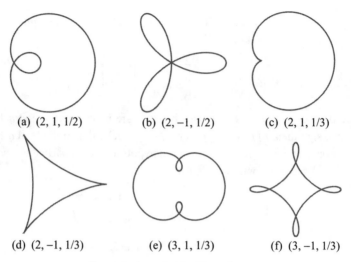

Figure 4. A trochoid gallery, (p, q, λ)

How Dürer did it

Dürer developed an instrument to draw trochoids, curves which he called *spiders* "because when they are fully drawn they resemble spiders" [39, p. D1v]. Figure 5 is a simplified version of the instrument, while Figure 14 is his instrument with four joints and telescoping rods. To orient the machine of Figure 5 with two joints and telescoping rods, position the main handle of the machine along the x-axis so that the hub labeled O is at the origin. Let B be the position of the second hub, and let P be the end of the rod extending out from B. Imagine some function relating θ and ϕ, where θ is the angle from the positive x-axis to the arm A from O to B and ϕ is the angle between the arm from B to P and the direction of arm A. The point labeled P traces out a curve. Because the arms are extensible to lengths a and b, this instrument can be used to construct a wide variety of trochoids whose parametrizations are

$$a \begin{bmatrix} \cos\theta \\ \sin\theta \end{bmatrix} + b \begin{bmatrix} \cos(\phi + \theta) \\ \sin(\phi + \theta) \end{bmatrix}. \tag{11}$$

Figure 6(a), from Dürer's book, is an epitrochoid corresponding to $\phi = \theta$, $a = 3$, and $b = 2$ in (11):

$$3 \begin{bmatrix} \cos\theta \\ \sin\theta \end{bmatrix} + 2 \begin{bmatrix} \cos 2\theta \\ \sin 2\theta \end{bmatrix}, \tag{12}$$

Dürer's Hypocycloid

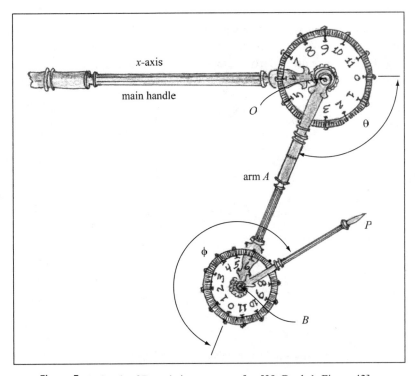

Figure 5. A sketch of Dürer's instrument, after [**39**, Book 1, Figure 43]

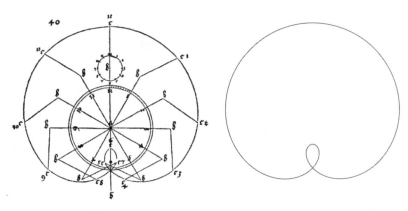

(a) Construction via Dürer's instrument
[**39**, Book1, Figure 40]

(b) Construction via a CAS

Figure 6. A spider from Dürer's treatise

which in terms of (7) is $(a+b)\mathcal{T}(1, 2, a/(a+b), b/(a+b), \theta)$. Dürer's convention for the direction of a positive angle was the clockwise direction. In 6(a), he shows the positions of the two arms of his machine at twelve points, at increments of thirty degrees. Figure 6(b) is the graph given by (12) rotated $\pi/2$ radians counterclockwise, which gives a good match to the curve in Figure 6(a).

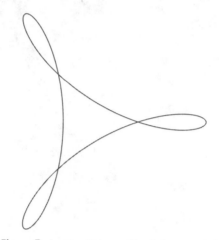

Figure 7. Another Dürer spider (a hypotrochoid)

Although no hypotrochoid graphic appears in Dürer's treatise, his machine could easily generate them. For example, (11) with $\phi = -3\theta$ and $a = 3/5$, $b = 2/5$ is

$$\frac{3}{5}\begin{bmatrix} \cos\theta \\ \sin\theta \end{bmatrix} + \frac{2}{5}\begin{bmatrix} \cos 2\theta \\ -\sin 2\theta \end{bmatrix},$$

which in terms of (7) is $\mathcal{T}(1, -2, 3/5, 2/5, \theta)$, and whose graph is Figure 7. With the values $\phi = -3\theta$ and $b = a/2$, Dürer's machine generates a hypocycloid whose graph looks like Figure 4(d). Exercise 4a shows a Dürer curve that resembles a hypotrochoid, and asks for the trochoid which best matches it.

Dürer's lamb

Is Dürer's lamb in Figure 1(a) resting on an arch of a hypocycloid? To investigate, we superimpose a circle C with center O onto a copy of Dürer's

Dürer's Hypocycloid

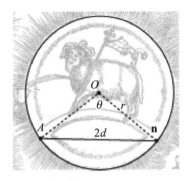
(a) A Dürer hypocyloid?

(b) The verdict

Figure 8. Superimposing the trochoid onto the cut

cut so that the circumference bounds the outer rim of the halo surrounding the lamb. Identify two endpoints A and B on C of the hypocycloid-like curve as shown in Figure 8(a). With a rule, approximate the radial distance r of the circle and d, half the distance between A and B. The measure θ of the central angle AOB is $2\sin^{-1}(d/r)$, which is $\theta \approx 1.88$ radians.

If the parameter t is viewed as a time variable, the trail left by the tack in the wheel that rolls around the circle is traced in time as given by (9). At those moments when the tack touches the circle (so forming a cusp), the tack's speed appears to halt, an event that occurs when $|\mathcal{Y}'(t)|^2 = 0$, where $|\mathcal{Y}'(t)|$ represents the magnitude of the vector $\mathcal{Y}'(t)$. Therefore

$$0 = |(-\sin t - \sin(\lambda t), \cos t - \cos(\lambda t)|^2 = 2 + 2\sin(t)\sin(\lambda t) - 2\cos(t)\cos(\lambda t).$$

Rearranging these terms gives

$$\cos((1+\lambda)t) = 1,$$

which has solutions at integer multiples of $t = 2\pi/(1+\lambda)$. Furthermore, since the first component of $\mathcal{Y}'(t)$ at $t = 2\pi/(1+\lambda)$ is 0, then $\sin(2\pi/(1+\lambda)) = -\sin(2\pi\lambda/(1+\lambda))$. Thus the sine of the angle spanning the first arch of the hypocycloid in Figure 3 is obtained by the quotient of the y-coordinate of $\mathcal{Y}(2\pi/(1+\lambda))$ and the length of $\mathcal{Y}(2\pi/(1+\lambda))$. That is, the sine of the spanning angle θ is

$$\sin\theta = \frac{\sin\left(\frac{2\pi}{1+\lambda}\right) - \frac{1}{\lambda}\sin\left(\frac{2\pi\lambda}{1+\lambda}\right)}{1 + 1/\lambda} = \sin\left(\frac{2\pi}{1+\lambda}\right). \tag{13}$$

Now identify $\angle AOB \approx 1.88$ with θ, and find λ. Since θ is between 0 and π, then

$$\theta = \frac{2\pi}{1+\lambda} \quad \text{or} \quad \theta = \pi - \frac{2\pi}{1+\lambda},$$

which means that $\lambda \approx 2.34$, the correct value, or $\lambda \approx 3.98$, a spurious value. With $\lambda \approx 2.34$, a graphic of the hypocycloid within a circle of radius $1 + 1/\lambda$ can be transported into an environment allowing graphics to be rotated, scaled, and superimposed, resulting in a graphic similar to Figure 8(b), showing that, no, Dürer most definitely was not thinking of a hypocycloid when designing "The Adoration of the Lamb." Instead, the arch is more like the arc of a circle.

Dürer's grapevine

Does Dürer's grapevine in Figure 1(b) follow the path of a limaçon $r = 1 + a\cos\theta$ for some positive number a? To investigate we remember that the limaçon's orientation in the xy-plane as generated by (10) will be that of the grapevine rotated clockwise by $90°$. To determine a good choice for the parameter a, we label the point of crossing in the limaçon as the origin O, and the low points of the outer and inner loops as A and B, respectively. Let C be the right-hand shoulder point of the limaçon, the point on the limaçon where a horizontal line falling from above would come to rest. Let D be the right-hand side point of the limaçon, where a vertical tangent line occurs. Let A_x, B_x, C_x, and D_x be the vertical displacements between O and the respective points A, B, C, and D.

Measure the distances A_x, B_x, C_x, and D_x. For example, on my scaled copy of "The Circumcision," $A_x \approx 4\frac{17}{32}$ inches and $B_x \approx 1\frac{15}{32}$ inches. The values for these quantities will vary according to the scaling used in reproducing Dürer's cut. However answers for the ratios B_x/A_x, A_x/C_x, and D_x/C_x should be invariant, that is, $B_x/A_x \approx 0.324$, $A_x/C_x \approx 7.75$, and $D_x/C_x \approx 2.85$.

For the limaçon, A corresponds to $\theta = 0$ and B corresponds to $\theta = \pi$, which means that A_x also has positive radial length $1 + a$ and B_x has positive length $a - 1$, (since a limaçon has an inner loop only when $a > 1$). Thus to find an a value for which the limaçon has a given B_x/A_x ratio, we solve

$$\frac{a-1}{1+a} = \frac{B_x}{A_x},$$

Dürer's Hypocycloid

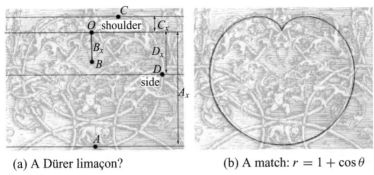

(a) A Dürer limaçon?   (b) A match: $r = 1 + \cos\theta$

Figure 9. Superimposing a limaçon onto the cut

giving $a \approx 1.685$. When a scaled, rotated version of the corresponding limaçon is superimposed upon Figure 1, its inner loop is much too wide although its outer loop matches very well, as the reader may verify in Exercise 6. However Exercise 3 shows that just because Dürer may have been trying to draw a certain curve, his final woodcut may not have been rendered accurately.

Can we find a better limaçon match? A look at Figures 1(b) and 9(a) shows that the grapevine loop, as it rises from the bottom of the woodcut, terminates slightly before the two vine ends cross at O. We find a second possible match by generating the limaçon corresponding to the A_x/C_x ratio, rather than the B_x/A_x ratio, and doubly check the result by using the D_x/C_x ratio.

Shoulder points of the limaçon occur at vertical tangent lines to $r = 1 + a\cos\theta$, which is where the derivative of $x = (1 + a\cos\theta)\cos\theta$ equals 0. Solving

$$x' = -a\sin\theta\cos\theta - (1 + a\cos\theta)\sin\theta = 0$$

for θ shows that the shoulder point C occurs at

$$\theta_C = \cos^{-1}\left(-\frac{1}{2a}\right),$$

which means that

$$C_x = |(1 + a\cos\theta_C)\cos\theta_C| = \frac{1}{4a}.$$

Since the top and bottom points of $r = 1 + a\cos\theta$—which we call the side points of the limaçon—are the points at which horizontal lines are tangent to the limaçon, then the side points occur where the derivative of

$y = (1 + a \cos \theta) \sin \theta$ equals 0. Similarly solving the appropriate equation shows that the side point D occurs at

$$\theta_D = \cos^{-1}\left(\frac{\sqrt{1 + 8a^2} - 1}{4a}\right),$$

which means that

$$D_x = (1 + a \cos \theta_D) \cos \theta_D = \frac{\left(\sqrt{1 + 8a^2} + 3\right)\left(\sqrt{1 + 8a^2} - 1\right)}{16a}.$$

Solving

$$4a + 4a^2 = \frac{A_x}{C_x} \approx 7.75$$

gives $a \approx 0.98$. Furthermore, solving

$$\frac{\left(\sqrt{1 + 8a^2} + 3\right)\left(\sqrt{1 + 8a^2} - 1\right)}{4} = \frac{D_x}{C_x} \approx 2.85$$

gives $a \approx 0.97$. That is, it appears as if the best-fitting limaçon for Dürer's cut is when $a \approx 1$, namely, a *cardioid*. In order to have been a perfect match for the cardioid, the ratios A_x/C_x and D_x/C_x would have to have been 8 and 3, respectively.

When a scaled, rotated version of $r = 1 + \cos \theta$ is superimposed upon the vine, the cardioid matches fairly well, as shown in Figure 9(b). That is, Dürer could have been thinking of what is now called a cardioid when designing "The Circumcision."

For other examples of a possible trochoid appearance in Dürer's artwork, see Exercise 7.

The hypocycloid meets the cycloid

A curve older than the hypocycloid is the cycloid that by tradition was discovered in 1450 by Nicholas of Cusa (1401–1464), a contemporary of Copernicus, who like Copernicus was also a cleric and an astronomer and who also thought that the earth circled the sun. However, it was Galileo who named the curve around 1599 while he was timing iron balls rolling down curved inclines. In 1658, Blaise Pascal, who by this date had more or less given up mathematics for theology, awoke one night with a toothache. Unable to sleep he began working on the geometrical properties of the cycloid.

Dürer's Hypocycloid

To his surprise the tooth ceased to ache. Regarding this as a divine intimation to proceed with the problem, he worked incessantly at it for eight days and completed a tolerably full account of the cycloid [**153**].

The cycloid, like the hypocycloid, is the path traced by a tack in the rim of a wheel as it rolls along level ground. We derive the parametric equations characterizing the cycloid in a nonstandard way by imagining the wheel of radius d having a tack in its rim as rolling around the inside of a very large circle of radius c as shown in Figure 10.

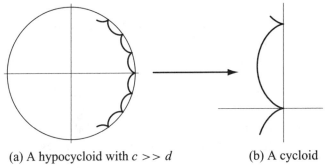

(a) A hypocycloid with $c \gg d$ (b) A cycloid

Figure 10. The cycloid as the limit of the hypocyloid

If $c \gg d$, then when t is near 0, $\cos(t) \approx 1$ and $\sin(t) \approx t$, which means that the hypocycloid characterization of (1) becomes

$$\mathcal{K}(t) \approx \left((c-d) + d\cos\left(\frac{c-d}{d}t\right), \ (c-d)t - d\sin\left(\frac{c-d}{d}t\right) \right),$$

near 0. Using the change of variable $s = (c-d)t/d$ and translating by $(-c, 0)$ (since $\mathcal{K}(0) = (c, 0)$) results in the approximation

$$d(\cos s - 1, \ s - \sin s), \tag{14}$$

which is indeed the parametric equations of a cycloid, albeit the road is the y-axis and the wheel of radius d rolls along its left-hand surface.

The hypocycloid as a differential equation

Voltaire had many pen-names. For a document to be printed according to appropriate channels in Voltaire's France, the manuscript had first to pass inspection by a royal censor. In fact, with respect to censorship and the new technology of printing starting in about 1450, the Catholic Church, partly in response to such books as written by Rabelais, decreed in 1543 that books could be printed and distributed only by permission of the Church. Following this ruling, in 1563 Charles IX (1550–1575) of France at age thirteen decreed that nothing could be printed without royal permission, a policy continued until the French Revolution. Meanwhile authors and publishers often tried circumventing this constrictive bottleneck. As a result many pamphlets and treatises were published anonymously or under pen-names.

Such practices pose a problem: how do we recognize Voltaire's work, or for that matter, that of any author?

The same is true for curves. Any given curve has many different names and characterizations. How do we know when any two are the same? Besides luck, we have a few guidelines. One way is to be familiar with how the curve arises dynamically as a rate of change of one variable with respect to another—that is, as a differential equation.

With this approach in mind, we seek a differential equation characterization of (9).

For (9), the largest radial value R is $R = 1 + 1/\lambda$ and the least radial value b is $b = |1 - 1/\lambda|$. Expanding and simplifying $x^2 + y^2$ gives

$$r^2 = x^2 + y^2 = \left(1 + \frac{1}{\lambda^2}\right) + \frac{2}{\lambda} \cos\left((1+\lambda)t\right). \tag{15}$$

By (15) and the double angle formulas $\sin^2 \psi = (1 - \cos(2\psi))/2$ and $\cos^2 \psi = (1 + \cos(2\psi))/2$,

$$\sqrt{R^2 - r^2} = \left|\frac{2}{\sqrt{\lambda}} \sin\left(\frac{1}{2}(1+\lambda)t\right)\right| \text{ and } \sqrt{r^2 - b^2}$$

$$= \left|\frac{2}{\sqrt{\lambda}} \cos\left(\frac{1}{2}(1+\lambda)t\right)\right|. \tag{16}$$

Since $\theta = \tan^{-1}(y/x)$ for $0 \le \theta < \pi/2$ and $\theta = \cot^{-1}(x/y)$ for $\pi/2 \le \theta < \pi$ and since $r^2 = x^2 + y^2$, then

$$\frac{d\theta}{dt} = \frac{xy' - yx'}{r^2} \text{ and } \frac{dr}{dt} = \frac{xx' + yy'}{r},$$

where $x' = dx/dt$ and $y' = dy/dt$. Therefore,

$$\frac{d\theta}{dr} = \frac{xy' - yx'}{r(xx' + yy')}, \tag{17}$$

where

$$xy' - yx' = \left(1 - \frac{1}{\lambda}\right)\left(1 - \cos\left((1+\lambda)t\right)\right) \tag{18}$$

and

$$xx' + yy' = -\left(1 + \frac{1}{\lambda}\right)\sin\left((1+\lambda)t\right). \tag{19}$$

By (15) through (19),

$$\frac{b\sqrt{R^2 - r^2}}{rR\sqrt{r^2 - b^2}} = \left|\left(\frac{\lambda - 1}{\lambda + 1}\right)\frac{\sin\left(\frac{1}{2}(1+\lambda)t\right)\cos\left(\frac{1}{2}(1+\lambda)t\right)}{r\cos^2\left(\frac{1}{2}(1+\lambda)t\right)}\right|$$

$$= \left|\frac{1}{r}\left(\frac{\lambda - 1}{\lambda + 1}\right)\left(\frac{\sin\left((1+\lambda)t\right)}{\cos\left((1+\lambda)t\right) + 1}\right)\right|$$

$$= \left|\frac{1}{r}\left(\frac{1 - \frac{1}{\lambda}}{1 + \frac{1}{\lambda}}\right)\left(\frac{1 - \cos\left((1+\lambda)t\right)}{\sin\left((1+\lambda)t\right)}\right)\right| = \left|\frac{d\theta}{dr}\right|.$$

That is, the differential equation that characterizes the hypocycloid is

$$\frac{d\theta}{dr} = \pm\frac{b\sqrt{R^2 - r^2}}{rR\sqrt{r^2 - b^2}}. \tag{20}$$

To see that any hypocycloid satisfies (20), let Γ be a general hypocycloid. There exists a standard hypocycloid γ for which Γ is a scaled copy of γ. That is, there is a positive number ϵ with $\Gamma = \epsilon\gamma$, which means that $\epsilon(x(t), y(t))$ is a parametrization of Γ. Then the radius $\hat{R}$ of the large circle (for the hypocyloid construction of Γ) is $\hat{R} = \epsilon R$, and the least distance $\hat{b}$ of Γ from O is $\hat{b} = \epsilon b$, and the radial distance $\hat{r}(t)$ of Γ from O at any time is $\hat{r} = \epsilon r$, which means that Γ satisfies (20) as well.

A famous riddle

Perhaps the greatest story of a mathematician posing a riddle for another is when Johann Bernoulli asked in *Acta Eruditorum*, June 1696,

Given two points A and B, find the curve that a point, moving from A to B in a vertical plane under its own gravity, must follow so that, starting from A with zero velocity, it reaches B in the shortest possible time.

Although tacitly directed to the "most acute mathematicians in the entire world," it was really a challenge to Isaac Newton. As the story goes, when shown the riddle after a day of working as Warden of the Mint, where he had been appointed to reform the coinage of the realm, Newton solved it before retiring for the night. Upon later seeing an anonymous solution in the proceedings of the Royal Society, Bernoulli identified the author, exclaiming, "I recognize the lion by his paw" [**17**, p. 115].

The solution to Bernoulli's problem is the cycloid. But if we change the riddle slightly, the answer is the hypocycloid:

Given two points A and B on the surface of a homogeneous earth, find the curve γ that a point mass, moving from A to B under its own gravity in a plane through earth's center, must follow so that, starting from A with zero velocity, it reaches B in the shortest possible time. (21)

When the radius of a homogeneous planet goes to ∞ while the arc-length between cities A and B along the planet's surface remains fixed, the arc along the earth's surface between A and B approaches a straight line, and for all practical purposes, the gravity at any point on a connecting (relatively shallow) curve through the planet between A and B is a constant. Thus, as was shown by the argument culminating in (14), solving this more general problem, also solves Bernoulli's problem.

Fortunately, Bernoulli's method of solution (for he solved the problem before posing it in print) to the original problem generalizes as well. The two key ideas in Bernoulli's approach are *Snell's law* and the *conservation of energy*. First of all, if a particle of mass m is at distance r from earth's center, then it has gravitational *potential* energy equal to

$$P(r) = \int_0^r m(ks)\,ds = \frac{1}{2}mkr^2, \qquad (22)$$

where s is a dummy variable representing distance from O, and $-ks$ is the gravitational acceleration on any particle s units from O with k being a positive constant. As a particle moves from the earth's surface ($r = R$) with initial *kinetic* energy 0, then its change in potential energy equals the change in kinetic energy. When the particle is at radial distance r, its change in kinetic energy is $\frac{1}{2}mv^2$ and its change in potential energy is $\frac{1}{2}mk(R^2 - r^2)$

Dürer's Hypocycloid

by (22), which means that the speed of the particle is

$$v = \sqrt{k}\sqrt{R^2 - r^2} \qquad (23)$$

at radial distance r.

Let γ be a curve in polar coordinates whose θ component is given by a function in terms of its radial component r. Let $P = (r, \theta)$ be a point on γ, and let ϕ be the angle between the ray through P and the tangent to γ at P in the direction in which arclength s increases.

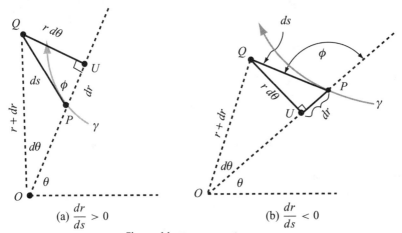

(a) $\dfrac{dr}{ds} > 0$ (b) $\dfrac{dr}{ds} < 0$

Figure 11. Tangent to the curve

Figures 11(a) and (b) each depict a right triangle PQU whose hypoteneuse PQ (along the tangent to γ at P) has approximate length ds. Since arc length is given by $ds/dr = \sqrt{1 + r^2(d\theta/dr)^2}$, we can construct a right triangle whose two other leg lengths are dr and $r\,d\theta$. To see how the geometry meshes with the algebra, the side opposite ϕ is approximately the arclength of a sector of a circle with radius $r + dr$ and central angle dr, namely, $(r + dr)d\theta$; since $dr\,d\theta$ is very small in comparison to $r\,d\theta$, a good approximation for the length of QU is $r\,d\theta$. Finally, side PU has length dr. In Figure 11(a), dr is positive and ϕ is acute, while in Figure 11(b), dr is negative and ϕ is obtuse.

By (23) and by Snell's law for a radially symmetric medium which says that $r \sin\phi/v$ is a constant (as shown in Item 12 of the appendix), then

$$\frac{r \sin\phi}{\sqrt{R^2 - r^2}} = K, \qquad (24)$$

where K is a constant.

With the geometry illustrated in Figure 11 that $\sin\phi = r\, d\theta/ds$, by (24)

$$\frac{d\theta}{ds} = \frac{K\sqrt{R^2 - r^2}}{r^2}. \tag{25}$$

Since $\cos\phi = dr/ds$, by (24),

$$\frac{dr}{ds} = \pm\sqrt{1 - \sin^2\phi} = \pm\frac{\sqrt{r^2 - K^2(R^2 - r^2)}}{r}. \tag{26}$$

Let b be the minimum radial coordinate on γ. Then $dr/ds = 0$ when $r = b$. By (26), $K = b/\sqrt{R^2 - b^2}$. So by (25) and (26), γ satisfies

$$\frac{d\theta}{dr} = \pm\frac{b\sqrt{R^2 - r^2}}{rR\sqrt{r^2 - b^2}},$$

the differential equation characterizing the hypocycloid as given by (20).

An Euler-Lagrange equation

In 1755, a nineteen year old Joseph Louis Lagrange wrote Euler a letter giving a way of solving problems like Bernoulli's riddle for Newton *without* having to use Snell's law. As some terminology, we say that a curve is *simple* if it does not intersect itself, that a parametrization is *smooth* if its derivatives of all orders exist, and that a parametrization is *regular* if the magnitude of its derivative is always positive. Here is an alternate way to find a regular curve γ which solves (21). As before, we utilize the principle of the conservation of energy so as to conclude that, if a point mass is at radial distance r from earth's center, its speed v along γ is given by (23). We say that the *transit time* for a particular arclength of γ is the time that it takes for the point mass to move along that portion of γ.

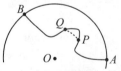
(a) A circular arc splice
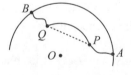
(b) A chordal splice
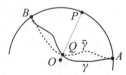
(c) A mirror image

Figure 12. Shorter time paths

If γ exists, we can conclude that it has the following three properties.

Dürer's Hypocycloid

Property 1. γ must have just one local extreme between A and B with respect to distance from O. If this were not so, consider Figure 12(a) where γ increases and then decreases its distance from O between points P and Q on γ that are equidistant from O. To find a path of shorter transit time, replace arc PQ of γ by the arc of a circle with center O and radius OP. This new arclength PQ is shorter than the old arclength PQ, and the speed along the new PQ is a constant that is no less than the variable speed along the old PQ. The splice-points at P and Q can be smoothened with negligible increase in transit time so that the new regular path has a shorter transit time than γ. What we have just shown is that γ is also simple so there are no places where γ crosses itself.

Property 2. γ's distance from O is ever changing as s increases. If this were not so, consider Figure 12(b) where the arclength along γ between P and Q is a circular arc concentric with O. Replace this arc with the line segment between P and Q. As before, the transit time along the new PQ is less than the transit time along the old PQ, and the splice-points can be smoothened so that the modified path has less transit time than does γ.

Property 3. Let P be the midpoint along the earth's surface between A and B. γ is symmetric with respect to the line through OP as shown in Figure 12(c). If this is not so, let $\bar{\gamma}$ be the mirror image of γ with respect to OP. Let Q be the point on γ and on $\bar{\gamma}$ and on OP closest to O. Let γ_1 be the portion of γ or $\bar{\gamma}$ from A to Q for which the transit time is least. Then γ can be replaced with the concatenation curve of γ_1 followed by the mirror image of γ_1. The splice point at Q can be made smooth with negligible increase in transit time so that the new regular path's transit time is no bigger than that of γ.

Let E be the point on γ nearest O by Property 1, and let b be the length of OE. By Property 3, we can conclude that E is the half-way point along γ. By Properties 1 and 2, there is a parametrization of γ from A to E giving θ in terms of r, denoted $\theta(r)$. Lagrange suggested that we consider a variation Θ of γ, (which led Euler to coin the technique as the *calculus of variations*),

$$\Theta(r, \epsilon) = \theta(r) + \epsilon \mu(r),$$

where ϵ is a real number and μ is any differentiable function for which $\mu(R) = 0 = \mu(b)$. Let

$$F(r, \Theta') = \frac{\sqrt{1 + (r\,\Theta')^2}}{\sqrt{R^2 - r^2}}, \qquad (27)$$

where $\Theta' = d\Theta/dr$. Since $\sqrt{1 + (r\Theta')^2}\,dr$ gives the infinitesimal arclength along Θ and $\sqrt{k}\sqrt{R^2 - r^2}$ gives the speed of the point mass at radial distance r then

$$M = \frac{2}{\sqrt{k}} \int_b^R F(r, \Theta')\,dr$$

is the time it takes the point mass to go from A to E to B along the path $\Theta(r, \epsilon)$, as we assume that the second half of θ, the part corresponding to γ going from E to B, mirrors the first half. By the chain rule (where r, Θ, and Θ are treated as independent variables),

$$\frac{\partial F}{\partial \epsilon} = \frac{\partial F}{\partial r}\frac{\partial r}{\partial \epsilon} + \frac{\partial F}{\partial \Theta}\frac{\partial \Theta}{\partial \epsilon} + \frac{\partial F}{\partial \Theta'}\frac{\partial \Theta'}{\partial \epsilon} = \frac{\partial F}{\partial r} \cdot 0$$
$$+ 0 \cdot \mu(r) + \frac{\partial F}{\partial \Theta'} \cdot \mu'(r) = \frac{\partial F}{\partial \Theta'} \cdot \mu'(r),$$

where $\mu' = d\mu/dr$. The minimum value of M occurs when

$$\left.\frac{\partial M}{\partial \epsilon}\right|_{\epsilon=0} = 0 = \left(\frac{\partial}{\partial \epsilon} \int_{r_0}^R F(r, \Theta')\,dr\right)\bigg|_{\epsilon=0}$$
$$= \int_{r_0}^R \left(\frac{\partial F(r, \Theta')}{\partial \epsilon}\bigg|_{\epsilon=0}\right) dr = \int_{r_0}^R \frac{\partial F(r, \theta')}{\partial \theta'} \mu'(r)\,dr,$$

where $\theta' = d\theta/dr$. Using integration by parts gives

$$\left.\frac{\partial M}{\partial \epsilon}\right|_{\epsilon=0} = 0 = \frac{\partial F}{\partial \theta'}\mu(r)\bigg]_{r_0}^R - \int_{r_0}^R \mu(r)\frac{\partial^2 F(r, \theta')}{\partial r\,\partial \theta'}\,dr$$
$$= -\int_{r_0}^R \mu(r)\frac{\partial^2 F(r, \theta')}{\partial r\,\partial \theta'}\,dr. \qquad (28)$$

The only way for the last integral of (28) can be zero for every arbitrary differentiable function μ for which $\mu(b) = 0 = \mu(R)$ is for

$$\frac{\partial^2 F(r, \theta')}{\partial r\,\partial \theta'} = 0.$$

Dürer's Hypocycloid

Integrating with respect to r gives the Euler-Lagrange equation,

$$\frac{\partial F(r, \theta')}{\partial \theta'} = K, \qquad (29)$$

where K is some constant.

Thus if there is a path of least transit time between A and B, by (27) and (29), there is a regular path of least transit time between A and B satisfying the equation

$$\frac{r^2 \theta'}{\sqrt{1 + r^2 \theta'^2}\sqrt{R^2 - r^2}} = K.$$

Solving this for θ' gives

$$\frac{d\theta}{dr} = \frac{K\sqrt{R^2 - r^2}}{r\sqrt{r^2 - K^2(R^2 - r^2)}}. \qquad (30)$$

When $r = b$, namely, when the point mass is at E, the low point on γ, then $dr/d\theta = 0$. By (30), $b^2 - K^2(R^2 - b^2) = 0$, which gives $K = b/\sqrt{R^2 - b^2}$. Substituting this expression for K in (30) gives (20), the differential equation whose solution is the hypocycloid. Thus—by way of several nifty convergence theorems from analysis which we have avoided mentioning specifically, such as being able to smooth a curve at a splice point or being able to interchange the order of integration and differentiation—the hypocycloid is indeed the curve of least transit time between A and B.

In this next chapter, we generalize (21) to a rotating earth and calculate the transit times for particles to move along the solution curves.

Exercises

1. (a) The epicycloid has equation

    $$\begin{aligned} E(t) &= (x(t), y(t)) \\ &= \left(\cos t + \frac{1}{\lambda}\cos(\lambda t),\ \sin t + \frac{1}{\lambda}\sin(\lambda t)\right). \end{aligned} \qquad (31)$$

    Find a formula analogous to (13) giving the spanning angle of one arch of the epicycloid.

    (b) Show that the limit of the epicycloid (as $\alpha \to \infty$) near 0 results in a cycloid parametrization as in (14).

2. (a) With respect to (9), if $\lambda = m/n$ where m and n are relatively prime positive integers, how many cusps does the hypocycloid have?
   (b) Find the arclength of one arch of the hypocycloid and of the epicycloid.

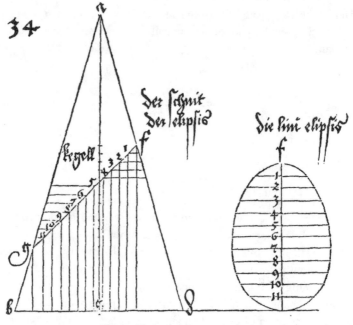

Figure 13. Dürer's ellipse [39]

3. Just because Dürer may have been trying to draw an abstract curve does not mean that the result will be a good approximation to it. For example, Figure 13 shows Dürer's attempt to draw an ellipse by transforming the curve of the intersection between a plane and a cone onto paper. Dürer called the result an egg-curve, "eierlinie." Obviously, something is amiss. Find the eccentricity of an ellipse which best approximates his curve, and superimpose this ellipse on the figure. See [65] for a discussion of Dürer's algorithm in constructing this conic section.

4. (a) Figure 14(a) is from Dürer's *The Art of Measurement*. The graphic contains two curves, one somewhat like a polar rose $r = \cos(k\theta)$ for some real number k, and the other like a hypotrochoid. Find trochoids which best approximate these curves.

Dürer's Hypocycloid

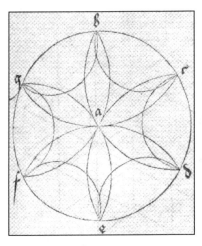
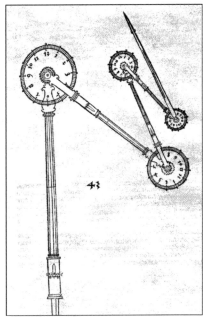

(a) Trochoid-like curves [**39**, book 2, portion of figure between figures 21 and 22]

(b) Dürer's machine [**39**, book 1, figure 43]

Figure 14. Dürer magic

    (b) Figure 14(b) is figure 43 from Dürer's treatise showing his machine for making curves. Generate some of the curves in the repertoire of this machine.

5. For Figure 1(a), demonstrate that the arc on which the lamb rests is probably that of a circle. How do the radii of the two circles compare (that of the arc and that of the circle encompassing the lamb)?

6. For Figure 1(b), generate the cardioid for the given ratio B_x/A_x on p. 172, and demonstrate that it is a poor match for Dürer's curve.

7. Figure 15 shows portions of two Dürer woodcuts that resemble epicycloids, one between two dog heads and the other beneath an open window. In each case find a λ value for which (31) generates an epicycloid best matching the indicated curve.

(a) *Triumphal Arch:* 1512, [**38**, cut 275]   (b) *Dürer's coat of arms:* 1523, [**38**, cut 327]

Figure 15. Dürer epicycloids?

8. Figure 16 is a woodcut from William Caxton's *The Mirrour of the World*, from 1481 that illustrates a passage in which Caxton explains why the world is round:

> God formed the world all round; for of all the forms that be, of what diverse manners they be, none are so full or receptive so much by nature as is the figure round that may hold within it so much in right quantity as does the round [**23**, pp. 58–59].

He goes on to show that figures with corners, when rotating, "take diverse places that the round seeks not." Inside the double circle, which

Figure 16. A Caxton hypocycloid?

represents the earth's form, are two shapes, a square and one that appears to be a hypocycloid. Superimpose a graph of the periodic hypocycloid with four cusps onto Caxton's curve. Do they match?

9. Find a differential equation analogous to (20) that characterizes the epicycloid as given by (31).

10. (a) Rewrite (14) in standard form so that it corresponds to the cycloid formed by a tack in a rim of a wheel of radius β rolling along (on the upward side of) the x-axis.

   (b) Use the result of Exercise 2b to find the arclength of one arch of the cycloid given by (14).

Vignette VIII: Maupertuis's Hole

Maupertuis championed the idea of digging a deep hole in the earth as we saw in Vignette I.

The idea is an old one, going back to at least the time of Hesiod who claimed that it would take nine days for an anvil to fall from earth's surface to the underworld. The idea was common knowledge by the late Middle Ages. It appeared in the first printed book with illustrations in the English language, William Caxton's *Mirrour of the World*, an encyclopedia of what everyone should know that appeared in 1481 and was a translation of a French manuscript of 1245, which in turn was a translation of an earlier text in Latin.

> And if the earth were pierced through in two places, of which that one hole were cut into the other like a cross, and four men stood right at the four heads of these two holes, one above and another beneath, and like-wise on both sides, and that each of them threw a stone into the hole, whether it were great or little, each stone should come into the middle of the earth without ever to be removed from thence. [**23**, p. 55]

Caxton illustrated these dynamics with the woodcut of Figure 1 and speculated what would happen if one stone was heavier than another.

About such a hole, Leonhard Euler remarked in his letters to a teen-ager, Princess Charlotte, whom he tutored for three years [**42**, p. 178]:

> You will remember how Voltaire used to laugh about a hole going to the center of the earth, mentioned by Maupertuis. But there is no harm in imagining it to see what would happen.

While most natural philosophers were content with mind experiments of a hole through the earth, Maupertuis wanted a real one.

Why?

As we have seen, Voltaire and Maupertuis were friends. While Voltaire was in exile to England, Maupertuis visited London in 1728 so as to learn

Figure 1. Holes through the center of the earth, a Caxton wood-cut of 1481

more about Newtonian mathematics, and was promptly named a member of the Royal Society. Perhaps Voltaire and Maupertuis first met in England. Maupertuis is said to have contracted syphilis there [**115**, p. 48], which at that time was basically incurable: its latter stages often involved intervals of dementia and clarity, which may help explain his more erratic behavior late in life. However Beeson [**16**, p. 264] cites a physician who concludes that although Maupertuis suffered from an unknown lung disease for the last twenty-five years of his life, there is only a "slight possibility" that his ailments were due to tertiary syphilis.

Back in France, Voltaire asked Maupertuis to tutor him so that he could understand Newton's ideas. In a letter of 1732, Voltaire thanks him using ecclesiastical metaphors.

> Your first letter baptized me in the Newtonian religion, your second gave me my confirmation. I thank you for our sacrament. [**140**, p. 83]

Voltaire was a cheerleader for Maupertuis during the polar expedition—despite the fact that Emily had had an affair with Maupertuis, when all that

Voltaire had asked from him was mathematics tutoring for her. In exultation of Maupertuis's triumph that Newton was correct and the Cassinis who had argued for a lemon-shaped earth were wrong, Voltaire called Maupertuis "the Flattener of the Earth and of the Cassinis, too," a title that Maupertuis perhaps wore too proudly as the years passed, and ultimately became a phrase of ridicule. Figure 2 is a self-commissioned portrait of Maupertuis in his Lapland garb posed in the act of flattening a globe. Of Maupertuis's published account of the expedition, Voltaire called it

> a story and a piece of physics more interesting than any novel. [**140**, p. 126]

In 1738, after reading Maupertuis's treatise on the shape of the earth, the crown prince Frederick, hoping to re-establish the Berlin Academy when he became king, asked in his ongoing exchange of letters with Voltaire whether Maupertuis was the man who could do so. Voltaire replied,

> A man like [Maupertuis] would establish in Berlin an academy of science that would outdo the Parisian one. [**140**, p. 181]

In course, Maupertuis met the king face to face, and returned the favor to Voltaire, arranging a private meeting between Voltaire and Frederick in 1740. In 1745, Maupertuis finally accepted the presidency of the Berlin Academy. The decision had been a soul-searching, radical move for a Frenchman because Prussia and France were oftentimes on opposing sides of the battlefield during much of the eighteenth century. Thus any French member of Frederick's court could not help but be considered a turncoat, a traitor, and a disappointment to France.

Maupertuis was a bold proponent of Newtonian thought. He is considered to be France's first Newtonian. Emily and Voltaire referred to him as "Sir Isaac Maupertuis" [**157**, p. 186]. As one who also loved optimization exercises, Maupertuis tended to minimize the ideas of Leibniz, adding more color to the already colorful controversy of ideas and reputations involving the rival claims with respect to the precedence and notation of Newton and Leibniz that yet continued long after after each giant had left the field of conversation.

Voltaire joined this court in 1750.

Why would he do so, as afterwards his already tenuous welcome in Paris was bound to be rescinded completely?

Voltaire was yet mourning Emily. He had been the royal historiographer for Louis XV for the past three years or so and his plays and poems were

Figure 2. Engraved portrait of Maupertuis by Jean Daullé of Tournières, courtesy of the Owen Gingerich collection

Vignette VIII: Maupertuis's Hole

currently achieving tepid reviews at best. Frederick had urged him to come to Prussia on numerous occasions over the past ten years. Voltaire was now 56 years old. It was now or never. Voltaire opted for a change. Perhaps he envisioned being Frederick's privy counselor on social issues. As some have speculated, Voltaire perhaps considered himself a secret agent, one who could send Louis XV timely news of Frederick's plans for European domination—after all, Voltaire had begun his career in the French diplomatic corps.

For a time all went well. But as Georges-Louis Leclerc, Comte de Buffon (1707–1788), a mutual friend of Maupertuis and Voltaire, anticipated [**140**, p. 305],

> Maupertuis and Voltaire are not made to live together in the same room.

Indeed, Voltaire was an alpha male. One biographer aptly called his account of Voltaire's life *Voltaire Almighty*. And Frederick's court was filled with alpha males. So recriminations from one to another richocheted thoughtlessly at times. Just as one instance: in response to a complaint that Frederick showed favoritism to Voltaire, the king replied [**101**, p. 222],

> I need [Voltaire] for another year at most. One squeezes an orange and throws away the peel.

During Voltaire's tenure at court, Maupertuis, as President of the Berlin Academy, proposed many ideas. In 1752, he published an open letter, ostensibly for the king, *A Letter for the Progress of Science,* containing a list of scientific investigations worthy of being pursued. Among his proposals were these [**84**, pp. 147–171].

- Continue looking for a northwest passage connecting the Atlantic to the Pacific. Explore Australia because its unique flora and fauna has developed in isolation from the rest of the world. Find and cultivate more spice islands besides those controlled by the current spice cartels.

- Find a practical way to determine longitude other than Cassini's method of looking at Jupiter's moons. Perhaps the key might lie with the *declination* (deviation from true north) of earth's magnetic field which Halley had mapped in 1700. Offer a prize for the development of reliable ship clocks.

- Offer a prize for advances in telescope design. Organize a multi-nation search for stars exhibiting *parallax*, the apparent movement of a nearer star against the backdrop of more distant stars as the earth moves along its orbit. Partition the sky into searchable regions by country. Continue looking for more moons in our solar system.

- Determine whether the aurora borealis is made from material escaping from the pole. Determine whether the southern hemisphere is shaped as the northern hemisphere.

- Use explosives so as to study the pyramids of Egypt. Dig a deep hole so as to study the earth's structure (see Vignette I for a translation of Maupertuis's paragraph on this proposal).

- Revive Latin as an international language by establishing a city in which it is the only authorized language. Raise a group of children in isolation from adults to determine the language they would develop.

- Test new medical procedures and remedies on criminals. For example, remove the kidney as a treatment for kidney stones. Probe the brains of the living to identify the location of mental functions. Use drugs such as opium to explore the mind. Determine whether the venom of scorpions, spiders, salamanders, serpents, and toads have medicinal value. Experiment with acupuncture.

- Pay physicians only if their treatments are effective. Allow doctors to specialize. Use statistics to determine the effectiveness of cures.

- Determine the limits of cross-breeding. Does the *jumar*—a cross between a donkey and a cow—exist? Grafting works with plants; does it work with animals? The crayfish regenerates a claw, the lizard its tail; can the human body regenerate a limb?

- Abandon all work on finding philosopher stones, perpetual motion machines, and algorithms for squaring the circle.

Upon hearing these open-ended proposals, as well as reacting to a dispute between Maupertuis and Samuel König, both of whom had been Emily's mathematics tutor years ago (and in fact König had been a two-year-long house guest of Emily and Voltaire [**147**, p. 4]), Voltaire wrote an essay ridiculing his long time friend, his rival in love, his mathematics tutor, and one whose scientific exploits he had admired. Printed by Voltaire for all to buy and read, *A Dissertation by Dr. Akakia* poked fun at the distinguished President of the Berlin Academy.

It is to this satire that Euler refers when he wrote about Voltaire's laughter at Maupertuis's idea of a hole toward earth's center. Unfairly or not, Voltaire wrote [**144**, vol. xxxvii, *Dr. Akakia*, pp. 197–198],

Vignette VIII: Maupertuis's Hole

> We must further inform [Maupertuis], that it will be extremely difficult to make, as he proposes, a hole that shall reach to the center of the earth—where he probably means to conceal himself from the disgrace to which the publication of such principles has exposed him. This hole could not be made without digging up about three or four hundred leagues of earth; a circumstance that might disorder the balance of Europe.

Voltaire sprinkled allusions to this image in his writings for years afterwards. For example, in an essay on rivers in his *Philosophical Dictionary* where he speaks of an ancient model of river networks penetrating the earth, as appears in the works of Hesiod, Plato, Virgil, and Ariosto, to name a few, he mentions a cavern [**144**, vol. xiii, p. 141]:

> In conformity to the physics of antiquity is the cavern (from which issue all the rivers of the earth) being in the very center of the earth. It is here that Maupertuis wanted to take a tour.

Voltaire does offer sound advice in *Dr. Akakia* [**144**, vol. xxxvii, pp. 197–198]:

> Let him [Maupertuis—and anyone in administration] be persuaded that all men of letters are equal, and we are sure he will gain by this equality. Let him never be so foolish as to insist that nothing should be printed without his order.

In this even-handed-at-first-glance counsel, the barb stings: Voltaire's appraisal is that Maupertuis is less than an average man of letters. In 1763, an early editor of Voltaire's work, Tobias Smollett, a satirist and novelist, prophesied that "such immortality as pertains to Maupertuis is due to [Voltaire's] exquisite satire" [**144**, vol. xxxvii, *Dr. Akakia*, p. 183]. Even today, the sting persists. For example, a reviewer of a recent biography of Maupertuis, with respect to Voltaire's phrase "that nothing be printed without [Maupertuis's] order," describes Euler as Maupertuis's hatchet man, using " 'glove-and-fist' operations [involving] sophistry and naked threats" to intimidate Leibnizians [**115**, p. 49]. In a recent comic book history of Euler, translated from German, two mid-eighteenth century pedestrians, on passing a book shop and seeing Maupertuis's name well down the list of scientific best-sellers, quip [**67**, p. 28],

> Since [Maupertuis] has proved that the poles are flat, his brain too has lost some of its sharpness.

As a last example of this controversy that fails to die, a current accomplished expositor of mathematics, referring to the *Dr. Akakia* satire, says that Voltaire is "a poison pen" and a "math-illiterate" [**92**, p. 134].

Although the king might laugh at Voltaire's manuscript and share in a private joke with Voltaire, a public printing was unacceptable. Public ridicule of an officer of the academy was tantamount to ridicule of Frederick. Voltaire was now in a hole of his own. His exit from the court was assured, whereupon [**101**, p. 223],

> I [Voltaire] resolved therefore to place the orange peel in safe keeping.

CHAPTER VIII

Newton's Other Ellipse

For a hole through the earth, how would a pebble fall therein?

Caxton imagined a straight line hole and said that a pebble would fall so as to reach earth's center with zero speed, and so remain there forever. The second century writer Plutarch had a different idea. He imagined that a boulder falling freely through a spherical earth would pass through earth's center and proceed to the other side, and then fall back, retracing its path, and fall again, and so on forever, somewhat like the motion of a balance scale or a swing [**107**, p. 65]. Much later, Galileo sharpened this image, using a cannonball rather than a rock in his thought experiment, and quantified simple harmonic motion [**47**, p. 227].

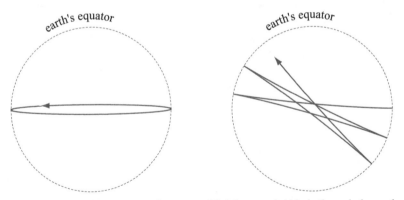

(a) An ellipse with respect to the stars (b) A hypocycloid hole through the earth

Figure 3. Pebble motion with respect to the stars and the earth

About fifty years later, Isaac Newton began the *Principia* with a discussion of Galileo's problem. He imagined a pebble falling without resistance through a homogeneously dense earth. With respect to the background of the fixed

197

stars, in corollary i of proposition 10 of book I of the *Principia* [**95**, p. 460], he showed that the path of the pebble lies in a plane and is an ellipse whose center coincides with earth's center. In contrast to the polar ellipse of Chapter IV, we refer to this instance of the ellipse as *Newton's other ellipse*. Along its natural parametrization in time, the pebble falls with simple harmonic motion in both its axial components, generating the ellipse of Figure 3(a). When the pebble is dropped at the equator, the path misses earth's center by about 376 km. But since the earth itself is turning, the hole of the pebble is quite different than an ellipse, and is in fact a hypocycloid as illustrated in Figure 3(b).

Lest the reader scoff at such motion as being impossible because of resistance, consider Stephen Hawking's thought experiment in [**63**, pp. 108–109]. As a potential power source, he proposes dropping a black hole with mass the size of a mountain from the surface of the earth. How to harness the energy of this motion we leave as a mystery. Yet because such a black hole's event horizon has radius about that of a nucleus of an atom, it rarely if ever strikes any particle of the earth as it falls back and forth through the earth. It truly is a free-falling pebble. And it is the path of this pebble through the earth that we wish to track.

In this chapter, we outline Newton's observation that a freely falling pebble in the earth follows an ellipse with respect to the stars. We then define elliptical precession, and demonstrate that any trochoid is a precessed ellipse. Along the way, we give the time lapse for a pebble to slide freely along an arch of any hypocycloid in the equatorial plane. And we end with an exploration of the surface generated by a pebble when dropped north of the equator: a hyperbolic surface of revolution, which gives rise to trochoidal lamp shade or flower vase designs.

The ellipse

For a homogeneously dense earth, Newton showed that its gravitational force function varies with radial distance r from earth's center.

He imagined the earth as composed of concentric spheres, and he showed that the net gravitational attraction by any hollow sphere on any body located anywhere inside the sphere is zero, and the net gravitational attraction by any sphere on any body located anywhere outside the sphere is exactly the same as that of a point of identical mass located at the center of the sphere. So as a body passes through the earth, the only part of the earth that attracts it is those spheres whose radii are less than or equal to the distance of the object from the center of the earth. Thus the mass of the earth acting on a body within the earth is proportional to r^3. Therefore, this attractive force on this body is proportional to r^3 and inversely proportional to r^2, and so the force $F(r)$ on the body is directly proportional to r, giving $F(r) = -kmr$, where k is some positive constant and m is the mass of the body. The associated gravitational acceleration on the body for our earth is thus $f(r) = -kr$, where $kR = g$ with R as earth's radius and $g \approx 9.8$ m/sec^2. For a careful derivation of this acceleration function see [**122**, ch. 2].

With this radially symmetric acceleration function f, the equations of motion, (III.16) and (III.18), for a freely falling pebble become

$$\frac{d^2 r}{dt^2} + kr = \frac{h^2}{r^3} \quad \text{and} \quad \frac{d^2 z}{d\theta^2} + z = \frac{k}{h^2 z^3}, \tag{1}$$

where θ is the polar angle to the pebble's position, $z = 1/r$, and h is the constant of angular momentum:

$$h = r^2 \left(\frac{d\theta}{dt} \right) = \frac{2\pi R^2}{Q} = \omega R^2, \tag{2}$$

with Q as earth's period and $\omega = 2\pi/Q$ as earth's angular frequency. The initial conditions for (1) are $r(0) = R = 1/z(0)$, $\left. \frac{dr}{dt} \right|_{t=0} = 0$, and $\left. \frac{dz}{d\theta} \right|_{t=0} = 0$. As shown in Exercise 1, the solutions to (1) are

$$r(t) = R \sqrt{\cos^2(\sqrt{k}\, t) + \frac{\omega^2}{k} \sin^2(\sqrt{k}\, t)} \tag{3}$$

and

$$r(\theta) = \frac{R}{\sqrt{\cos^2 \theta + \frac{k}{\omega^2} \sin^2 \theta}}. \tag{4}$$

The polar plot of (4),

$$\begin{bmatrix} X(\theta) \\ Y(\theta) \end{bmatrix} = r(\theta) \begin{bmatrix} \cos \theta \\ \sin \theta \end{bmatrix}, \tag{5}$$

is an ellipse of semi-axial lengths R and $\omega R/\sqrt{k}$ because $X^2/R^2 + Y^2/\left(\omega R/\sqrt{k}\right)^2 = 1$. Furthermore, if the pebble is dropped from rest at the earth's surface on the equator, then the pebble's furthest distance, its *apogee*, from earth's center O is R, its nearest distance, its *perigee*, is

$$\frac{2\pi R}{Q\sqrt{k}} = \frac{\omega R}{\sqrt{k}}, \qquad (6)$$

and its period is

$$T = \frac{2\pi}{\sqrt{k}}. \qquad (7)$$

An elegant form of (3), which incorporates position as well as distance from the origin, is the parametrization

$$P(t) = \begin{bmatrix} x(t) \\ y(t) \end{bmatrix} = R \begin{bmatrix} \cos(\sqrt{k}\,t) \\ \frac{\omega}{\sqrt{k}} \sin(\sqrt{k}\,t) \end{bmatrix}, \qquad (8)$$

as outlined in Exercise 2. Newton's first theorem in the *Principia* [**95**, p. 444] along with the aforementioned corollary i of proposition 10 is equivalent to (8), although he fails to quantify the semi-minor axis and the angular frequency. While (8) is an ellipse with respect to the fixed stars, in the next section we demonstrate that the hole it traces through the earth is a hypocycloid.

Toward that end, we adopt a standardized form of (8). Let

$$\mathcal{E}(b,\, t) = \begin{bmatrix} \cos t \\ b \sin t \end{bmatrix}$$

where $|b| \leq 1$. We say that $\mathcal{E}(b, t)$ is a *normalized* parametrization in t of an ellipse because with respect to t it parametrizes an ellipse with semi-major axial length 1 and semi-minor axial length $|b|$. When b is positive the ellipse is traced counterclockwise in time, and when b is negative, clockwise. Thus (8) is $R\,\mathcal{E}(\omega/\sqrt{k},\, \sqrt{k}\,t)$.

Precessing the ellipse

To *precess* a parametrization, we need a dynamic rotation matrix, so-called because the angle of precession changes with respect to time. Let $\mathcal{Q}(\alpha, t)$ be the 2×2 matrix which when multiplied with a two-dimensional vector will rotate it αt radians counterclockwise about the origin O,

$$\mathcal{Q}(\alpha,\, t) = \begin{bmatrix} \cos(\alpha t) & -\sin(\alpha t) \\ \sin(\alpha t) & \cos(\alpha t) \end{bmatrix}.$$

Since the earth rotates ωt radians counterclockwise in a time lapse of t, then with respect to the earth, the falling pebble is at $\mathcal{Q}(-\omega, t)P(t)$; that is, at time t, the pebble must be rotated ωt radians clockwise from its position with respect to the stars so as to obtain its position with respect to the earth. The graph of this hole through the earth parametrized in time by $\mathcal{Q}(-\omega, t)P(t)$ is shown in Figure 3(b).

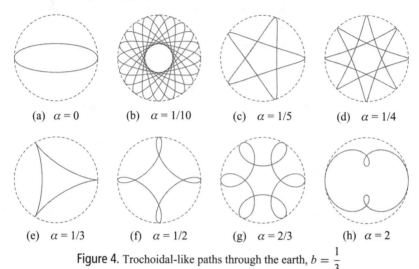

Figure 4. Trochoidal-like paths through the earth, $b = \dfrac{1}{3}$

We say that

$$\mathcal{H}(\alpha, b, t) = \mathcal{Q}(-\alpha, t)\mathcal{E}(b, t) = \begin{bmatrix} x(t) \\ y(t) \end{bmatrix}$$
$$= \begin{bmatrix} \cos\alpha t \cos t + b \sin\alpha t \sin t \\ -\sin\alpha t \cos t + b \cos\alpha t \sin t \end{bmatrix} \quad (9)$$

is a *parametric precession* of $\mathcal{E}(b, t)$ in t by the angular frequency α, where $|b| \leq 1$, and that its graph is a *precession of the ellipse* parametrized by $\mathcal{E}(b, t)$.

To imagine the dynamics of (9) consider this mind experiment (using distance and time units so that $R = 1$ and $\sqrt{k} = 1$). Suppose that the Norse god Thor has the supernatural ability both to change the angular frequency of earth at will and to manipulate Hawking's black hole. Varying William Caxton's suggestion in the woodcut of Figure 1 about throwing a stone down a hole from the equator, Thor, on a stationary earth, throws the black hole at a tangent to what will be the equator. Provided Thor's toss is not overly fast, the black hole follows an ellipse through the earth as given by $\mathcal{E}(b, t)$ for some b with $|b| \leq 1$. Next, Thor imparts to the earth an angular frequency of α. We assume that this rotation rate does not change the shape of the earth. This change in the motion of the earth has no effect upon the motion of the black hole in any way, but it does change the path of the hole through the earth so that the black hole's position at time t is $\mathcal{H}(\alpha, b, t)$. Figure 4 shows the graphs of some of these holes through the earth, as engineered by Thor. They look like members of the trochoid family.

We write the normalized parametrization of an ellipse in trochoid form from Chapter VII as

$$\mathcal{E}(b, t) = \begin{bmatrix} \cos t \\ b \sin t \end{bmatrix} = \frac{1+b}{2}\begin{bmatrix} \cos t \\ \sin t \end{bmatrix} + \frac{1-b}{2}\begin{bmatrix} \cos(-t) \\ \sin(-t) \end{bmatrix}$$

$$= T\left(1, -1, \frac{1+b}{2}, \frac{1-b}{2}, t\right).$$

By the addition identities for cosine and sine,

$$\mathcal{Q}(\alpha, t)\begin{bmatrix} \cos \beta t \\ \sin \beta t \end{bmatrix} = \begin{bmatrix} \cos(\alpha+\beta)t \\ \sin(\alpha+\beta)t \end{bmatrix} = \mathcal{Q}(\alpha+\beta, t)\begin{bmatrix} 1 \\ 0 \end{bmatrix}.$$

Therefore,

$$\mathcal{H}(\alpha, b, t) = \mathcal{Q}(-\alpha, t)\mathcal{E}(b, t) = \mathcal{Q}(-\alpha, t)\begin{bmatrix} \cos t \\ b \sin t \end{bmatrix}$$

$$= \frac{1+b}{2}\mathcal{Q}(-\alpha, t)\begin{bmatrix} \cos t \\ \sin t \end{bmatrix} + \frac{1-b}{2}\mathcal{Q}(-\alpha, t)\begin{bmatrix} \cos(-t) \\ \sin(-t) \end{bmatrix}$$

$$= T(-\alpha+1, -\alpha-1, \frac{1+b}{2}, \frac{1-b}{2}, t),$$

which means that $\mathcal{H}(\alpha, b, t)$ parametrizes a trochoid.

To reverse the argument, consider a trochoid E parametrized by $T(p, q, \lambda, \mu, t)$. Solving $\lambda = (1+b)/2$ and $\mu = (1-b)/2$ gives $b = \lambda - \mu$, which means that $|b| < 1$. Solving $p = 1 - \alpha$ and $q = -(1+\alpha)$ gives $\alpha = -(p+q)/2$. So E is parametrized by $\mathcal{H}(-(p+q)/2, \lambda - \mu, t)$. That is, $T(p, q, \lambda, \mu, t)$ and $\mathcal{H}(-(p+q)/2, \lambda - \mu, t)$ are the same parameterization. Therefore, any trochoid is also a precessed ellipse, and can be, courtesy of Thor, perceived as the hole followed by a black hole through the earth.

Furthermore, when $\alpha = b$, then $\mathcal{H}(b, b, t) = T(-b+1, -b-1, (1+b)/2, (1-b)/2, t)$ which matches the form of (VII.4) with $p = 1 - b$ and $q = 1 + b$; thus

$$\mathcal{H}(b, b, t) \tag{10}$$

parametrizes a hypocycloid.

Newton's Other Ellipse

Finally, we show that $Q(-\omega, t)P(t)$ is a hypocycloid. Since $P(t) = R\mathcal{E}(\omega/\sqrt{k}, \sqrt{k}\,t)$, then

$$Q(-\omega, t)P(t) = R\,Q(-\omega, t)\mathcal{E}\left(\frac{\omega}{\sqrt{k}}, \sqrt{k}\,t\right) = R\mathcal{H}(\frac{\omega}{\sqrt{k}}, \frac{\omega}{\sqrt{k}}, \sqrt{k}\,t),$$

which traces the same curve as $R\mathcal{H}(\omega/\sqrt{k}, \omega/\sqrt{k}, t)$, a hypocycloid by (10). That is, when a pebble is dropped from the equator of a rotating, homogeneous spherical earth, its path through the earth is a hypocycloid.

Transit times

In keeping with the classic riddle posed by Bernoulli for Newton from Chapter VII, we ask for the time for a pebble to slide without resistance along any given hypocycloid through the earth, for any given angular rotation rate. The key in calculating this transit time is the relationship between the polar angle θ and time t for the pebble's elliptical path through the earth. If we write θ as a function in terms of t, then, with ω as earth's angular frequency and $r(t)$ as in (3),

$$r(t)\left(\cos(\theta(t) - \omega t), \sin(\theta(t) - \omega t)\right) \tag{11}$$

is a parametrization of the same hypocycloid that $Q(-\omega, t)P(t)$ parametrizes. And in fact they are the same parametrization, although they look different. To derive a relation between θ and t, we combine (2) and (3) as

$$\frac{d\theta}{dt} = \frac{h}{R^2 \cos^2(\sqrt{k}\,t) + \frac{\omega^2 R^2}{k} \sin^2(\sqrt{k}\,t)}. \tag{12}$$

As outlined in Exercise 3, the solution to (12) with initial condition $\theta = 0$ when $t = 0$, using standard trigonometric identities and integration techniques, is

$$\theta(t) = \sqrt{k}\,t + \tan^{-1}\left(\frac{(\frac{\omega}{\sqrt{k}} - 1)\sin(\sqrt{k}\,t)\cos(\sqrt{k}\,t)}{1 + (\frac{\omega}{\sqrt{k}} - 1)\sin^2(\sqrt{k}\,t)}\right) \quad \text{for } t \geq 0. \tag{13}$$

The hypocycloid that a pebble follows as it falls freely through the earth varies when earth's rotation rate varies. That is, we allow a pebble at A, at polar distance R from O and at polar angle 0, to fall freely through a rotating earth of period Q for any $Q > 2\pi/\sqrt{k}$, where $k = g/R$ with $g = 9.8$ m/sec^2. If Q is less than $2\pi/\sqrt{k} \approx 1.42$ hours, the pebble at A will immediately veer into space because centripetal acceleration exceeds gravitational acceleration. For each $Q > 2\pi/\sqrt{k}$, the pebble resurfaces at some point $B(Q)$. We refer to the path between A and $B(Q)$ followed by the pebble as the hypocycloid *associated* with Q or associated with the rotational frequency $\Omega = 2\pi/Q$. For example, for $Q = \infty$, the associated hypocycloid is a line whose other endpoint is A's antipodal point. As Q decreases, $B(Q)$ moves further west toward A, as suggested by Figure 5 where Q values are given in hours. The hypocycloid whose first arch endpoints are A and $B(24)$ is the same hypocycloid as in Figure 3(b), and $B(1.5)$ is not far from A. For any given point B_0 on earth's equator between $B(\infty)$ west to A, there is a unique hypocycloid associated with a unique period Q_0 for which $B_0 = B(Q_0)$. As we will show, the hypocycloid associated with Q_0 is the path between A and B_0 of least transit time for all $Q > 2\pi/\sqrt{k}$.

To determine $B(Q)$ for a given Q, let b be the minimum distance from O to the hypocycloid associated with Q, which occurs when $t = \pi/(2\sqrt{k})$. Thus by (3),

$$b = r\left(\frac{\pi}{2\sqrt{k}}\right) = \sqrt{\frac{h^2}{kR^2}} = \frac{2\pi R}{\sqrt{k}\,Q} = \frac{R\Omega}{\sqrt{k}}. \tag{14}$$

Furthermore, by (11) and (13), the central angle between A and $B(Q)$ is

$$\theta(\pi/\sqrt{k}) - 2\pi(\pi/\sqrt{k})/Q = \pi(1 - \Omega/\sqrt{k})$$

so that the great circle distance between A and $B(Q)$, denoted $d(A, B)$, is

$$d(A, B) = \pi R\left(1 - \frac{\Omega}{\sqrt{k}}\right). \tag{15}$$

The speed of a pebble sliding freely (without resistance) along a tunnel through a rotating earth is determined by its change in kinetic and potential energy. When a pebble of mass m begins sliding along a tunnel at the earth's surface, its speed $\frac{ds}{dt}$ along the tunnel is 0, where s denotes arclength along the tunnel. At radial distance r, the kinetic energy of the pebble sliding along the tunnel is $\frac{1}{2}m(\frac{ds}{dt})^2$. The potential energy of the pebble at radial distance r in a nonrotating earth is

$$\int_0^r mk\rho\, d\rho = \frac{1}{2}mkr^2, \tag{16}$$

Newton's Other Ellipse

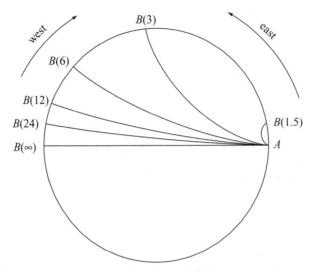

Figure 5. Hypocycloid arches with respect to earth period (in hours)

where ρ is a dummy distance variable. The pebble's rotational energy at radial distance r is

$$\frac{1}{2}m\omega^2 r^2, \tag{17}$$

which the pebble inherits as being part of the rotating earth. This rotational energy can be viewed as negative potential energy. For example, if ω gets too large, that is if the earth rotates too quickly, the centripetal force outwards will equal the gravitational force inwards, which means that there will be no (positive) potential energy at radial distance R, and so the pebble cannot lose potential energy by falling and will never begin to slide down the tunnel from earth's surface. Thus the net potential energy of the pebble at radial distance r with respect to its behavior in the tunnel can be taken as

$$\frac{1}{2}mr^2(k-\omega^2), \tag{18}$$

the difference of (16) and (17). Therefore the change in kinetic energy with respect to the rotating earth as the pebble falls from radial distance R to r is $\frac{1}{2}m(\frac{ds}{dt})^2$ while the change in potential energy is $-\frac{1}{2}m(k-\omega^2)(R^2-r^2)$ by (18). The law of the conservation of energy says that the sum of the energy changes in kinetic and potential energy is 0, so

$$\left(\frac{ds}{dt}\right)^2 = (k - \omega^2)(R^2 - r^2), \tag{19}$$

which we can rewrite as

$$dt = \frac{ds}{\sqrt{(k-\omega^2)(R^2-r^2)}} = \frac{1}{\sqrt{k-\omega^2}}\sqrt{\frac{1+r^2(\frac{d\theta}{dr})^2}{R^2-r^2}}\,dr, \tag{20}$$

since arc length ds along a polar curve is

$$ds = \sqrt{1 + r^2\left(\frac{d\theta}{dr}\right)^2}\,dr.$$

Let σ be a parameter so that $r(\sigma)(\cos\theta(\sigma), \sin\theta(\sigma))$ is a parametrization of a given tunnel. By the chain rule, (20) becomes

$$dt = \frac{1}{\sqrt{k-\omega^2}}\sqrt{\frac{(\frac{dr}{d\sigma})^2 + r^2(\sigma)(\frac{d\theta}{d\sigma})^2}{R^2 - r^2(\sigma)}}\,d\sigma. \tag{21}$$

When the tunnel is a hypocycloid associated with rotational frequency Ω, the second radical factor in (21) simplifies to a constant,

$$\sqrt{\frac{(\frac{dr}{d\sigma})^2 + r^2(\sigma)(\frac{d\theta}{d\sigma})^2}{R^2 - r^2(\sigma)}} = \frac{\sqrt{kR^4 - h^2}}{R^2}$$

$$= \frac{\sqrt{kR^4 - (2\pi R^2/Q)^2}}{R^2} = \sqrt{k - \Omega^2}, \tag{22}$$

as shown in Exercise 4. We emphasize that the parameter σ in (22) is the same as time t only when Ω is ω. Therefore, by (14), (21), and (22) the total time it takes for a pebble to transit the hypocycloid associated with frequency Ω when the earth's frequency is ω and where b is the distance of the hypocycloid from O is

$$2\int_0^{\frac{\pi}{2\sqrt{k}}} \frac{1}{\sqrt{k-\omega^2}}\sqrt{k-\Omega^2}\,d\sigma = \sqrt{\frac{k-\Omega^2}{k-\omega^2}}\frac{\pi}{\sqrt{k}} = \pi\sqrt{\frac{1-(b/R)^2}{k-\omega^2}}. \tag{23}$$

When $\omega = 0$, (23) agrees with Lin and Lyness [80].

As an example, let Γ be the hypocycloid corresponding to the earth period $Q = 24$ hrs, so that $\Omega = \pi/12$ rad/hr. We determine the transit time for Γ

in a stationary earth ($\omega = 0$) and in a rapidly spinning earth ($\omega = \pi$ rad/hr). By (23), the respective transit times are 42.24 min and 59.58 min. That is, the transit time is more than 17 minutes longer in the rapidly spinning earth than in the stationary earth, which makes sense because a faster earth spin reduces the effective acceleration of gravity within the earth, which in turn means a slower speed for a pebble sliding along a tunnel.

As a second example, let Γ be the hypocycloid arch between A and B: equatorial points at Quito, Ecuador and at Entebbe, Uganda. Quito lies at longitude 78.5° W and Entebbe lies at longitude 32.5° E. We find the transit time along Γ. Since the great circle distance between Quito and Entebbe is $D = 111°\pi R/180° \approx 12398.8$ km, then the radius of a circle rolling around on the inside of the equator with circumference D has radius $q = D/(2\pi) \approx 1973.33$ km. A fixed point on this circle traces Γ whose minimum distance to O is $b = R - 2q \approx 2453.33$ km. In a non-rotating earth, by (23), the transit time along this hypocycloid is 39.08 minutes. By (14), the associated period for Γ is $Q \approx 3.679$ hours, corresponding to angular frequency $\Omega \approx 1.708$ radians/hr. So a pebble falling freely from Quito will arrive at Entebbe 42.31 minutes later. By (23) with $\omega = \pi/12$ radians/hr, the transit time along the hypocycloid is 39.15 minutes which is 4.2 seconds slower than for a stationary earth.

Lastly, we point out that the solution to the brachistochrone problem of Chapter VII for a rotating earth, given that A and B are on the equator, is a hypocycloid arch with endpoints A and B. To see why this is true, Bernoulli's modified derivation from Chapter VII uses a conservation of energy argument resulting in the pebble's speed along the tunnel being $ds/dt = \sqrt{k}\sqrt{R^2 - r^2}$; that is, ds/dt is proportional to $\sqrt{R^2 - r^2}$. For a rotating earth, the only change in the entire derivation is that $ds/dt = \sqrt{k - \omega^2}\sqrt{R^2 - r^2}$, which again means that ds/dt is proportional to $\sqrt{R^2 - r^2}$, and so the path of least transit time is a hypocycloid when the earth rotates.

Tunnels starting north of the equator

What kind of tunnel results when the pebble is dropped from northern latitudes on a rotating earth?

Let us drop a pebble at latitude ψ at point D in Figure 6. As before, with respect to the fixed stars the pebble's path is an ellipse. Since the pebble's initial tangential speed at latitude ψ is less by a factor of $\cos \psi$ in comparison to the initial tangential speed of a particle at the earth's equator, then the semi-minor axial length for the ellipse originating at latitude ψ will be $\cos \psi$ of

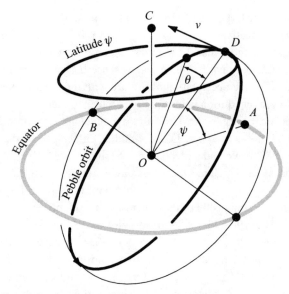

Figure 6. Ellipse of motion for a pebble dropped at latitude ψ

the semi-minor axial length for the ellipse originating at the equator. That is, in comparison to (8), a parametrization for the ellipse originating at latitude ψ is

$$R \begin{bmatrix} \cos(\sqrt{k}t) \\ \frac{\omega}{\sqrt{k}} \cos \psi \sin(\sqrt{k}t) \end{bmatrix}.$$

To this parametrization, append a third component with value 0, and call the new parametrization $P(\psi, t)$:

$$P(\psi, t) = R \begin{bmatrix} \cos(\sqrt{k}t) \\ \frac{\omega}{\sqrt{k}} \cos \psi \sin(\sqrt{k}t) \\ 0 \end{bmatrix}.$$

Generalizing the precession approach we used in two dimensions, we let

$$\mathcal{E}(b, \psi, t) = \begin{bmatrix} \cos t \\ b \cos \psi \sin t \\ 0 \end{bmatrix}$$

where $|b| \leq 1$. Let $L(\psi)$ be the matrix that when multiplied with a vector rotates it about the y-axis by angle ψ, counterclockwise with respect to the positive y direction:

$$L(\psi) = \begin{bmatrix} \cos\psi & 0 & -\sin\psi \\ 0 & 1 & 0 \\ \sin\psi & 0 & \cos\psi \end{bmatrix}.$$

Thus, $L(\psi)\mathcal{E}(b, \psi, t)$ is a rotation of the ellipse $\mathcal{E}(b, \psi, t)$ into alignment with the trajectory's natural position with respect to the equatorial plane being the xy-plane and the drop point (at $t = 0$) being at latitude ψ, as shown in Figure 6. Let $\mathcal{Q}(\alpha, t)$ be the matrix that when multiplied with a vector rotates it about the z-axis by αt radians, counterclockwise with respect to the positive z direction:

$$\mathcal{Q}(\alpha, t) = \begin{bmatrix} \cos(\alpha t) & -\sin(\alpha t) & 0 \\ \sin(\alpha t) & \cos(\alpha t) & 0 \\ 0 & 0 & 1 \end{bmatrix}.$$

Thus, $\mathcal{F}(\alpha, b, \psi, t) = \mathcal{Q}(-\alpha, t)L(\psi)\mathcal{E}(\alpha, b, \psi, t)$ is a precessed parametrization of $\mathcal{E}(b, \psi, t)$, and $R\mathcal{F}(\omega/\sqrt{k}, \omega/\sqrt{k}, \psi, \sqrt{k}t) = \mathcal{Q}(-\omega, t)L(\psi)P(\psi, t)$ is the path of the pebble with respect to the earth.

As an example, we drop a pebble at Torino, Italy, 45° N, 7.5° E, which we call point A. At what longitude will the pebble resurface? With ω as one rotation per day,

$$R\mathcal{F}(\omega/\sqrt{k}, \omega/\sqrt{k}, \pi/4, \pi/\sqrt{k})$$

gives $(-4448.6, 830.8, -4525.5)$ which means that the resurface point is about 169.4° east of Torino, giving the resurface point as 45° S and 177° E, which is about 300 miles southeast off the coast from Wellington, New Zealand. However, if the earth rotates at, say, once every two hours (and the earth maintains its spherical shape), then the resurface point B is at 45° S and 60.5° E, putting it well off the southeast coast of Africa, out where the Indian Ocean merges with Antarctic waters. Figure 7 shows the geometry of these tunnels.

The tunnels are not planar, even from A to B. We demonstrate this result for the case when dropping a black-hole at latitude $\psi = \pi/4$. Without loss of generality, we may define a time unit so that $1/\sqrt{k}$ is unit time, and a distance unit so that R is unit distance. Under such a scheme, the valid values for ω are between 0 and 1 radian per unit time. (For any higher value of ω, a black-hole dropped at the equator would fly off into space.) Let $A = \mathcal{F}(\omega, \omega, 0, \psi)$, $B = \mathcal{F}(\omega, \omega, \pi, \psi)$, $C = \mathcal{F}(\omega, \omega, \pi/2, \psi)$, and $D = \mathcal{F}(\omega, \omega, \pi/4, \psi)$.

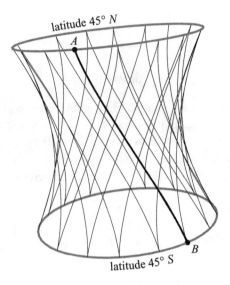

Figure 7. A wicker basket, $\omega = \pi$ radians/hr

Let U be the cross product of $C - A$ and $B - A$, let V be the cross product of $D - A$ and $B - A$, and let ϵ be the magnitude of the difference between the unit vectors in the U and V directions. If the curve given by $\mathcal{F}(\omega, \omega, t, \psi)$ is planar then U and V point in the same direction, which means that ϵ should be close to 0. However the graph given in Figure 8 shows that ϵ is nowhere near 0 as ω goes from 0 to 1 (except at $\omega = 0$). Thus the curve from A to B is non-planar for any ω with $0 < \omega < 1$.

While the tunnels given by $\mathcal{F}(\omega, \omega, t, \psi)$ form a network resembling a wicker basket or a vase of reeds as in Figure 7, their projection onto the xy-plane is a hypocycloid. To see this result, the first two coordinates of $\mathcal{F}(t, \psi)$ are

$$R \cos \psi \begin{bmatrix} \cos(\omega t)\cos(\sqrt{k}\,t) + \frac{\omega}{\sqrt{k}} \sin(\omega t)\sin(\sqrt{k}\,t) \\ -\sin(\omega t)\cos(\sqrt{k}\,t) + \frac{\omega}{\sqrt{k}} \cos(\omega t)\sin(\sqrt{k}\,t) \end{bmatrix}. \quad (24)$$

With $\tau = \sqrt{k}\,t$, (24) becomes a scaling of $\mathcal{H}(\omega/\sqrt{k}, \omega/\sqrt{k}, \tau)$, a hypocycloid by (10).

These surfaces also allow us to see the trochoid family sprout into three dimensions. For simplicity, let $\sqrt{k} = 1 = R$. For almost all values of α, the graph of $\mathcal{F}(\alpha, b, \psi, t)$ is non-periodic, which means that the graph of $\mathcal{F}$ in time t generates a two-dimensional surface in three-space. Rather than

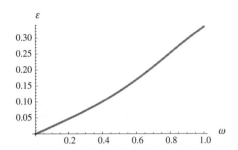

Figure 8. The non-planarity of non-equatorial tunnels

tracing out the surface in the single dimension of time t as given by $\mathcal{F}$, an easier way is to use a two-dimensional approach:

$$\mathcal{G}(s, t) = \begin{bmatrix} \cos(\alpha t + s) & \sin(\alpha t + s) & 0 \\ -\sin(\alpha t + s) & \cos(\alpha t + s) & 0 \\ 0 & 0 & 1 \end{bmatrix} L(\psi)\mathcal{E}(b, \psi, t),$$

for any given α, b, and ψ. That is, $\mathcal{G}$ precesses the ellipse $L(\psi)\mathcal{E}(b, \psi, t)$ by an additional offset parameter s, and in essence allows every point on earth's surface at latitude ψ to be a simultaneous drop point. Figure 9 shows the surface of $\mathcal{G}$ when pebbles are dropped from latitude 45° N. The contour lines on these surfaces correspond to a pebble's path through the earth.

Furthermore, the projection of these tunnels onto the xz-plane is a shape whose left-hand and right-hand boundaries are almost always a hyperbola as is explored in Exercise 6b.

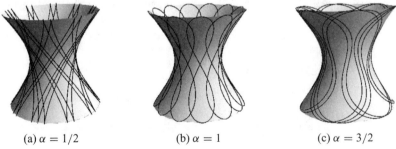

(a) $\alpha = 1/2$ (b) $\alpha = 1$ (c) $\alpha = 3/2$

Figure 9. Trochoidal lampshades or flower vases, $\psi = \pi/4$ and $b = 1/2$

In Figure 9(a), the contours are the tunnels corresponding to dropping a pebble on a rotating earth. In Figures 9(b) and 9(c), we need Thor to speed up the earth's rotation after the pebble has been dropped. When projected onto the xy-plane, Figure 9(b) is a hypotrochoid, and Figure 9(c) is an epitrochoid, Exercise 6a. Such curves might make pleasing rib forms for lampshades or designs in cut crystal vases.

Exercises

1. The two differential equations of (1) are of the form

$$\frac{d^2 w}{du^2} + \alpha^2 w = \frac{\beta^2}{w^3}, \tag{25}$$

where α and β are constants. Since we wish to solve (25) corresponding to the initial conditions, $r(0) = R = 1/z(0)$, $\left.\frac{dr}{dt}\right|_{t=0} = 0$ and $\left.\frac{dz}{d\theta}\right|_{t=0} = 0$, the initial conditions for (25) are $w(0) = \delta$ for some $\delta > 0$ and $\left.\frac{dw}{du}\right|_{t=0} = 0$.

(a) Let $\rho = \frac{dw}{du}$. Show that (25) can be written in terms of ρ and w as

$$\rho \, d\rho = \left(\frac{\beta^2}{w^3} - \alpha^2 w\right) dw, \tag{26}$$

where $\rho = 0$ when $w = \delta$.

(b) Integrate (26) and show that ρ is given by

$$\frac{dw}{du} = \rho = \pm \sqrt{\beta^2 \left(\frac{1}{\delta^2} - \frac{1}{w^2}\right) - \alpha^2 (w^2 - \delta^2)}.$$

(c) Use the substitution $q = w^2 - \delta^2$ to show that (b) can be written as

$$\frac{\delta}{2\sqrt{\beta^2 q - \alpha^2 \delta^2 q (q + \delta^2)}} dq = \pm du,$$

where $q = 0$ when $u = 0$.

(d) Integrate (c) to show that

$$\sin^{-1}\left(\frac{2\alpha^2 \delta^2}{\beta^2 - \alpha^2 \delta^4}\left(w^2 - \frac{\beta^2 + \alpha^2 \delta^4}{2\alpha^2 \delta^2}\right)\right) = \pm 2\alpha u - \frac{\pi}{2}.$$

(e) Show that (d) simplifies as

$$w^2 = \frac{\beta^2 + \alpha^2 \delta^4}{2\alpha^2 \delta^2} - \frac{\beta^2 - \alpha^2 \delta^4}{2\alpha^2 \delta^2} \cos(2\alpha u).$$

(f) Use the double angle formula for cosine to show that (e) simplifies as

$$w = \sqrt{\delta^2 \cos^2(\alpha u) + \left(\frac{\beta}{\alpha\delta}\right)^2 \sin^2(\alpha u)}.$$

(g) Finally, show that (f) gives (3) and (4).

2. Complete the outline below to demonstrate that (8) is a valid parametrization of the pebble's motion as it falls through the earth.

   (a) Demonstrate that (5): $r(\theta)(\cos\theta, \sin\theta)$ and (8): $R\left(\cos(\sqrt{k}\,t), \frac{\omega}{\sqrt{k}}\sin(\sqrt{k}\,t)\right)$ represent the same ellipse.

   (b) Let $A(t)$ be the area swept out by the position vector of (5) from time 0 to time t. Use (2) to show that $\Delta A = \frac{h}{2}\Delta t$.

   (c) Define the angle $v(t)$ and the length $s(t)$ so that $\tan(v) = y/x$ and $s(t) = \sqrt{x^2 + y^2}$. Show that

   $$s^2\, dv = h\, dt.$$

   (d) Let $\mathcal{A}$ be the area swept out by the position vector of (8) from time 0 to time t. Use part (c) to show that $\Delta \mathcal{A} = \frac{h}{2}\Delta t$.

   (e) Now use (b) and (d) to argue that the arclengths generated by the two parametrizations over any time period are the same, and so conclude that (8) faithfully describes the motion of the falling pebble.

3. Complete the outline below to solve (12) given that $\theta = 0$ when $t = 0$. Start by letting

$$\rho(t) = \frac{1}{a^2 \cos^2(\omega t) + b^2 \sin^2(\omega t)}$$

where a, b, ω are constants, and letting $\Theta(t)$ be that function for which $\frac{d\Theta}{dt} = \rho(t)$ with $\Theta(0) = 0$.

   (a) Explain why $\Theta(t)$ increases without bound for $t \geq 0$.
   (b) Show that

   $$\int \rho(t)\, dt = \frac{1}{ab\omega} \tan^{-1}\left(\frac{b}{a}\tan(\omega t)\right) + C,$$

where C is a constant of integration, and thus conclude that

$$\Theta(t) = \frac{1}{ab\omega} \tan^{-1}\left(\frac{b}{a} \tan(\omega t)\right), \quad \text{for } 0 \le t \le \frac{\pi}{2\omega},$$

$$\text{and } \Theta\left(\frac{\pi}{2\omega}\right) = \frac{\pi}{2ab\omega}.$$

(c) Justify the recursive relation

$$\Theta(t) = \frac{\pi}{ab\omega} - \Theta(\frac{\pi}{\omega} - t), \quad \text{for} \quad \frac{\pi}{2\omega} < t \le \frac{\pi}{\omega}.$$

(d) For $t > \pi/\omega$, let $n(t) = \lfloor \omega t/\pi \rfloor$ and let $q(t) = t - \pi n(t)/\omega$, the integral number of times that π/ω divides t and the remainder, respectively. Then

$$\Theta(t) = \frac{n(t)\pi}{ab\omega} + \Theta(q(t)) \quad \text{for} \quad t > \frac{\pi}{\omega}.$$

Thus $\theta(t) = h\Theta(t)$ for all t, with $a = R$, $b = \dfrac{h}{R\sqrt{k}}$, and $\omega = \sqrt{k}$. Justify the equation

$$\theta(t) = \tan^{-1}\left(\frac{h}{R^2\sqrt{k}} \tan(\sqrt{k}t)\right), \quad \text{for } 0 < t < \frac{\pi}{2\sqrt{k}}. \quad (27)$$

(e) Let $\tan \alpha = c \tan \beta$. Show that

$$\alpha = \beta + \tan^{-1}\left(\frac{(c-1)\sin \beta \cos \beta}{1 + (c-1)\sin^2 \beta}\right).$$

(f) Substitute $\alpha = \theta$, $\beta = \sqrt{k}t$, and $c = h/(R^2\sqrt{k})$ in (e) so as to give the desired result. Explain why this formula is valid for all t.

4. (a) Show that the derivatives of (3) and (13) simplify as

$$r'(t) = \frac{\sqrt{k}(-R^2 + h^2/(kR^2)) \sin(\sqrt{k}t) \cos(\sqrt{k}t)}{r}$$

$$\text{and } \theta'(t) = \frac{h(1 - r^2/R^2)}{r^2}.$$

(Hint: To find $\theta'(t)$ use (27) rather than (13).)

(b) Justify each step in the sequence of identities:

$$\frac{(\frac{dr}{dt})^2 + r^2(t)(\frac{d\theta}{dt})^2}{R^2 - r^2(t)}$$

$$= \frac{k(h^2/(kR^2) - R^2)^2 \sin^2(\sqrt{k}\,t)\cos^2(\sqrt{k}\,t) + h^2(R^2 - r^2)^2/R^4}{r^2(R^2 - r^2)}$$

$$= \frac{\frac{1}{k}(h^2 - kR^4)^2 \sin^2(\sqrt{k}\,t)\cos^2(\sqrt{k}\,t) + h^2(R^2 - h^2/(kR^2))^2 \sin^4(\sqrt{k}\,t)}{r^2 R^4 (R^2 - h^2/(kR^2))\sin^2(\sqrt{k}\,t)}$$

$$= \frac{\frac{1}{k}(kR^4 - h^2)\cos^2(\sqrt{k}\,t) + h^2(kR^4 - h^2)\sin^2(\sqrt{k}\,t)/(k^2 R^4)}{r^2 R^2/k}$$

$$= \frac{(kR^4 - h^2)}{R^4} \frac{(R^2 \cos^2(\sqrt{k}\,t)) + h^2/(kR^2)\sin^2(\sqrt{k}t)}{r^2} = \frac{kR^4 - h^2}{R^4}.$$

5. (a) Find values of α and b for which $\mathcal{H}(\alpha, b, t)$ parametrizes the polar rose $r = \cos(3\theta)$.

   (b) Just as $\mathcal{H}(b, b, t)$ is a hypocycloid for b where $0 < b < 1$, perhaps there are α and b values for which $\mathcal{H}(\alpha, b, t)$ is an epicycloid. Look for them. (Try $\alpha = 2b$.)

   (c) Show that $\mathcal{H}(1/3, 1/3, t)$ is the same parametrization as $\mathcal{T}(1, 2, 2/3, 1/3, 2t/3)$.

6. (a) Show that the curves Γ of Figure 7 when projected onto the xy-plane are trochoids with respect to a circle of radius $R/\sqrt{2}$.

   (b) Show that the $y = 0$ cross-section of the graph of $\mathcal{G}(s, t)$ for any $\alpha, b,$ and ψ with $0 < b < 1$ and $0 < \psi < \pi/2$ is a hyperbola.

7. Take A as Entebbe, Uganda and B as Pontianak, Borneo (longitude 109° E) along the equator.

   (a) For the hypocycloid between A and B, find b, the distance between O and the hypocycloid.

   (b) Find the rotation Ω corresponding to the hypocycloid arch between A and B.

   (c) What is the transit time for the hypocycloid at the rotation rate of part (b)?

   (d) Find the transit time for this hypocycloid when the earth rotates once per day.

8. Find a rotation rate Ω corresponding to the hypocycloid arch whose endpoints are at New Orleans, Louisiana (30° N and 90° W) and Durban,

South Africa (30° S and 32° E). Estimate the transit time for this tunnel if the earth rotates once per day.

9. Experiment with different paths from the north pole to the equator for a rotation rate of 2 hours. (Assume that the earth remains a homogenous spherical ball at this rotation rate.) What is the planar path of least transit time? In particular, compare your path against a straight line and an arch of a hypocycloid.

10. On p. 193, Maupertuis, as a leader of a scientific community, proposed a number of promising research areas. Imagine yourself in a similar position in today's world. Name ten items on your list.

Vignette IX: The Man in the Moon

In writing Micromégas, Voltaire drew upon an old literary tradition of life on other worlds.

When viewed from the earth, the shadowy features on the full moon often prompt suggestive images of a face: the man in the moon. Our nearest neighbor in space, it seems, graphically declares the fanciful notion that the universe is full of people. Plutarch wrote an essay in the first century AD, "Concerning the face that appears in the man in the moon," speculating upon the causes of the face. He went on to suggest that people might live there.

One hundred years earlier, Lucretius (94–49, BC), in *The Way Things Are* [**82**, book ii, lines 1052–1078, p. 82], voiced that compelling possibility, not only on the moon but elsewhere.

> Out beyond our world
> There are, elsewhere, other assemblages
> Of matter, making other worlds. Oh, ours
> Is not the only one in air's embrace.
>
> Let's admit—
> We really have to—there are other worlds,
> More than one race of men.

Echoing these sentiments, Christiaan Huygens, a Dutch mathematician-astronomer, in his posthumous book *Cosmotheoros* of 1698, gave us a veritable writers' guide to designing aliens.

> A man that is of Copernicus's opinion, that earth is carried round the sun like the other planets, cannot but sometimes have a fancy that it is not improbable that the rest of the planets have their inhabitants too [**70**, pp. 1–2].

He goes on to say that intelligent creatures probably inhabit planets of other star systems. Such people, like earthlings, would have five senses, walk upright, and have hands. They think like us, but may not look like us. Perhaps they have

> a neck four times as long, and great round saucer eyes five or six times as big [**70**, p. 77].

Writers going back to the Pythagoreans spoke of life existing beyond planet earth. In the legend of Icarus and Daedalus, the heroes fashion wings of feathers and wax and fly too close to the sun, as shown in the woodcut of Figure 1 which comes from a 1493 printing of *The Mirror of True Rhetoric*, leaf xi. Since the air up above was breatheable—according to the ancients—then it should be an easy matter for people to live in heavenly lands—if only they could get there.

Figure 1. The fall of Icarus (1493), woodcut attributed to Dürer [**38**, cut 84]

For example, Lucian of Samosata (129–188 AD), sometimes called "the Voltaire of the Greeks" [**154**, p. 73], describes a voyage from the earth to the moon in a ship caught up in a waterspout; the sailors encounter men on the moon, and are enlisted to fight against winged men from the sun, all of which is told in the satirical story *The True History*.

Vignette IX: The Man in the Moon

Almost all poets and philosophers have probably speculated upon alien life, if only to accentuate the issues that we face on our little planet. As C. S. Lewis says [**76**, p. 59],

> Is any man such a dull clod that he can look at the moon without asking himself what it would be like to walk [there]?

My favorite among these musings is from Washington Irving (1783–1859), an American storyteller and historian who wrote often about early America and enjoyed looking at events from alternate points of view. To describe how the American Indians felt upon being displaced by European settlers with their gunpowder, booklearning, and culture, he imagined how we would feel if we were displaced by moon-men.

Figure 2. Jefferson and Napoleon at the court of the Man in the Moon

In his thought-experiment in *Knickerbocker's History of New York* [**71**], a party of exploring moon-men philosophers land on earth in 1808. They were pea-green, had one eye below their shoulders, and had tails. These features, they thought, were far more beautiful and natural than the skin color, the two eyes, the silly head, and the tailless body of the earth men they met, as suggested in Figure 2. The moon men had ray guns and force fields and great machines by which they pacified protests from earth's disorganized armies. As specimens to show their king, the Man in the Moon, they kidnap President Thomas Jefferson, Napoleon of France, and a few other notables.

At the palace court, the Man in the Moon is unimpressed by Jefferson's jibberish and Napolean's uniform. He decrees that earth men must abandon their backward beliefs and primitive customs. All must accept a new culture. Everyone must become a Lunatic. Any earthman who refused could go live in the swamps of South America. The rest of the world now belonged to the Man in the Moon.

CHAPTER IX

Maupertuis's Pursuit Problem

In 1638, Francis Godwin's story, *The Man in the Moone*, was published. In it, an astronaut harnesses a wedge of 25 swans, as in Figure 3(a) [**31**, plate facing p. 118], and flies to the moon. The swans fly at a constant rate and always head toward the moon, so their trajectory from the earth to the moon is not a straight line, but a pursuit curve as shown in Figure 4. This was studied by Maupertuis and is referred to as the *pursuit problem of Maupertuis*. In Godwin's story, the flight to the moon takes twelve days, whereas the return trip, which follows a straight line, takes eight days. Was Godwin accurate?

We use a little calculus to determine the consistency of these flight times. We also explore some popular fancies regarding motion from the literature of yesteryear.

A little history

Let's put Godwin's estimate into historical context, and compare it with some others. In each of the following narratives, unless otherwise stated, the tacit assumption is that the return trip is as long as the outbound trip.

Lucian, in *The True History*, written in 174 AD, has a ship, initially lofted by a waterspout, sail to the moon in eight days. In 1300, Dante travels through the solar system in *The Divine Comedy*. Together with his guide Beatrice, they leave earth from a mountaintop, and fall upwards. To describe their speed, Beatrice says to Dante,

For lightning ne'er ran, as thou hast hither, [*Paradisio*, canto i].

In a matter of minutes, they arrive on the moon. Around 1610, Johannes Kepler wrote *The Dream*, in which his hero is transported by a force field to the moon in four hours. Around 1630, Galileo calculated that a ball would fall from the moon to the earth in 3 hours, 22 minutes, and 4 seconds, [**47**, p. 224]. In 1640, John Wilkins, the first secretary of the Royal Society, anticipated that a manned moon trip would last 180 days. In 1656, in *Voyage to the Moon*,

(a) via a wedge of swans (b) via a flock of wild birds
Figure 3. Space travel

Cyrano de Bergerac, famous for a long nose and swordsmanship, initially launching a fictional version of himself in a winged contraption powered by fireworks as depicted in Figure 5, travels there and back, each leg lasting several days. In 1835, in *The Adventures of Hans Pfaall*, Edgar Allan Poe describes a two week long moon voyage in a balloon inflated with gas "37.4 times lighter than hydrogen." In 1870, in *Round the Moon*, Jules Verne describes a round trip projectile shot to the moon, each leg lasting four days. In 1901, in *The First Men in the Moon*, H. G. Wells describes an anti-gravity moon flight of a few days with a return trip lasting several weeks. In 1968, Apollo 8, the first manned craft to circle the moon, reached the moon in 3.5 days, and returned in 2.5 days.

Godwin (1562–1633), an Anglican bishop, is known primarily as a church historian. He wrote his moon story around 1599, and circulated it privately among friends since the story builds on Copernican theory and publication would have caused controversy. He chose a Spanish background for his astronaut, Domingo Gonsales, so as to be in the spirit of the Spanish exploration of America by Columbus a hundred years earlier. After he died, a friend

Maupertuis's Pursuit Problem

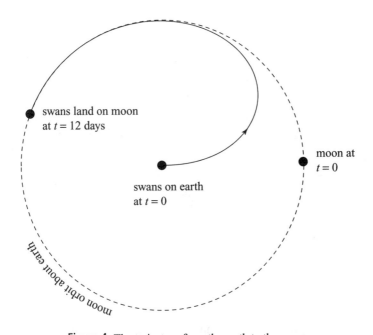

Figure 4. The trajectory from the earth to the moon

published the story anonymously. It went through many printings in the next two centuries and inspired similar stories. For example, in *The Little Prince*, a classic children's story, the hero leaves his planet for earth by harnessing a flock of birds, as shown in Figure 3(b) [33, p. 2].

Just as in Dante's story, Domingo and the swans start their journey from a mountaintop. For the first few moonward miles, the birds strain at transporting Domingo. But thereafter the earth's pull of gravity vanishes. The birds have an easy time until just before landing on the moon, when gravity again becomes an observable phenomenon.

Domingo's trajectory to the moon is the solution of a pursuit problem, a puzzle with a long tradition going back to Zeno's mythic race of Achilles overtaking the tortoise. The earliest Latin version of the problem dates to the court of Charlemagne, where the scholar Alcuin collected some mathematical conundrums to sharpen young minds. The 26th of these reads,

> There is a field which is 150 feet long. At one end stood a dog, at the other a hare. The dog advanced to chase the hare. Whereas the dog went nine feet per stride, the hare went seven. How many feet and how many

Figure 5. Cyrano's craft

leaps did the dog take in pursuing the fleeing hare until it was caught? [**130**]

Later versions of this recreational problem vary the nature of the pursued and the pursuer. Figure 6 shows Reynard the Fox chasing a hare, from the epic satirical poem of 1483 [**8**, plate facing p. 94]. As the hare later testifies in court of Reynard's treachery,

Figure 6. Reynard the Fox in pursuit of the rabbit, drawn by J. J. Mora (1901)

> Thanks to my legs, I got away,
> Or I should not have liv'd this day;
> But still I lost part of my ear,
> And was almost half dead with fear.
> Whoever ventures o'er the plain,
> Risks, by him [Reynard] basely to be slain.

A popular renaissance version includes couriers overtaking one another on roads between various Italian cities [**132**, p. 70]. The first to apply Newton's calculus to pursuit problems was Pierre Bouguer in 1732, who imagined a privateer overtaking a merchant ship on the high seas [**9**, p. 91]. More recent versions have a fighter plane overtaking a bomber, a missile overtaking a jet, and the space shuttle overtaking the Hubble Telescope. In the next section, we solve Bouguer's pursuit problem wherein a fox chases a rabbit, a standard example in many differential equations texts.

Maupertuis generalized the pursuit problem in 1735 so that the pursued can follow paths other than straight lines. Indeed, the pursuit problem of Maupertuis commonly has the pursued moving along the arc of a circle. See Exercise 6 for a 1748 version of the problem involving a spider and a fly from the conundrum section of the *Ladies Diary*. More recently, versions of Maupertuis's problem appeared several times in early volumes of the *American Mathematical Monthly* as

> A dog at the center of a circular pond makes straight for a duck which is swimming along the edge of the pond. If the rate of swimming of the dog is to the rate of swimming of the duck as n is to 1, determine the equation of the curve of pursuit and the distance the dog swims to catch the duck.

Summarizing their limited success at solving this problem in the first volume of the *Monthly* in 1894, one of four solvers concluded that the problem solution "transcends the present limits of mathematical genius" [**9**]. In 1921, Hathaway [**62**] and Morley [**89**] gave detailed approximate solutions to the dog-and-duck problem. We use a similar approach in presenting a solution to Godwin's problem, and add a new twist in that, contrary to the terrestrial problem, the celestial one involves both the pursued and the pursuer rotating about a common center in good Copernican fashion.

An interlude: Bouguer's Solution

Imagine a rabbit being chased by a fox. The rabbit, initially at the origin, runs along the y-axis at constant speed a while pursued by a fox, initially out along the x-axis at $(c, 0)$, who always runs towards the rabbit at constant speed b.

At any time t the rabbit's position is $(0, at)$. Let $(x(t), y(t))$ denote the fox's position at time t, as illustrated in Figure 7(a). We seek a representation for the fox's path towards the rabbit as y in terms of x. The slope of the line of sight from the fox to the rabbit is $(y - at)/x$. This is also the direction in which the fox runs, a direction that is tangent to the path of the fox, whose slope is dy/dx. That is,

$$\frac{dy}{dx} = \frac{y - at}{x}. \tag{1}$$

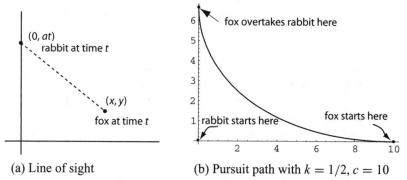

(a) Line of sight (b) Pursuit path with $k = 1/2, c = 10$

Figure 7. Bouguer's pursuit problem

Let $s(t)$ be the arclength along the fox's path towards the rabbit starting from $(x(0), y(0)) = (c, 0)$. Since the fox runs at constant speed b,

$$\frac{ds}{dt} = \sqrt{\left(\frac{dx}{dt}\right)^2 + \left(\frac{dy}{dx}\right)^2} = b. \tag{2}$$

To obtain y in terms of x, we combine (1) and (2), eliminating s and t. By the chain rule,

$$dt/dx = (dt/ds)(ds/dx) = (1/b)(ds/dx).$$

Since dt/dx is negative (as time goes on, the fox gets closer to the y-axis) and since dt/ds is positive (as time goes on, the distance run by the fox increases), then $ds/dx = -\sqrt{1 + (dy/dx)^2}$. With $y' = dy/dx$, (1) can be written as $xy' = y - at$, which upon differentiating with respect to x and simplifying gives

$$xy'' = -a\frac{dt}{dx} = \frac{a}{b}\sqrt{1+(y')^2}. \tag{3}$$

Since (3) involves x, y', and y'' but not y, we use the method of reduction in order, replacing a second-order equation with an equivalent first-order one. Let $p = y'$, so that $dp/dx = p' = y''$, and (3) becomes

$$\frac{1}{\sqrt{1+p^2}}\,dp = \frac{a}{bx}\,dx.$$

Integrating and using the initial condition $y'(c) = p(c) = 0$ (since the initial line of sight from the fox to the rabbit is along the x-axis), we get

$$\sinh^{-1} p = \frac{a}{b}\ln\left(\frac{x}{c}\right).$$

Rewriting this as $y' = \sinh(\ln(x/c)^{a/b})$ and remembering that $\sinh(z) = (e^z - e^{-z})/2$ gives

$$y' = \frac{1}{2}\left((x/c)^k - (x/c)^{-k}\right),$$

when $k \neq 1$ where $k = a/b > 0$. Integrating again, with initial condition $y(c) = 0$, gives

$$y(x) = \frac{1}{2}\left(\frac{c}{k+1}(x/c)^{k+1} + \frac{c}{k-1}(x/c)^{1-k}\right) + \frac{ck}{1-k^2}. \tag{4}$$

Figure 7(b) depicts the fox's path given by (4) when the fox runs twice as fast as the rabbit. If the rabbit and fox run at the same speed the pursuit curve is called a *tractrix*,. If the fox runs faster than the rabbit (corresponding to $0 < k < 1$), the fox eventually overtakes the rabbit. Much information about pursuit curves can be found in Nahin [93] and on the web (for example, http://mathworld.wolfram.com and the National Curve Bank site at http://curvebank.calstatela.edu/pursuit/pursuit.htm).

An earth-moon model of the journey

To model Domingo's flight, position the earth at the origin, O. Assume that the moon orbits the earth counterclockwise in a circle with period $p = 27.3$ days and radius $D = 1$ *lunar unit* (l.u.), where 1 l.u. $\approx 384{,}000$ km. We take p as the *sidereal* period (the period with respect to the fixed background of

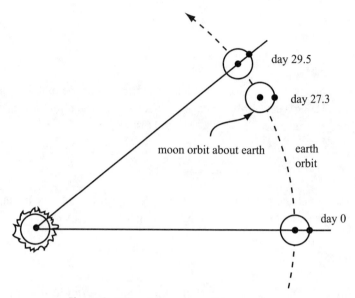

Figure 8. Sidereal versus synodic moon periods

stars) of 27.3 days rather than the *synodic* period (the period with respect to the sun) of 29.5 days. Figure 8 shows the geometry of these two terms. (In 29.5 days, the moon goes about 1.08 times around a circle of one lunar unit.)

Assume that at time $t = 0$, the moon is at $(1, 0)$. The center of the moon at any time is given by

$$M(t) = \left(\cos \frac{2\pi t}{p}, \sin \frac{2\pi t}{p}\right). \tag{5}$$

In lunar units, the earth's radius is $R \approx 0.0167$, and the moon's radius is $r \approx 0.00453$, since the earth and moon radii are about 6400 km and 1738 km, respectively. Let $X(t) = (x(t), y(t))$ denote Domingo's position along his trajectory at time t. Assume that at $t = 0$, Domingo and his wedge of swans are on the earth's surface at $(R, 0)$. Let c be the constant speed at which the swans fly. Since the swans always fly in the direction of the moon $M(t) - X(t)$ at speed c, then

$$X'(t) = c \frac{M(t) - X(t)}{||M(t) - X(t)||}, \tag{6}$$

as long as $||M(t) - X(t)|| > r$. Thus, one way to generate Domingo's pursuit curve to the moon is to solve (6) along with the initial condition $X(0) = (R, 0)$.

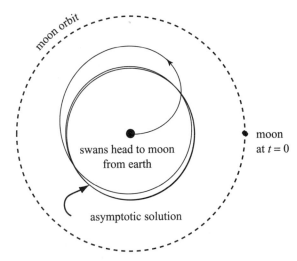

Figure 9. A hopeless quest, $c = 0.15$ lunar units/day

As long as the swans' cruising speed c exceeds the speed k of the moon ($k = 0.230$ l.u./day) the swans will ultimately reach the moon. However, if $c \le k$, their trajectory will asymptotically approach a circle with center O and radius $\rho = c\,p/(2\pi)$. For example, as shown in Figure 9 with $c = 0.15$, the swans initially get about three-fourths of the way to moon orbit before falling back into a circular orbit about half-way to the moon. In particular, the asymptotic solution to (6) for any initial condition with $c < k$ (except when the swans are already on the moon) is

$$X(t) = \rho\left(\cos\left(\frac{2\pi t}{p} + \theta\right), \sin\left(\frac{2\pi t}{p} + \theta\right)\right), \tag{7}$$

where θ is the phase angle $\cos^{-1}(\rho/D)$. As can be verified, (7) satisfies (6).

To see why (7) is a solution, see Figure 10. The swans at s see the moon at m. Since the periods of the moon and the swans are the same, the tangent lines to the inner circle at the swans' position always go through the moon. Thus the phase angle between (5) and (7) is θ for all t.

Since (6) cannot be solved analytically, we use a computer algebra system as discussed in Exercise 4. The graphs of Figures 4 and 9 show the pursuit

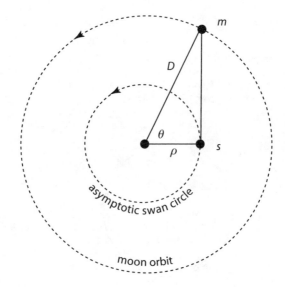

Figure 10. Finding the phase angle

curves corresponding to c values of 0.15 and $c = 0.252$, respectively. With $c = 0.15$, the swans reach the moon's surface in 12.02 days and return home in 3.88 days, rather than Godwin's estimate of 8 days. In general, the duration of the return flight is $(1 - R - r)/c$. Furthermore, as c increases from k, the ratios of the time lengths of there and back again decrease from ∞ to 1, as illustrated in Figure 11. This graph makes intuitive sense, because as the swans' speed increases, their trajectory to the moon will converge to a straight line, whereas if their speed is close to k, their trajectory spirals in multiple loops that converge towards the circle of the moon's orbit.

A sun-earth-moon model of the journey

Since Godwin describes himself as a Copernican—"I joyne in opinion with Copernicus" [**103**, p. 21]—let's see if adding the sun into the model affects the journey lengths to the moon and back. This time, let the sun be at the origin. Assume that the earth moves around the sun in a circle with period $q = 365$ days and radius $Q = 390.625$ l.u., since the earth is about 150,000,000 km from the sun. Then the center of the earth is at

$$G(t) = Q\left(\cos\frac{2\pi t}{q}, \sin\frac{2\pi t}{q}\right).$$

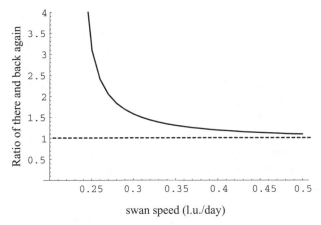

Figure 11. The ratio of travel times to the moon and back again

In order to explore more fully the possible range of trajectories to the moon and back in the sun-earth-moon model, we let the center of the moon be at

$$\mu(t, T) = G(t) + M(t + T), \tag{8}$$

where $0 \leq T \leq p$. In this model, the speed K of the earth with respect to the sun is $K = 2\pi Q/q \approx 6.72$ l.u./day, while the speed of the moon with respect to the sun is non-constant, fluctuating slightly above and below K. As long as the swans fly faster than $K + k$ with respect to the sun, they will reach the moon. When the trajectory of Domingo from the earth to the moon is plotted using (8) rather than (5), the graph appears to be an arc of the earth's orbit about the sun because Q is much larger than D. Therefore,

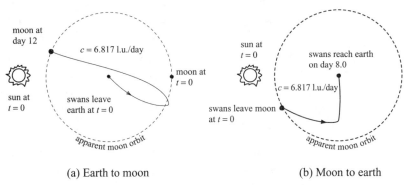

(a) Earth to moon   (b) Moon to earth

Figure 12. Apparent trajectories

rather than plotting the points generated with the sun at O, we graph the points as they would appear from earth by subtracting $G(t)$ from the point generated at time t. Figure 12(a) shows a 12-day apparent trajectory from earth when the swans fly at $c = 6.817$ l.u./day leaving the earth at midnight, when the sun, earth, and moon, initially in that order, are on the non-negative x-axis, and the swans are at position $(Q + R, 0)$.

At first glance, this trajectory may appear to be counter-intuitive since the curve goes down and to the right rather than up and to the right, as might be expected. But let's consider the first iteration. Because of the alignment of the earth and the moon, the y-coordinate of Domingo's position remains at 0. Meanwhile, the earth has moved so that its y-coordinate at time Δt is positive. It follows that the y-coordinate of Domingo's apparent position is negative rather than positive. The apparent trajectory to the earth is shown in Figure 12(b), which corresponds to the earth's center being initially at $(Q, 0)$, the moon at phase $T = -11.27$ days, the swans at polar angle $\theta = \pi/3$ with respect to the moon, and $c = 6.817$ l.u./day. This time, Domingo's return trip lasts 7.99 days. So Godwin's guess was correct, at least for this choice of parameters.

Exercises

1. Solve Alcuin's dog-and-hare pursuit problem on p. 223.

2. Solve Bouguer's pursuit problem when $k = 1$. Contrast your formula for $y(x)$ with (4).

3. In the solution to Bouguer's pursuit problem illustrated in Figure 7(b),

    (a) allow the rabbit to run at 1 meter/sec. How long does it take the fox to catch the rabbit (where the fox starts at 10 meters out along the x-axis)?

    (b) As Bouguer pointed out, a better solution for the fox to catch the rabbit is to run directly to where the rabbit will be. That is, the fox should run along a straight line. If the rabbit of part (a) continues to run up the y-axis at 1 meter/sec, how much time has the fox saved in taking the straight-line path?

4. (a) For the earth-moon model, we use the following *Mathematica* code, suggested by Wagon [149], to generate an interpolating function up to at most an arbitrary default value of 30 days for the swans to fly to the moon (in the earth-moon) model.

```
R=0.0167; r=0.00453; p=27.3; c=0.252;
M[t_?NumericQ]:={Cos[2π t/p],Sin[2π t/p]};
swan1=NDSolve[{X'[t]==c(M[t]-X[t])/Norm[M[t]-X[t]],
    X[0]=={R,0}}, X[t], {t,0,30}][[1,1,2]]
```

When executing the swan1 command above, an interpolating function for $0 \le t < 12.229$ is generated, as the swans reach the center of the moon in about 12.229 days. Figures 4 and 9 were generated by plotting swan1 for c values of 0.15 and $c = 0.252$, respectively. Generate the pursuit curve corresponding to $c = 0.2$.

(b) To determine the time for the swans to reach the surface of the moon with $c = 0.252$, we solve the equation swan1=r, which in *Mathematica* can be achieved by the code

```
FindRoot[Norm[M[t]-swan1]==r, {t, 1}]
```

which gives a flight time of 12.02 days and a return flight of 3.88 days. Determine the flight times of there-and-back-again corresponding to $c = 0.2$.

(c) To implement the dynamics of the sun-earth-moon model, we modify swan to be swan2 with three parameters, where θ is the initial polar angle of the swans' position with respect to the earth, T is a phase shift of the moon, and c is the swans' speed. For various values of θ and T, the swans may initially be unable to see the moon, such as when $\theta = 0$ and $T = p/2$.

```
Q=390.625; q=365; θ = 0; T=0; R=0.0167;
r=0.00453; p=27.3; c=6.817;
G[t_?NumericQ]:=Q{Cos[2π t/q],Sin[2π t/q]};
μ[t_?NumericQ]:=G[t]+M[t+T];
swan2=NDSolve[{X'[t]==c(μ[t]-X[t])/Norm[μ[t]-X[t]],
    X[0]=={Q+R Cos[θ],R Sin[θ]}}, X[t], {t,0,30}][[1,1,2]]
```

Plotting the vector expression swan2 - G[t] gives Figure 12(a), with a flight time of 11.98 days.

When Domingo returns to earth, the swans fly always toward the earth. However, unlike in the earth-moon model of the last section, now the earth is moving. A slight modification of swan2 enables us to generate return trajectories.

```
Q=390.625; q=365; θ = π/3; T=-11.27; R=0.0167;
r=0.00453; p=27.3; c=6.817;
```

```
swan3=NDSolve[{X'[t]==c(G[t]-X[t])/Norm[G[t]-X[t]],
X[0]==μ[0]+r{Cos[θ],Sin[θ]}}, X[t], {t,0,30}][[1,1,2]]
```

Generate, if possible, pursuit curves in this model for which the flight from the earth to the moon is 10 days, and the return trip is 5 days.

5. Another way to attack Godwin's consistency problem of 12 versus 8 is to focus on the number of swans in the wedge. In the story, two swans die on the moon, so that Domingo's wedge has 23 swans on the return trip. Develop a model between wedge speed and swans in the wedge, so that in the earth-moon model, 12 and 8 days are consistent times for the trip to the moon and back.

6. Solve this pursuit problem from a 1748 issue of the *Ladies Diary*, as posed by John Ash.

   A spider, at one corner of a semi-circular pane of glass, gives uniform and direct chase to a fly, moving uniformly along the (semi-circular) curve before him as shown in Figure 13. The fly is initially at 30° from the spider, and is captured at the opposite corner. What is the ratio of both uniform motions?

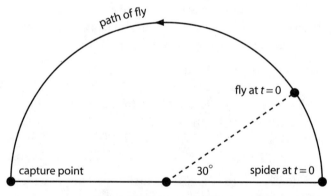

Figure 13. Spider stalking a fly

In due course, the editors of the journal reported that a solution was received but, it "being so operose [tedious]," they declined to publish it.

7. (a) Using a Mercator projection map, solve this pursuit problem from a 1749 issue of the *Gentleman's Diary*, as posed by Thomas Perryam.

Maupertuis's Pursuit Problem

A captain of a privateer seeing a merchant ship at SSE sailing due west, continually bears directly upon her. On coming upon the ship, it appeared that the merchant ship sailed 30 miles during the chase, whereas the privateer was 21 miles from t when the chase began. Supposing each ship's motion as uniform during the chase, find the distance sailed by the privateer, the difference in longitude, and the difference in latitude of the privateer from start to finish.

(b) Resolve Exercise 7a if the privateer is initially at 50° N, 10° W. Assume that the earth is a sphere of radius 6400 km. (This is a 3-dimensional pursuit problem.)

8. In the sun-earth-moon model, for a given positive number λ, find trajectory pairs so that the ratio of flight times to the earth and the moon is λ. For a given value of c, find the optimal λ values.

9. Imagine that the Little Prince's asteroid B-612 pictured in Figure 3(b) is in the asteroid belt of our solar system. Plot trajectories of the Little Prince's flight to the earth and home again for various speeds c and configurations of the earth and B-612.

10. Position n snails at the vertices of a regular n-gon. Each snail crawls directly toward its nearest counterclockwise neighbor at unit speed. Draw the pattern of the snail trails for various positive integers n.

Vignette X: Voltaire and the Almighty

With respect to theology, where does Voltaire stand? Most would classify him as a Deist, or in Voltaire's terminology, a Theist. Here is his definition [**144**, vol. xiv, *Theism*, p. 80]:

> What is a true theist? It is he who says to God: "I adore and serve You," and [to all men]: "I love you."

He differentiates Theism from both Islam and Christianity using anecdotes [**144**, vol. xiv, *Theism*, p. 81]:

> [A Theist] doubts, perhaps, that Mohamed made a journey to the moon and put half of it in his pocket; he is tempted not to believe the story of St. Amable, whose hat and gloves were carried by a ray of the sun from Auvergne to Rome.

To be more blunt, he generalizes [**144**, vol. xiv, *Theism*, p. 80],

> Theism is good sense not yet instructed by revelation; other religions are good sense perverted by superstition.

Of course, Voltaire was being flippant about revelation and superstition in these last two quotes. In a manner that appears to be more serious, he outlines Christian faith [**144**, vol. viii, *Faith*, pp. 328–332]:

> The faith which [Christians] have for things that they do not understand is founded upon that which they do understand; they have grounds of credibility. Jesus Christ performed miracles in Galilee; we ought therefore to believe all that he said. In order to know what he said we must consult the Church. The Church has declared the books which announce Jesus Christ to us to be authentic. We ought therefore to believe those books. We ought to submit our reason to it, not with infantile and blind credulity, but with a docile faith, such as reason would authorize. Such is Christian faith.

If this entry on faith from his *Philosophical Dictionary* was all that we had from Voltaire's pen, we might be tempted to say that Voltaire is espousing orthodox Christianity. However, when we read much more than this passage, we might suspect that Voltaire means almost the opposite of what he writes, namely, that whenever he says "ought" or "must" in this passage he really means "ought not necessarily" or "must not necessarily." At least, that's how the church in Voltaire's day primarily decided to interpret him.

For example, here is Voltaire speaking about believing that Rabelais's Gargantua was a historical figure [**144**, vol. ix, *Gargantua*, p. 144–146]:

> If no writer, with the exception of Rabelais, has mentioned the prodigies of Gargantua, at least no historian has contradicted him. Believe then, I repeat, in Gargantua; if you possess the slightest portion of avarice, ambition, or knavery, it is the wisest part you can adopt.

These last two passages on faith and Gargantua are parallel. Clearly, Voltaire is being deliberately silly in arguing for the historical existence of Gargantua. He concludes however that anyone who says that they believe the argument is lying and is therefore probably a knave of a human. Now replace Rabelais and Gargantua with the Church and Jesus, respectively, along with the accompanying conclusion about knavery—and it is easy to see why the Church sometimes had difficulty with Voltaire.

But Voltaire was not a complete skeptic. Here's an entry, in my opinion, in his *Philosophical Dictionary* when he seems to be genuine with respect to faith—without any irony, sarcasm, or multiple meanings: in a long imaginative entry on religion, Voltaire has a dialogue with the leading religious leaders of the past, the last of whom is Jesus, whereupon the dialogue concludes with these two lines [**144**, vol. xiii, *Religion*, p. 73]:

> Jesus: When I was in the world, I never made any difference between the Jew and the Samaritan [the non-Jew].
> Voltaire: Well, if it be so, I take you for my only master.

Nevertheless, Voltaire's writings are full of skepticism. In their fifteen years together, Emily and Voltaire discussed many things. Besides Newton and his ideas and Voltaire's poetry and plays, they enjoyed critiquing scripture and ecclesiastical history and tradition, taking delight in exposing inconsistencies and superstition. Almost every other entry in *The Philosophical Dictionary*, although written long after Emily had died, is a rendition of their discussion on these matters. As just one of many examples of this skepticism: Voltaire considers a proverb of Solomon (Proverbs 23:31),

Vignette X: Voltaire and the Almighty

> Look not upon the wine when it is red, when it gives its color in the glass.

Voltaire comments [**144**, vol. xiii, *Solomon*, p. 243],

> I doubt very much whether there were any drinking glasses in the time of Solomon; it is a very recent invention; all antiquity drank from cups of wood or metal.

That is, Voltaire suggests that not all of the proverbs attributed to Solomon were written by him or were even written in the days of Solomon. However, glass vessels were first made in about 1600 BC in Egypt and Mesopotamia. Moreover, an ancient glass-making factory from the days of Ramses the Great (c. 1303–1213 BC) has been well studied [**96**]. Apparently the ancients pressed and rolled hot glass. The much more recent technology of blown glass dates to Syria in the first century BC. Nevertheless in this instance and in many more like unto it in the *Philosophical Dictionary*, Voltaire raises open-ended questions. He delights in ferreting out why things are the way they are. For him, blind faith is no faith at all, but delusion.

It is in the spirit of Voltaireian inquisitiveness that we include this next chapter on the biblical value of π. We also raise a question. How do we handle measurements from other eras? We cannot return and ask that the measurements be taken according to our specifications. Pliny the Elder wrestled with this issue when he compiled his encylopedic *Natural History*. After chronicling at length the astronomical measurements of the past, he says [**106**, book ii, section xxi, p. 229],

> [Some of] these figures [computations] are really unascertained and impossible to disentangle.

Nevertheless, he included them in his narrative because they were interesting.

Finally, with respect to Voltaire's theology, here is his short list of what to believe [**144**, vol. xii, *Necessary*, p. 57].

> Distrust the inventions of charlatans; worship God; be an honest man; and believe that two and two make four.

CHAPTER **X**

Solomon's π

Of the beginnings of π, it is often said that its biblical value is 3, for *I Kings 7:23* reads

> Solomon made a molten sea, ten cubits from the one brim to the other: it was round all about, and its height was five cubits: and a line of thirty cubits did compass it round about.

This bronze basin is known as Solomon's Sea, and is illustrated in Figure 1. The text seems to say that for a circumference of 30, a circle's diameter is 10, implying that π is 3. To delineate this strand of the lore about π, we present various perspectives on this puzzle, presenting them roughly in order

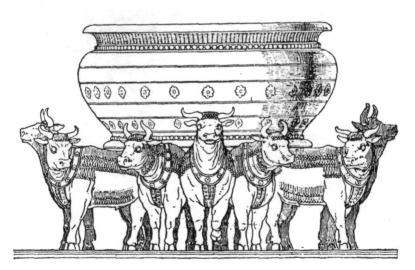

Figure 1. Solomon's Sea [**14**, plate 87.]

of increasing complexity, although not necessarily in order of increasing credibility.

I. Noise

We call those who measured Solomon's Sea *surveyors*, those who wrote the historical books of the Bible *chroniclers*, and those who copied or translated the books *scribes*. It is possible that the surveyors measured wrongly, that the chroniclers recorded information imperfectly, or that later scribes transcribed erroneously—miscues that we categorize as noise.

The biblical books of *Kings* and *Chronicles* are parallel texts. Scholars have tried to harmonize the apparent noise of their differing accounts, which of course includes Solomon's Sea. Payne [**100**] accounts for textual anomalies as instances of "accidental corruption by a later scribe" either through a "mistaken reading" of word form or through ambiguous, "unclear" numerical expressions; "rounding numbers" and "hyperbolically" inflating numbers so as to make a point; and "different methods of reckoning." As a simple example of noise, Herzog [**66**] points out that the Greek Septuagint translation renders the Sea's circumference 33 cubits in *Kings*, while rendering it 30 cubits in *II Chronicles 4:2*.

As a more serious example, consider the disagreement on the capacity of Solomon's Sea, where *I Kings 7:26* puts it at 2000 baths (where a bath is somewhere between 4.5 and 10 gallons) while *II Chronicles 4:5* puts it at 3000 baths. Rabbinic scholars in the *Talmud, Erubin 14b*, dating to about 500 AD, explain this difference by rendering the 2000 baths of the Kings passage as liquid measure and the 3000 baths of the Chronicles passage as dry measure, and say that the dry measure would include a heap above the brim, being one third of the total measure. Wylie [**155**] attributes this difference in volume measurement to a confusion about the Sea's structure: he says that whereas the Sea probably had a hemispherical basin, the *Chronicles* chronicler assumed that the Sea had a cylindrical basin, and so recorded its corresponding volume. A cylinder circumscribed about a hemisphere—where the two share the same base—has 1.5 times the volume of the hemisphere, so the 3000 baths of *Chronicles* would be 1.5 that of the 2000 baths of *Kings*.

A highly whimsical and more recent example of surveyor noise in the context of calculating π's value comes from Dudley [**35**]. He stumbled across a compilation of mid-nineteenth century approximations for π (taken from apparently sincere attempts at squaring the circle), as shown in Table 1.

For a moment, imagine that π's value is linear in time and that the given data is indicative of π's variable value. The line of best fit through the data, $p(t)$, gives the approximate value of π in year t, where t is the Gregorian year:

$$p(t) = 3.1239827671 + 0.0000157082t. \tag{1}$$

Table 1. Novice attempts at the first five decimal digits of π, 1832–1879

$t: \pi$	$t: \pi$	$t: \pi$	$t: \pi$	$t: \pi$
1832: 3.06250	1845: 3.16667	1855: 3.15532	1865: 3.16049	1874: 3.15208
1833: 3.20222	1846: 3.17480	1858: 3.20000	1866: 3.24000	1874: 3.14270
1833: 3.16483	1848: 3.20000	1859: 3.14159	1868: 3.14214	1874: 3.15300
1835: 3.20000	1848: 3.12500	1860: 3.12500	1868: 3.14159	1875: 3.14270
1836: 3.12500	1849: 3.14159	1860: 3.14241	1869: 3.12500	1875: 3.15333
1837: 3.23077	1850: 3.14159	1862: 3.14159	1871: 3.15470	1876: 3.13397
1841: 3.12019	1851: 3.14286	1862: 3.14214	1871: 3.15544	1878: 3.20000
1843: 3.04862	1853: 3.12381	1862: 3.20000	1872: 3.16667	1878: 3.13514
1844: 3.17778	1854: 3.17124	1863: 3.14063	1873: 3.14286	1879: 3.14286

Dudley makes some amusing extrapolations, such as when π would have value 3.2, or when it had value π, 3, or 0. For example, by (1), π's decimal expansion would be 3 in the year 7893 BC, perhaps the year when Adam and Eve found themselves in the Garden, or when early man first drew the sun and moon on cave walls and discovered that the proportion of circumference to diameter is close to three.

II. Tradition

The ancients had many different rules for π, some of whose natural interpretations implicitly define π as 3. Castellanos [22] and Gupta [58] cite various documents which demonstrate that the ancient Babylonians, Egyptians, Indians, and Chinese had such a rule. Of course, we focus on the Jewish tradition of π being 3 because the Bible is readily available, and the image of a large bronze basin being the center of some controversy makes a great story. One such Hebrew rule is found in the *Mishnah*, a compilation of Jewish traditions, dating to the second century AD. In *Mishnah, Erubin 1:5*, we read,

> Whatsoever is three handbreadths in circumference is one handbreadth in width.

Zuckermann, a nineteenth century German scholar, points out that the rabbis who compiled the *Mishnah* "were aware of more exact values [of π], but accepted the value of 3 as a workable number for religious purposes" [**44**, p. 23]. By way of illustration of religious purposes, consider the following passage

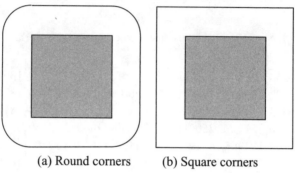

(a) Round corners (b) Square corners

Figure 2. Levite gardens about the cities of refuge

from *Erubin 14b* from the *Talmud*, which is an expansive commentary on the *Mishnah*. The above *Mishnah* rule is given in the following equivalent form:

> But consider: By how much does a square exceed that of a circle? By a quarter.

This rule is to be interpreted in the following way: Take a square of side length 2 and inscribe a circle within it; the area of this square is 4; removing $\frac{1}{4}$ of this area from 4 leaves 3, the approximate area of the circle and the implicit, practical Talmudic value of π. One of the early applications for this rule, and in fact an application which may have led to the formulation of this rule (see *Erubin 56b-57a*), is the problem as described in *Numbers 35:4-5* in the time of Joshua: cities measuring 2000 cubits from north to south and 2000 cubits from east to west, with 1000 cubits outward from the walls roundabout, were to be given to the Levite tribe—were the corners to be round or square, as indicated in Figure 2? The difference in area is worth discussing, at least if you were a Levite.

This same problem was a lively issue in resolving the problem of how far one is allowed to walk on the sabbath; *Erubin 4:8* says that

> [one could] travel within two thousand cubits in any direction as [though he was within] a circle [while] the Sages say: As [though he was within] a square, so that he wins the benefit of the corners.

This passage may be one of the oldest instances of a non-Pythagorean measure for distance.

Thus, in light of the above examples, even if the surveyors had measured the Sea as a $31\frac{1}{2}$ cubit circumference and a 10 cubit diameter, for example, it is possible that these values may have been adjusted to harmonize with the tradition of π being 3.

It is also possible that this tradition of implicitly identifying 3 with π arose from a practice of rounding to the nearest integer. And therefore, as Meeus [86] points out, if the diameter of the Sea lay between 9.5 and 10.5 cubits, and the circumference lay between 29.5 and 30.5 cubits, then the biblical value of π is between the bounds of $29.5/10.5 \approx 2.81$ and $30.5/9.5 \approx 3.21$, thereby resolving any measurement anomaly. However *Exodus 37:1* gives the measurements of the ark of the covenant as $2\frac{1}{2}$ by $1\frac{1}{2}$ by $1\frac{1}{2}$ cubits. Adjusting Meeus's argument to round to the nearest half leaves the biblical value of π between 2.90 and 3.10, not nearly so satisfactory.

One shortcoming of these kinds of arguments is that since a great deal of thought and effort went into casting the Sea, one might expect those dimensions to be recorded accurately.

III. The Hidden Key

Perhaps a hidden key exists to unlock the meaning of this passage. Posamentier [109] relates the story of an 18th century Polish rabbi, Elijah of Vilnah, who observed in the Masoretic text, the Hebrew Bible, that the word "line" in the parallel *Kings* and *Chronicles* texts of this passage are spelled קוה and קו, respectively. (Others attribute this gematria reasoning to Rabbi Matityahu Hakohen Munk [30].) The extra ה is the key. How is it used? Take the ratio of the sums of the standard numeric values of the Hebrew letters (ק $\equiv$ 100, ו $\equiv$ 6, ה $\equiv$ 5) for each of these words, obtaining $111/106$; multiply by 3—the apparent value of π—and obtain $\pi \approx 333/106 \approx 3.141509$, a value agreeing with π to four decimal digits. Stern [135] comes to the same conclusion independently by examining only the *Kings* passage, observing that the word "line" while written as קוה is pronounced only as קו since ה is silent.

A natural question with respect to this method is, why add, divide, and multiply the letters of the words? Perhaps an even more basic question is, why all the mystery in the first place? Furthermore, H. W. Guggenheimer in his *Mathematical Review* note on [135] seriously doubts that the use of letters as numerals predates Alexandrian times; if such is the case, the chronicler did not know the key. Moreover, even if this remarkable approximation to π is more than coincidence, this explanation does not resolve the obvious measurement discrepancy—the 30 cubit circumference and the 10 cubit diameter.

Finally, Deakin [30] points out that if deity truly is at work in this phenomenon of scripture revealing an accurate approximation of π, a much better fraction not far from 333/106 would most definitely have been selected instead. The basic idea is from ancient Egypt, where the custom of dealing with a fractional quantity was to write it as the sum of unitary fractions: fractions with numerator one and denominator a positive integer. Thus an Egyptian would write 5/6 as $1/2 + 1/3$. The notion of continued fractions is a natural development of this bias. That is, the number denoted as the sequence of positive integers (except that the first may be 0) $[a_0, a_1, a_2, \ldots]$ is

$$[a_0, a_1, a_2, a_3, \ldots] = a_0 + \cfrac{1}{a_1 + \cfrac{1}{a_2 + \cfrac{1}{a_3 + \cdots}}}.$$

Each positive rational number can be expressed as a terminating sequence $[a_0, a_1, \ldots, a_n]$ for some n, while every irrational number including π has a unique infinite sequence. Its continued fraction, from almost any text on number theory, starts with

$$\pi = [3, 7, 15, 1, 292, 1, 1, 1, 2, 1, 3, \cdots],$$

so that π is the limit of the progression

$$3 \longrightarrow 3 + \frac{1}{7} \longrightarrow 3 + \cfrac{1}{7 + \frac{1}{15}} \longrightarrow 3 + \cfrac{1}{7 + \cfrac{1}{15 + \frac{1}{1}}} \longrightarrow \cdots$$

written more familiarly as

$$3 \longrightarrow \frac{22}{7} \longrightarrow \frac{333}{106} \longrightarrow \frac{355}{113} \longrightarrow \frac{103993}{33102} \longrightarrow \cdots.$$

That is, God might prefer 355/113 (along with a suitable key so as to reveal the secret to man) rather than 333/106 as representative of π, for three reasons:

- The denominators 113 and 106 are very close; that is, the fraction 355/113 is marginally more complicated than 333/106.

- The fraction 355/113 correct to ten decimals is 3.141592920, giving 6 digit agreement with π rather than four, a hundred times better!

- In the Indo-Arabic number system, the fraction 355/113 is easily remembered—the digits of the fraction when following an S pattern from below form the sequence 113355.

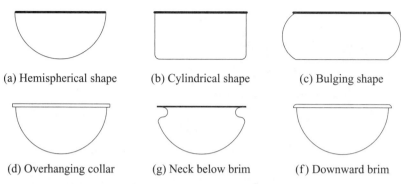

(a) Hemispherical shape (b) Cylindrical shape (c) Bulging shape

(d) Overhanging collar (g) Neck below brim (f) Downward brim

Figure 3. Possible Sea Profiles

IV. The Inside Story

The *Talmud*, *Erubin 14a* maintains that the 30 cubit measurement was the inside circumference of the Sea. Such a measurement when made compatible with $\pi \approx 3.14$ and a 10 cubit outside diameter means that the thickness of the Sea is about four inches, the approximate width of a man's hand, which is how *I Kings 7:26* describes it. That is, if t is the thickness, then the inside diameter is $10 - 2t$ and so $30 = \pi(10 - 2t)$, which means that $t \approx 0.225$ cubits; since a cubit is approximately 18 inches, then $t \approx 4$ inches. Rabbi Nehemiah, in the *Mishnat ha-Middot*, the earliest extant Hebrew work on geometry, dating to about 150 AD, outlines this same approach [**15**, pp. 75–76].

Measuring the inside circumference of a basin with a line is tricky however. One way to approximate this measure is to "walk" a cubit stick around the inside of the opening, so tracing out an inscribed 30-gon of sorts. Along these lines, Zuckermann proposed a dodecagonal shape for the Sea's opening [**44**, p. 51]; see Figure 4(d). Both of these models are in agreement with the *Talmud*'s conclusion in *Erubin 14a*.

A tradition which the *Talmud* may have used as justification for its explanation is described in *Mishnah*, *Kelim 18:1*:

> The School of Shammai say: A chest should be measured on the inside [to determine its capacity]. And the School of Hillel say: On the outside.

Since the diameter measure is clearly an outside measurement from the *Kings* passage, and since there is some ambiguity in the measurement of the circumference, the *Talmud* adopted the former tradition rather than the latter for that measurement, even though the English translation, "a line of thirty cubits did compass it round about," suggests an outside measurement.

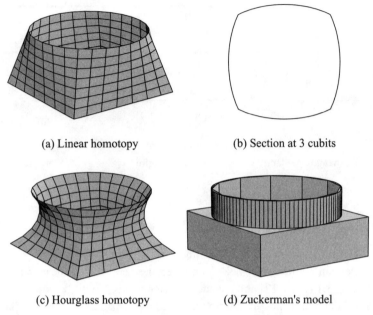

(a) Linear homotopy (b) Section at 3 cubits

(c) Hourglass homotopy (d) Zuckerman's model

Figure 4. Some "square-round" models

V. The Protruding Brim

A natural model for the Sea's shape is a hemispherical bowl whose girth is greatest at the brim so that the Sea has a somewhat circular profile as in Figure 3(a). Josephus says as much. In *Erubin 14b*, Rami bar Ezekiel says that the Sea was square from its base to three cubits up while round at the brim to two cubits down. Another interpretation is that the cross-sections from the base to the rim follow a homotopy of a square transforming into a circle as is done linearly in Figure 4(a); a more elegant rendering is the hourglass transformation of Figure 4(c); in these models, the juncture which Rami bar Ezekiel alludes to is illustrated by Figure 4(b), the cross-sectional shape at height three cubits, above which the cross-sections are more round and below which the cross-sections are more square. Zuckermann interprets this passage literally, so that the top (two cubits) is cylindrical and the bottom (three cubits) is prismatic, as in Figure 4(d), [**44**, p. 51]. Zuidhof [**156**] proposes a cylindrical body, and thus a rectangular profile. Payne [**100**, p. 122] maintains that the Sea had a "considerable bulge to accommodate even (the) two thousand baths (of *I Kings 7:26*)." So the shape of the Sea is quite unresolved. But *I Kings 7:24* describes that beneath the brim of the

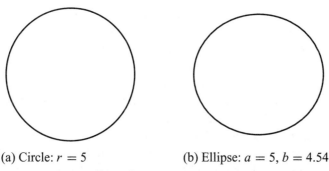

(a) Circle: $r = 5$ (b) Ellipse: $a = 5, b = 4.54$

Figure 5. The Sea from above

Sea were two rows of *knops*—grape-like, decorative knobs—forming a kind of collar, so that the upper part of the Sea's silhouette looked something like the upper part of Figure 3(d); perhaps the circumference measurement was taken just beneath this collar, as Steveson [136] and Zuidhof [156] suggest, or was taken as the measurement around the neck of Figure 3(e) or around the waist of Figure 4(c).

Another explanation is that the brim of the Sea overhung its crest as in Figure 3(f), so that the length of a cord strung "from one brim to the other" would be greater than the actual diameter. If this extra downward curve of the Sea's lip gives an extra four inches or so on each side, the measurement anomaly is resolved.

VI. The Premature Conic

As suggested by Read [111], suppose that the brim's contour is merely round or oval shaped or an ellipse, so that the diameter—the major axis—is 10 cubits. To find the minor axis, $2b$, where the ellipse in parametric polar coordinates is $x = 5\sin(\theta)$ and $y = b\cos(\theta)$, write the integral expression for arc length, and equate it to a perimeter of 30 cubits, resulting in the equation

$$4\int_0^{\frac{\pi}{2}} \sqrt{25\cos^2(\theta) + b^2\sin^2(\theta)}\,d\theta = 30,$$

which when solved gives $b \approx 4.54$. That is, the minor axis of such an ellipse is about an inch more than 9 cubits. To model the Sea's opening by other ovals, the integral formula in [88] may be useful.

Although ellipses were not defined until Menaechmus (c. 380–320 BC), ovals were certainly familiar to the ancients. So if one wished to design a round object with perimeter 30, long diameter 10, and short diameter an integer, then the ellipse of Figure 5 (or an oval very close to it) is what will most likely be designed by trial and error.

Steveson [136] demurs, saying that the twelve symmetrically placed oxen upon which the Sea sat *(I Kings 7:25)* supports a circular shape. Three of these oxen faced north, three west, three south, and three east in the counterclockwise convention—and from which the *Talmud, Yoma 58b* says, "Hence you are taught that all the turns you make [in the Temple] must be to the right." In such a tradition that each direction is of equal importance, an oval opening might be viewed as improper.

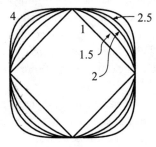 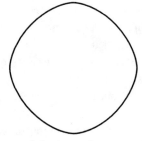

(a) A family of pseudo-circles   (b) With $p \approx 1.7$, circumference to long diameter: 3

Figure 6. Pseudo-circles

A family of curves that has a more proper four-fold symmetry in keeping with the four natural directions and the four oxen teams are the pseudo-circles,

$$|x|^p + |y|^p = r^p, \tag{2}$$

where r is the pseudo-radius and p is any positive number. For example, Figure 6(a) shows various family members, for $p = 1$, 1.5, 2, 2.5, 4. By the integral arc-length formula, the circumference $C(p)$ of a pseudo-circle is the intimidating looking integral

$$C(p) = 8r \int_0^{(\frac{1}{2})^{\frac{1}{p}}} \sqrt{1 + z^{2p-2}(1-z^p)^{\frac{2(1-p)}{p}}}\, dz,$$

where $z = x/r$. Using a computer algebra system for computing $C(p)/(2r)$ gives a ratio of 3.03 when $p = 1.7$. That is, Figure 6(b) is a good candidate for the shape of the Sea's rim.

VII. The Double Standard

Since the cubit is approximately the length of a forearm from elbow to finger tip, about 1.5 feet, an idea that resolves the ratio dilemma is for a taller craftsman to measure the circumference and a shorter craftsman to measure the diameter.

Is there any merit to this argument?

There were at least three different cubit lengths in use in biblical times. The second temple (dating to no earlier than 500 BC) housed a bureau of standards, as we would call it, within its eastern gate, referred to as the Castle of Ŝuŝan [**72**, p. 121]. *Mishnah*, *Kelim 17:9* describes the relationship between three of these units.

> And there were two (standard) cubits in the castle of Ŝuŝan, one on the north-eastern corner, and the other on the south-eastern corner. The one on the north-eastern corner exceeded that of Moses by half a fingerbreadth, [while] the one on the south-eastern corner exceeded the other by half a fingerbreadth, so that the latter exceeded that of Moses by a fingerbreadth. And why did they prescribe one large and one small? Only [for this reason]: that the craftsmen might take [material] according to the small [cubit] and return [their finished work] according to the large [cubit], so that they might not be guilty of trespass [of Temple property].

In conventional units, two standard cubits were used in the time of Solomon and the first temple, the cubit of Moses (M) of length 42.8 cm, and the large cubit (L) of length 44.6 cm; a third standard, the small cubit (S) of length 43.7 cm came much later, according to Kaufman [**72**].

This *Kelim* passage describes a curious measurement tradition in the days of the second temple. That is, temple craftsmen took materials of wood or stone from the temple in terms of the *profane* (ordinary) S cubit, worked with those materials outside the temple (as the sound of hammer and chisel was forbidden on the temple site), and returned the finished items in terms of the *holy* L cubit, installing them inside the temple. This measurement rule seems austere for craftsmen, because actual lengths of finished products are usually less than the actual lengths of the raw materials used. It looks like double jeopardy! Apparently, temple personnel held craftsmen to a very strict accounting. An editorial footnote for this passage of the *Talmud* (Soncino Press, 1948) summarizes this point:

> [these rules made] sure that they [the workmen] neither appropriated any material that belonged to the Temple nor received payment for labour they had not performed.

Kelim 17:10 goes on to point out that all measurements of the second temple itself were in terms of the S cubit except for the measurements of "the Golden Altar and the horns and the Circuit and the Base [of the Altar]." The editorial notes go on to say that these most holy and inner things of the temple appear to have been measured in terms of the M cubit.

In view of such measurement traditions in the days of the second temple, it is reasonable to imagine similar ones in the days of the first temple. In particular, since M was an older standard than L, profane or ordinary objects were probably measured with L while holy objects were probably measured with M. It is therefore possible that as a meaningful gesture, since this basin's function was to cleanse, rendering the profane into the holy, the engineers of the Sea may have ceremoniously designed the Sea so that the outside—the circumference—was in terms of L, and that the inside—the diameter—was in terms of M. Such a conjecture results in $\pi \approx 3.12$, (where $\pi(10)(42.8) \approx 30(44.6)$).

Furthermore, this value of π is independent of the stated cubit's lengths of 42.8 cm and 44.6 cm. Let l and m be the lengths of L and M respectively. Since each cubit is 24 fingerbreadths long, and since this *Kelim* passage asserts that the L cubit exceeds the M cubit by a fingerbreadth, then since M is an older unit than L, a natural interpretation is that $l = (25/24)m$. If so, a circumference of 30 of the L cubits and a diameter of 10 of the M cubits yields the relation

$$30\left(\frac{25}{24}\right)m = 10\pi m,$$

which gives $\pi \approx 3.125 = 25/8$, which is a Babylonian approximation for π in vogue during Solomon's day [15, pp. 21–22]. With such a close approximation to π, it is natural to wonder whether the L cubit was initially defined so that 3 of its cubits would encompass a circle of diameter 1 of the cubits of Moses.

Finally, there are mathematical variants of this double standard idea for making π evaluate to 3.

Andersen [5] and Chasse [25] suggest using non-Euclidean elliptical geometry. As such a good model is the surface of a sphere or globe. Identify the rim of the Sea as the circle of latitude at 60°N on a globe, so that the diameter measure will follow a great circle arc across the north pole. The ratio of circumference to this curved diameter is indeed 3. This reasoning is reminiscent of Figure 3(f), wherein a line measuring the diameter may drape across a curved surface.

R. Euler [43] and Adler [3] use a pseudo-metric to measure circumference and diameter of the pseudo-circles (2) where pseudo-distance $d(X, Y)$ between two points in the plane is

$$d(X, Y) = \left((x_1 - y_1)^p + (x_2 - y_2)^p \right)^{\frac{1}{p}},$$

with $X = (x_1, x_2)$, $Y = (y_1, y_2)$, and p a positive number. As they point out, the least possible value for the ratio of circumference to diameter for pseudo-circles is π. However if the circumference of the pseudo-circle is measured in the pseudo-metric while the diameter is measured with the usual Euclidean metric, then for the pseudo-circle with $p \approx 2.37$, the ratio of circumference to diameter taken along the line $y = x$ is 3.

Norwood [97] explains that by the Lorentz-Fitzgerald contraction principle of relativity theory, a circle of radius 1 meter spinning at 10 million rpm will make the proportion of circumference to diameter to be 3. He explains further in [98] that his measuring stick is stationary and outside the spinning system, so that the measuring stick's length remains invariant during the measurement process.

Concluding Remarks

Which explanation is correct? Since the Sea is reported as broken and carted away by the conquering Babylonians in *Jeremiah 52:17* in about 586 BC, there are no irrefutable answers. Each of the arguments has some merit. And it may very well be that the true story lies in a combination of these explanations. Whatever the explanation for this puzzle, what I find most interesting is that the chroniclers somehow decided that the diameter and girth measurements of Solomon's Sea were sufficiently striking to include in their narrative. It is almost as if they saw "as through a glass darkly" the abstract π, and could not but help to record in passing this particular instance of a most curious geometric relationship.

Exercises

1. (a) Use the following series, discovered by James Gregory in 1671, to calculate two decimal place accuracy for π by hand by evaluating at $x = 1$:

$$\tan^{-1} x = x - \frac{x^3}{3} + \frac{x^5}{5} - \frac{x^7}{7} + \cdots, \tag{3}$$

that is, $\pi = 4(1 - 1/3 + 1/5 - 1/7 + 1/9 - 1/11 + \cdots)$.

   (b) Calculate π to four decimal place accuracy by hand using (3) with $x = 1/\sqrt{3}$, as was first done by Abraham Sharp (1651–1742) who used this idea to compute 72 digits of π [**15**, p. 144].

2. (a) From Table 1, draw a scattergram of the data. Then calculate the correlation coefficient,

$$\rho = \frac{\sum_{i=1}^{n}(t_i - \bar{t})(p_i - \bar{p})}{\sigma_t \sigma_y},$$

where n is the number of data items, $\bar{x}$ is the average value of the variable x, and σ_x is the standard deviation of the variable x, where $\sigma_x^2 = \sum_{i=1}^{n}(x_i - \bar{x})^2/n$. The correlation coefficient varies from -1 to 1 in general; when there is a strong, direct relationship between two variables, the correlation coefficient has magnitude close to 1; when the relationship is weak, it is close to 0. What can you conclude about the relationship between t and p?

   (b) For the data of Table 1, generate the quadratic $q(t)$ of best fit, and use it to find the two years in which $q(t)$ is π.

3. Find the difference in areas for the region given to the Levites outside the cities of refuge as described on p. 244. Figure 2 shows circular versus rectangular corners.

4. The curious rule for travel on the sabbath as given on p. 244 seems to say that the distance $D(X, Y)$ between two points X and Y is

$$D(X, Y) = \max\{|x_1 - y_1|, |x_2 - y_2|\},$$

where $X = (x_1, x_2)$ and $Y = (y_1, y_2)$. Is this a metric? Prove your answer.

5. As Meeus [86] noted, if measurements are rounded to the nearest integer, the ratio of circumference to diameter ranges from 2.81 to 3.21 on p. 245. Find the range for the ratio of circumference to diameter if the custom is to round to the nearest quarter.

6. Find the fifth rational approximation to π using the continued fraction representation on p. 246. To how many decimal places does this fraction agree with π?

7. Find the regular n-gon of radius 5 for which the perimeter is closest to 30.

8. For the pseudo-circle and the Euclidean metric, there is a number $p > 2$ for which $C(p)/(2R)$ is also 3, where R is the long diameter. Find it.

9. For the explanation in section (III), devise a key—perhaps similar in spirit to the key described for the number 333/106—which suggestively leads one to conclude that a good approximation for π is the number 355/113.

10. On p. 253, find the value of r for which the circumference of the pseudo-circle is 30 with $p = 2.37$.

Vignette XI: Laputa and Gargantua

The Micromégas story owes much to Jonathan Swift and François Rabelais.

Figure 1. Gulliver espying Laputa, sketch courtesy of Jack Boyles (2009)

During Voltaire's exile to England, Jonathan Swift and Voltaire spent three months together on the country estate of an English nobleman. Since Swift's *Gulliver's Travels* had just recently been printed, they probably discussed giants and political/social lampooning in general. Like the Micromégas story, one of Gulliver's adventures was about a community of scientists. On the island country of Laputa, supposedly off the coast of Japan, the Laputians held court on a levitated loadstone. When rebellion broke out in any part of the kingdom, the floating rock could be manouvered so as to hover over that part of his kingdom, to block out the sun and, if necessary, to be lowered to crush buildings and any imagined rebels within. Figure 1 is a sketch of Gulliver and the floating island; and Figure 2 is a photo demonstrating that some diamagnetic materials, being repelled by a magnetic field and attracted by a gravitational field, will levitate in equilibrium in space. In this case, a thin rectangular slice of pyrolithic graphite hovers about three millimeters above some nickel-cadmium magnets.

Figure 2. Pyrolithic graphite floating in a magnetic field, courtesy of Jared Newton

The Laputian court consisted primarily of Swiftian scientists: people obsessed with the discovery and theory of new objects and phenomena to the exclusion of concern for the daily affairs of their own lives and the general welfare of the peasants in the cities below. Of course, Swift was poking fun at the imbalance that can occur between doing research for research's sake and ignoring societal problems to the detriment of society itself; he was mocking man's tendency in a new age of enlightenment to make reason alone the foundation of society and governance, as is further explored much later in such stories as Aldous Huxley's *Brave New World* and George Orwell's *1984*. C. S. Lewis termed this extreme rational outlook as *scientism*, which he describes as

Vignette XI: Laputa and Gargantua

the belief that the supreme moral end is the perpetuation of our own species, and that this is to be pursued even if, in the process of being fitted for survival our species has to be stripped of all those things for which we value it—of pity, of happiness, and of freedom. [**76**, pp. 71–72]

Martin Luther, two centuries before Swift and Voltaire, broadly lashed out at this kind of outlook, describing reason as a whore.

But the devil's bride, reason, the lovely whore comes in and wants to be wise, and what she says, she thinks, is the Holy Spirit. Who can be of any help then? Neither jurist, physician, nor king, nor emperor; for she [reason] is the foremost whore the devil has. [**83**, vol. 51, p. 374, Martin Luther's Last Sermon in Wittenberg : Second Sunday in Epiphany, January 17, 1546]

From this tension between reason and tradition, or reason and faith, Swift accordingly coined his brave new world, for *la puta* is Spanish for *the whore*.

Voltaire wrote his story with elements of this Swiftian spirit, albeit with less coarseness and directness. Indeed, in the only instance in *Micromégas* where he refers to his friend by name, Voltaire borrows some of Swift's bawdy humor. Doing so is also an indirect allusion to Rabelais, for Voltaire describes Swift as "the English Rabelais" [**144**, vol. xii, *Prior, Butler, and Swift*, p.312] and [**144**, vol. xxxix, *Dean Swift*, p. 90]:

Rabelais was in every respect superior to his age, though Swift is infinitely [superior] to Rabelais. Our curate of Meudon [Rabelais], in his extravagant and unintelligible book, has exhibited extreme gayety and equally great impertinence. He has lavished at once erudition, coarseness, and ennui. [Yet Rabelais] is a drunken philosopher, who only wrote in moments of intoxication.

In his poem *The Temple of Taste*, about an ideal library containing only the best of the world's literature, Voltaire acknowledges Rabelais' craftsmanship as a writer, albeit with a proviso [**144**, vol. xxxvi, *The Temple of Taste*, p. 64]:

The work of Rabelais is to be seen there [in the Temple of Taste], reduced to less than half a quarter of its bulk.

In his youth, François Rabelais had been a monk, first with the Franciscans, who, as part of an anti-intellect campaign meant to reform their order, confiscated the books from his cell. Not long thereafter, he joined the Benedictines, where he served as secretary to a bishop. In mid-life, he became a medical doctor and also worked in the printing trade: proof-reading, editing classics, and translating manuscripts. In the summer of 1532, an anonymously written book of folk-tales entitled *The Great and Inestimable Chronicles of the Huge Giant Gargantua* appeared, being a compilation of traditional local stories about a giant who carries friends around in his coat pockets. The book was popular and sold well. Knowing he could write a better story, and hoping to cash in on the Gargantua legends too, Rabelais wrote a story about the son of Gargantua, Pantagruel, whose name means "always thirsty" [**154**, p. 82]. As 1532 had been a year of great drought, he hoped that the Pantagruel name would add to the book's allure.

It did all that and more.

Four subsequent books about Pantagruel and his father Gargantua appeared as the years passed, although the last volume is but attributed to Rabelais. Curiously, nowhere in his writings does Rabelais give Pantagruel's exact height, although at times he appears to be anywhere from seven feet to a mile in stature, as is pointed out at greater length in Vignette II. Pantagruel assumed the appropriate size for whatever the occasion, with no apologies given to the reader. Throughout the narrative, Rabelais' giants speak of wine, war, and love with wit and scatalogical musings rivaling those of Jonathan Swift. In fact, his first book was banned by the Sorbonne—the Parisian University established in 1257 that took upon itself the duty of judging literary work—as being obscene and immoral. Rabelais also sprinkled his work with talk of utopia, as Sir Thomas More's 1516 *Utopia* was still immensely popular, a book that raised natural questions about just and sensible societies. Rabelais was a scholar who knew five languages and who was versed in all knowledge. Wine for him was a metaphor for wisdom and knowledge. His characters drink from bottomless cups; and he means for that symbolism to be applied both literally and figuratively.

At the end of his first volume in the series (which is placed second in his collected works), Rabelais outlines Pantagruel's adventures yet to be chronicled. One of these is for Pantagruel to visit the moon [**110**, p. 217]. However Rabelais never got around to telling the story. So, in the Rabelaisian tradition of continuing what another author began, we therefore take up the tale.

This story raises a question. How can we measure things that are currently beyond society's technological ability to do so, in this case the composition of the moon using pre-twentieth century technology?

Of course, it cannot be done.

Vignette XI: Laputa and Gargantua

But one can tell stories about outlandish-as-yet-ideas, sketch diagrams of fanciful machines, and dream dreams. For example, de Fontenelle in his 1686 mind experiment tour of the solar system predicted

The art of flying will be perfected, and some day we will go to the moon. [**32**, p. 34]

Da Vinci sketched exotic machines, including the monstrous crossbow of Figure 3. Such yesterday's musings lead to tomorrow's exploits.

CHAPTER **XI**

Moon Pie

Pantagruel was reading. He was puzzling over the privately circulated notes of the canon Copernicus, who had odd ideas about celestial motion, that the earth rotated in space and moved about the sun.

Meanwhile, his friends were extolling their skill at the crossbow.

"Yes, I could knock an apple from atop a man's head at fifty paces, God rest his soul," said Eusthenes while looking askance at a wench serving overfull tankards.

"That deed's been done already," said Panurge through the suds of his ale, "on the shores of Lake Geneva," and then as his own boast took shape, his eyes sparkled. "But I could strike the hat nailed to the post at which the bailiff had demanded all to bow—from across that lake." As if that weren't embellishment enough, he added, "In the moonlight."

At the mention of the moon, Pantagruel's concentration was broken, and the babblings of his friends, having logjammed in a queue in some part of his brain, became intelligible. Not to be outdone, he laughed. "William Tell! I don't think so. But it could be hit."

"The apple or the hat?" they chimed, clearly expecting an exhibition of gargantuan proportions, for although Pantagruel had faults, boasting was not among them.

"No. That's nothing. I can shoot the moon."

"How will you know if you hit it?" asked Eusthenes.

"You mean, which is easier to say to the cripple, 'Your sins are forgiven' or 'Rise and walk'?" Pantagruel lived to discuss theology, metaphysics, philosophy, alchemy, mysticism, and any ology or ism, and interjected rambling discourse at every opportunity, even if there was none.

Ignoring Pantagruel's witticism, Panurge refined the objection. "That's right, how can we see it?"

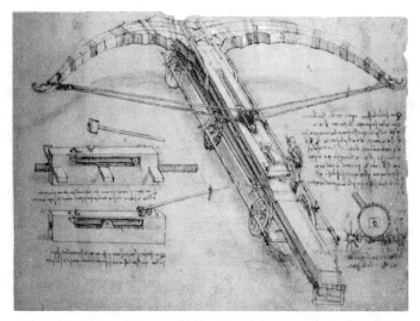

Figure 3. *Ballista*, a large crossbow, da Vinci, c. 1485 [**53**, p. 18]

"Coconuts," murmured Pantagruel, his mind racing through some calculations, "and dye." To their disbelief, he elaborated, "Big coconuts." More skepticism. "The dye will spread over the face of the moon."

From his coat pocket, he extracted four specimens. "These are from a place called Con-go, from trees taller than me," he said. Each coconut was larger than the inn on whose veranda this discussion was taking place.

A full moon illuminated their faces, mocking the giant's words.

Stirred to action, Pantagruel produced an awl, and plunged it into each shell, so creating orifices about the diameter of a wine cask. Down each of these he poured concoctions of powders and elixirs like a master alchemist. One after the other, he pressed a thumb against the openings, and shook them. The sloshing of coconut milk sounded like surf crashing against rocks. Everyone in the inn and its environs gathered at the noise.

"Volunteers?" the giant looked around.

In some regions of the world, an up and down nod of the head means yes, in others, no. Signals can be confusing. But anyone observing the hand waves and head nods of the four men Pantagruel selected might swear to a negative meaning.

Onto each volunteer Pantagruel poured a drop from a different coconut. Of course, Pantagruel's drop was roughly a half dozen gallons. The first man, with an oversized belly and spindly legs, turned blue. A second man, tall and bald, his hat left inside on a table in the confusion, turned green. A third, who had had white, wild eyebrows matching a greasy once-white tunic, turned red. The fourth, the stocky, short innkeeper, with a deep frown on his face at seeing his domain splashed with sticky dye, turned bright yellow.

Satisfied with the hues, Pantagruel corked each nut with a cask of burgundy, and set the nuts onto the muddy street, lining them as artillery shells to be fired.

Pantagruel stood, bent over, and retrieved his crossbow. It had been leaning against a cathedral belfry several streets away. He began explaining. "Shooting the moon is like shooting ducks on the wing," he explained. "If I aim where the moon is now, by the time the coconut reaches its target, the moon will have moved along. Instead, I must shoot ahead of the moon. I have four shots and four colors."

The crowd roared its approval.

"Each will be shot at a different angle. One should make it. The ancients are not in complete agreement as to the moon's distance from us." He enumerated various authorities, and the crowd quieted. Pantagruel sat on the ground and stretched the bow, now modified to take a coconut.

Before he fired, his friends seized the moment. "Bets! Who will bet on a color?"

"Blue!" rose one cheer, to be drowned out by "Green!" which in turn was silenced by "Red!" Few chose yellow. The innkeeper's dour, moldy-yellow countenance checked would-be partisans.

Taking aim while calling out a color and a series of numbers for Panurge to write in a logbook, Pantagruel released the bowstring. Its snap raised a breeze that banged shutters, snuffed candles, and fluttered skirts and wigs.

All eyes focused on the heavens. As one, the crowd held its breath hoping to see a color splash across the moon.

Pantagruel broke the tension. "Two days," he said. "In two days, one of the coconuts will strike the target."

A moan greeted his words. Horse races lasted minutes. This experiment would take time. The only one genuinely happy was the innkeeper, who recognized a business opportunity. Everyone in Paris would be watching. Many would drink his spirits, sleep on his mats, dine at his tables, and add copper to his coffers. Life was yellow, but good. Profit was in the air.

Paris hummed with expectation. People sported blue sashes, green hats, red tunics, and yellow vests, and took to calling each other Blue Bellies, Green Beans, Bloody Reds, and Yellow Jackets. Bets were placed not only on color, but time of impact and size of stain. Skeptics bet on nothing happening. The king lobbied for royal white to be added as a color. He finally bet on red.

Meanwhile, the queen sent Pantagruel a request bound with a lace handkerchief.

Most gracious philosopher and defender of honour,
    Greetings!
Of you I ask a boon. I know not if it be possible. No other would attempt what you do. The lady across the channel and I have wagered. Some say the moon is more real than this earth, that a chunk of it is heavier than a chunk of common rock. Others say less. Old women's tales say the moon is of cheese. I am not an old woman! I say the moon is like the earth. The lady wagers for heavier. Who can decide? How can we know? Can you, O mighty man? I offer my finest crown jewel as prize for answering this riddle. To this pledge I set my seal. My hope is in your hands.

Your humble servant,
Stewardess of all France

The giant sniffed the scented lace. The queen's puzzle kept him awake. In the morning, his friends breakfasted while tallying the betting books. Panurge inquired, "How are you, my friend?"

The giant outlined the queen's challenge, and then proposed a solution.

Eusthenes was doubtful. "Why go to that trouble? Just find a suitable rock and say it came from the moon."

"Yes," echoed Panurge. "For example, think of the True Cross. In villages from here to the holy land, each parish has a piece of the one True Cross. Should those pieces be amassed as one, the cordage would stoke the king's hearths for a year and a day. If they are said to be genuine, why not a moon rock? We can find a bishop to authenticate it."

"You have identified the crux of the matter," Pantagruel said. He was silent for a minute, and then continued, "We need a savant recognized by both the English and the French to observe as we go. Otherwise, we have nothing, even though I retrieve tons of rock."

"Hmm. Will Paracelsus do?" suggested Panurge, "Otherwise known as Theophrastus Bombastus von Hohhenheim, the famous Swiss physician."

"You know him?"

Eusthenes looked up from his ledgers. "He's a Yellow Jacket, down for 50 sous."

The giant clapped his hands, "Good, he will do nicely. Now to get my hair cut. I'll need a troupe of weavers."

Before the day was out, barbers went to work. Pantagruel had initially wanted comely maidens. But the mademoiselles had little luck. Each hair was more stout than their wrists. Fortunately, teams of lumberjacks from the Pyrenees had been pressed into the French army, and their commandant gave leave for them to shave the giant, as the commandant was a Blue Belly, in for twenty sous.

Pantagruel's thought was to harpoon the moon. He knew how far it was, a distance equal to circumnavigating the globe ten times. Rather than using a barb at the tip of the harpoon, he planned on using a hollow shaft, which should fill with rock upon impact. When pulling on the harpoon tether, Pantagruel hoped to retrieve a core sample for the queen. Nothing could be simpler.

He needed rope, lots of it, tough rope, strong enough to withstand a tug so as to loosen an embedded shaft. Pantagruel's hair from his head to his toes was to be woven into multistrand lengths with an inner core of leather air bladders. News of this enterprise filtered through Paris. The fever pitch about the coconuts escalated. As is often the case, the story told from one to another soon differed from the truth. Some said that the giant intended to pull the moon to earth. Crowds of confused men and women used this excuse to overrun vegetable and bread markets. Some set fires. Others launched fireworks.

One more day to wait.

Late that afternoon, Pantagruel fired a test shot. Taking aim at a distant hill, he threw a makeshift harpoon. It snaked with deceptive speed, coil after coil vanishing into the sky. And then it all stopped. The rope flopped to the ground. Paracelsus, who had agreed to observe, had spent the afternoon reaching the target area. He climbed over the debris heaped up by the impact. At his command, a flare soared into the sky. Pantagruel yanked on the rope, dislodging the head of the harpoon, and rewound the tether. Villagers jumped aboard the shaft for brief joy rides through grain fields and across rivers and lakes. Some cows and sheep were discomfited. Flower beds and huts were swept away. No one got overly upset. This was a gala time.

The queen agreed to underwrite expenses. Her cousins, King Ferdinand and Queen Isabella had financed Columbus's adventures not long ago, and thereafter reaped the lands of North and South America. Now it was France's turn. Maybe this venture would lead to the moon and its wealth, whatever that might be. The harpoon could serve as a temporary flag pole.

The next morning, Pantagruel rented a mountain in the Alps to serve as a spindle. When hurled, with the harpoon going away from the earth, the coils should unwind smoothly. Panurge arranged for the rope to be transported east. Another army of artisans was hired to splice the lengths together on site and wind the rope about the mountain.

As evening of the second day approached, speeches were given, bands played, troupes danced and sang in the steets. Few could avoid staring at the moon. Chants for blue, red, green, and yellow punctuated the proceedings.

At midnight a tinge of red spread near the western boundary of the moon.

Drums boomed. Cheers echoed through the city, and not only from the Bloody Reds. Everyone likes a winner.

A minute later, a similar tinge of yellow spread along the eastern rim of the moon.

Two coconuts had hit the target. Two colors had won.

The innkeeper smiled broadly, "Drinks are on the house this night!" He knew how to celebrate.

Pantagruel and his friends waved and bowed.

Afterwards, long into the wee hours of the morning at Rainbow Inn, as it was now called, Eusthenes and Panurge settled bets. With two winning colors, their share was less than anticipated, but more than enough to buy ale far into the future.

Four months later, the rope enterprise ran as clock-work. The spindle was filling. It was time for a second test shot. From atop a nearby peak, Pantagruel fiddled with a larger crossbow. A rope draped from his perch across to Spindle Mountain.

At noon, great bell horns sounded hollowly. After a melancholy cadence, Pantagruel, released a bolt toward the sun overhead. It was the first day of summer. The twang reverberated across the valleys. The accompanying breeze bent mountain flowers and grasses. The air was sweet.

The mountain sprouted vocal cords. The rope sang. Coil after coil whirred heavenward, louder when the rope unwound from the near side of the mountain, and fainter when it unwound from the far side.

Within an hour, something was wrong.

The rope was following the sun.

Paracelsus was the first to notice, "This cannot be. Who will believe this?"

"Why not?" asked Eusthenes, mesmerized by the coils slipping off the mountain.

Pantagruel understood. A shadow of the idea had haunted him since the coconuts had struck the moon. To him, the filament winding into the sky seemed almost scriptlike: a string of letters, tumbling dreamily into parsable words and meaning. He danced atop the mountain, yelling, "Glory be, Copernicus is right!" His celebration was with the same legendary abandon as Archimedes who, when first understanding the idea of specific gravity, had run naked through the streets shouting, "Eureka!" Whereas Archimedes had been in a bath tub at the moment of discovery, Pantagruel had been relieving himself. His codpiece was undone, flapping in the wind. His arms, too, flapped at the elbows like a big bird. His feet and knees were akimbo in a wild jig. His chortling echoed from every mountain face.

After the excitement, Pantagruel sped across to Spindle Mountain. If the test continued, the rope would wrap about the earth. He grabbed an unwinding loop and yanked.

This time, it took longer to rewind. Five thousand miles.

What to do?

"Spindle Mountain won't work," said Paracelsus.

"We could give up," agreed Eusthenes.

Panurge added more discouraging news. "Rumors say that our experiments are hexing crops and jinxing the weather. The Sahara sends dry winds, and the arctic sends cold. They say we are to blame."

Where to go? Pantagruel and his friends had sailed the seven seas. Although not the first to round the Cape or circumnavigate the world, they were not far behind da Gama and Magellan. They too had sought a Northwest Passage and weathered the Horn.

As Pantagruel thought about a better place, he brightened. "The poles. The arctic is no good. It is probably all ice. But the antarctic—that might be good."

Logistics were now a problem. Spindle Mountain contained 150,000 miles of rope. How to get it to the South Pole?

By August, Panurge had recommissioned their old ship, the Thalemege, stowed it with supplies and gear for a trek to the bottom of the world, and assembled a crew. Eusthenes had moved the rope factory to Lisbon. The last 100,000 miles were to be paid out into the ocean in a continuous strand. The Thalemege would carry one end to the antarctic. Once there, Pantagruel could pull it ashore.

Meanwhile, the giant sectioned the rope into lengths of 5000 miles. These he launched in a series of 30 harpoon shots. To onlookers, the streaming strands of black hair looked like a strange breed of birds in continuous single file migrating south. To those who knew nothing of the expedition, it was an omen of doom. Endless filaments in the sky for what purpose? The bonds that held the world together were coming undone.

By December, Pantagruel and his crew stood on the shores of what is now called the Wedell Sea. The penguins and sea elephants ignored them. Fortunately this was the time of the midnight sun. The strands had been retrieved and spliced together. The Portuguese strand too was being reeled in. Another long-time friend of the giant, Friar John, had assumed Lisbon oversight duties.

"Peace to you, my brothers," he had saluted them as they left. "Your exploits will tickle the world."

Pantagruel located an ideal spot for the moonshot: a mostly empty, still simmering caldera from an old volcano blast. Within it, glacial ice had formed a great cave. What nature had begun, Pantagruel and his crew perfected. Here the rope was safe from the weather.

By the new year the moon shot seemed like lunacy.

Everyone was cold.

"Remind me again why we are doing this," Eusthenes said.

"It's not really for a woman's honor or a gemstone, is it?" observed Paracelsus.

"Perchance it is a phallic symbol," bantered Panurge. "Every sea elephant bull works hard at maintaining its harem on the beach. Maybe this gambit is Pantagruel's release. A harpoon extending from the earth to penetrate the moon—any elephant or blue whale or prehistoric monster, for that matter, would be jealous."

Pantagruel responded. A simple denial would encourage more psychoanalysis, as people would later call it. "The moon riddle is a puzzle I think we can solve. And yes, honor is at stake. The lady is our queen. But no, not for any gem. If she gave one to me, as a Frenchman, I would but give it back. And yes, to honor again. It is one thing to boast, and another to do. Let us do, and then talk, and," he eyed his friends, "and drink—to France, and," he looked at Paracelsus, "to Switzerland, and to all men everywhere."

The next day was cloudless and windless, ideal conditions. At noon, the team sang an Ave Maria. Paracelsus gave a short speech: was the moon light or heavy? In a few days, they would know.

The sun was low on the horizon, as was the moon. Pantagruel checked the settings one last time, and fired the bolt, not upwards, but off to the horizon. The harpoon faded from view, and the rope flowed furiously from the cave.

It rubbed against the opening. Ice began melting. Jets of steam streamed into the atmosphere. Great clouds formed. The rope disappeared into a fog.

Snow began falling.

The storm became a white-out. The launch team took shelter, and bedded down. Nothing more could be done.

Two days later, the unending whir of the unwinding rope was silent. The company emerged from their tents, and waded through drifts of snow. The moon rope was limp. Pantagruel picked it up and coiled the slack.

He braced himself and yanked.

Nothing gave.

He pulled harder.

Teeth clenched, the veins in his forearms and neck pulsed. The ground trembled. For a long minute he strained.

Everyone held their breath as if they too were at the windlass, hauling at an anchor so as to journey back to warm inns and tart ale.

Panatagruel sighed and let go.

"If I pull harder, the line will snap. The harpoon is stuck."

"We've snagged heaven, and cannot keep it," said Panurge.

Paracelsus agreed, "A noble effort."

As if to jeer their failure, some circling birds perched on the rope and cawed.

Panurge waved and shouted.

They fluttered off and perched yet higher, and cawed again.

Eusthenes whispered, "An aerie without bound."

Not one to refuse a suggestion, be it ever so unintentional, Pantagruel said, "I'll try it."

Securing the loose end of the rope about an iceberg, he gripped the line, and tested his weight against it. Climbing hand over hand, he hauled himself skyward. During the first ten miles, the rope stretched so that his feet touched the ground. He continued for another ten before retreating to tell of his plan.

"Madness," muttered Paracelsus who saw a thousand reasons why it was foolhardy.

"I can climb it in twelve days," Pantagruel said. "I cannot ask anyone to accompany me. Nor will I choose volunteers," he quipped.

"Is there ale on the moon?" asked Panurge.

"Are there women on the moon?" added Eusthenes.

Pantagruel shrugged, "We will bring provisions, my friends. As to the ladies, who knows? I plan to see for myself."

Paracelsus promised to man the earthward side of the ropes.

Figure 4. ...the earth receded and the climb grew easier....
sketch by Trudie Simoson

Before they parted, the giant secured the crossbow to the rope, and outlined his plan. "If we fail to return by day twenty-seven, release the rope from the earth. We will find a way home. Good luck."

As Pantagruel shimmied up what now looked impossibly thin, Panurge and Eustenes peered from Pantagruel's side pockets and waved.

"To the moon or bust," they yelled. "Utopia! We never found it on earth." They hooted and laughed.

The remnant of the crew echoed their cheer, but with less conviction. Paracelsus was certain that the climb was to the death.

As Pantagruel continued, the earth receded and the climb grew easier. Each time Pantagruel closed his fist for the next handhold, air from the bladders within the rope escaped, allowing them to breath.

Pantagruel relieved himself periodically. He aimed toward the sun because that direction was warmer than when facing the stars. The streams of liquid coalesced into great iceballs. These became sun-grazing comets. In future years, when scientists analyzed core samples from them, they were perplexed. The best they could conclude was that these comets tested positive for kidney stones.

Meanwhile, the spirits of Pantagruel and his friends were high. They were following the rope to the moon. Every now and then the giant pulled forward on the rope, increasing their speed. It was mostly effortless. Eusthenes and Panurge had secured themselves by tether to the giant's jacket. At times they clung to Pantagruel's shoulder and conversed, reminiscing about past adventures, the women they had known, their favorite ales, and speculating as to what might lie ahead.

"Some say that Eden is on the moon," said Pantagruel, "and that the garden is as beautiful as at creation. Others say that the far side of the moon is home to the Elysian Fields where the spirits of the just lodge. Still others say that Pythagoras was from the moon, that his thighs were of gold, and that he is up there now. Animals and flowers are larger and more perfect than on earth. A nibble of moon food satisfies hunger for a week."

"Your nibble or mine?" asked Eusthenes.

"It looks empty to me," said Panurge.

The moon filled the sky. Its face was scarred. Deep shadows, jagged mountains, and empty plains lay before them. Far behind, the bluish-white earth was but a beacon.

"It looks like hell, not Eden," agreed Pantagruel. "One more day, and we will know."

Nearing the moon was like being in a canoe careering towards a waterfall. Pantagruel braked their progress periodically, and then continuously.

Whilst two hundred feet from the surface, Pantagruel jumped from the rope onto a level space. They landed with a thud. But the harpoon was not to be seen. It was buried. In fact, its impact had brought a mountain down.

"Welcome to Selenia," called a voice. "Come in out of the sun." A figure emerged from the shadows. She was slender: a tall human figure clad in a toga, sandals on her feet. "I believe that you are the ones who have tumbled our catacombs," the figure eyed the giant up and down, "but you are welcome. Come, few visitors arrive from the sky."

Pantagruel surveyed the area. It had been a thoroughfare. On one side was a boulder heap, a jumble of pillars and blocks. On the other were dark archways leading into the moon.

One passageway was big enough for the giant's bulk.

Once inside, the walls glowed luminescently. The brilliance outside had masked the softness inside. Further in led further down, and the ceiling height grew. Pantagruel no longer crouched. Soon the passageway emptied into a great cavern dwarfing the space within Notre Dame Cathedral. Here they were invited to sit and talk.

Pantagruel promptly fell asleep. After twelve days without sleep, he could no longer keep his eyes open. Between the giant's snores, his friends had many questions.

"Who are you people? Are we dead? What do you eat? Is this a dream? Do you know Pythagoras?"

Selenia's emissary beckoned to a table, "Here is refreshment. And yes, all of your questions will be answered."

"Why didn't you cut the rope? You knew not whether we would be friend or foe. Why wait for an enemy to come upon you?"

The guide smiled, "You first sent up a keg of wine. Anyone having taste like that cannot be all bad."

Their talk continued for hours. The food was strange but the drink was not unlike the ale at Rainbow Inn.

They too fell asleep.

Three days later, the three felt as usual. Each answer they heard prompted more questions. A state tour was promised to all Selenia and the far side of the moon. Their adventures there are an altogether different story, which means that we ought to finish this one first.

After thirty days, the link between earth and the moon was severed. Paracelsus had waited a few extra days. Pantagruel retrieved the cord and the crossbow. The Selenites helped with a solar-powered winch.

Then the giant selected various rocks. Into a chest, he packed a dense black opal-like rock, a porous pumice-like stone, and a third specimen that was an edible fungus. The moon caverns were filled with it; much of the Selenite diet was derived from it. He added letters to the chest and then sealed it.

To one end of the rope he afixed the chest. With little ceremony, he fired a bolt earthward. Two days later it struck the Pacific along the Tropic of Cancer. In thirteen days the rope coiled itself around the earth, and the chest splashed down.

By good fortune, Paracelsus and the crew spotted the loops of cable. Since it had been marked by color dyes along its length, the doctor was able to determine within ten miles where the end of the rope might be. Within another month, they retrieved the chest.

Standing on the quarterdeck, Paracelsus read the letter addressed to the crew.

Comrades,

Bonjour! I hope all is well with you and the ship, and that the chest has been recovered intact. Sad to say, we have no definitive answer for the queen's riddle. Although everything here seems lighter—indeed my weight seems to be about one sixth as on earth—you will find the moon opal is heavier than earth granite, but the pumice is lighter. Furthermore, parts of the moon, although not cheese, are altogether edible. The third sample is a delightful truffle, found in lunar caves. So everyone was right in a way, including the old women.

The people here have kindly invited us to tour their lands. Seeing that we have come this far, we have elected to stay and have an adventure. The French standard has been unfurled. But flags are useless here as the lunar surface has no breeze whatsover. We have no timetable for a return date. Greet our friends. Also in the chest is a sack of gems for the crew: mementos, treasure, or dust collectors. Perhaps they can be swapped for a pint.

>May fair winds bring you home safely,
>Pantagruel and company

The chest was duly presented to the queen, and Paracelsus's recounting of the tale charmed the court. Few gave credence to Pantagruel's demonstration that the earth moved. No one could repeat the experiment. The opal was cut. Its proceeds underwrote new uniforms for the king's officers. The pumice was placed in storage and later tossed into a quarry, whereupon it was crushed, and made into cement. The truffle was served at state dinners for years to come. It was always a favorite. The menu listed it as "moon pie, a delicacy from one of the king's domains."

Exercises

In these exercises, assume that Pantagruel's harpoon is 100 meters long with density 100 kg/meter. Assume that the earth's radius is $R = 6400$ km, and that it is $Q = 375{,}000$ km from the earth to the moon, which is also the length of the rope.

1. Pliny the Elder says that in India there are "trees so lofty that it is not possible to shoot an arrow over them" [**106**, book vii, section ii, p. 519]. Find a lower bound for the height of such a tree.

2. (a) Determine the speed with which Pantagruel must hurl the coconuts so that they reach the moon in two days.

(b) If the cross sectional area of the rope is 100 cm^2, determine the mass and volume of a length of rope that extends from the earth to the moon. Make some reasonable assumptions about the rope's density.

3. On p. 265 Pantagruel reminds the crowd that the ancients were not in complete agreement about the distance between the earth and the moon. Pliny enumerates some of these guesses [**106**, book ii, sections xix, cxii–cxiii, pp. 227, 369–373]. He says that

- Pythagoras inferred that the distance D of the moon from the earth was $D = 15,750$ (Roman) miles from the earth, that of the sun from the moon was $2D$, and that of the zodiac from the sun was $3D$.

- Another authority, perhaps Aristarchus, said that the distance from the sun to the moon was $19D$.

- Eratosthenes concluded that the earth's radius was 42,000 stades, or about 5250 Roman miles. To this, Pliny adds, perhaps for earth's atmosphere, 12,000 stades, so that he takes the radius R of an atmosphered earth as $R = 6750$ Roman miles.

Let C be the distance from earth's center to the zodiac, the limits of the cosmos in a geocentric system. Pliny concludes that the ratio R to C is 1/96. How does he get this value?

4. (a) Spindle Mountain has a base diameter of 15 km and height above its base of 5 km. Wrap Q kilometers of rope about the mountain. How many coils of rope are there? Use the result of Exercise 2b. Assume the mountain is a cone.

(b) The ancients used the phrases "the hairs on a head," "the sand in the sea," and "the stars in the sky" to refer to large numbers. How many hairs are in a full head of hair, approximately? If Pantagruel had shoulder length hair and the hairs were all laid end to end, would it reach the moon?

5. Fix the center of the earth in space, yet allow the earth to turn upon its axis once each day. Imagine that Pantagruel has hurled the harpoon, attached to the rope, straight up at 1 km/sec while standing at the equator.

(a) Determine the curve of the rope from the earth's surface to the harpoon for the first quarter of a day.

(b) On p. 269, Pantagruel concludes that Copernicus's heliocentric system is correct, that the earth rotates. Why does he not ascribe the phenomenon to something like Descartes's vortices?

6. Assume that the earth is fixed in space. At the south pole, Pantagruel hurls the harpoon, attached to the rope, at a speed of 10 km/sec. Will the harpoon reach the moon? If so, at what speed?

7. Paracelsus thought of many reasons why Pantagruel could not climb to the moon. List some of these reasons.

8. How large must a circular stain on the moon be so as to be visible from the earth?

9. (a) Pantagruel throws the harpoon from the moon to the earth whereupon it strikes the earth near Oahu Island in the Pacific and coils around the earth. Where does the treasure chest land?

   (b) Paracelsus and the crew find the coiled rope south of the Canary Islands, at longitude 20° W. From the markings on the rope, they are at marker 360,000 km. How many coils of rope are there? At what longitude line is the treasure chest?

10. (a) One of the more troubling scenes in Rabelais's works is when a flock of sheep together with their shepherds are tossed into the sea and left to drown in Chapter VIII of Book IV [**110**, pp. 422–426]. Rabelais gives no comment on the action, except to describe it. What message, if anything, was Rabelais trying to convey to his readers in this passage?

    (b) Explain why Rabelais's Pantagruel stories (especially book iv), *Gulliver's Travels*, and *Micromégas* may be viewed as a retelling of Homer's *Odyssey*.

Vignette XII: A Last Curtain Call

In mid-life, Voltaire received a letter asking him to describe himself. He responded in verse, giving an answer which characterizes his entire life. The selected, translated lines below from his poem give a candid self-appraisal: his shortcomings, his desires, his guiding principle, and a restlessness as a fellow human being.

> High praise you bestow on me,
> And finish with desire to know me;
> You'll praise me less when I am known;
> But what I am I'll freely own.
>
> Sometimes I to science soar
> So of nature may explore,
> Following Newton through the sky
> I to find natural causes try.
>
> I read philosophers profound,
> Who nature by their reason found;
> I see Clairaut, Maupertuis, rise
> By calculation to the skies;
>
> And I indeed too often find
> Such studies but perplex my mind.
> Obscure researches set apart,
> I study next the human heart.
>
> A friend to man, I strive to show
> How he to love himself may know.
> Yet notwithstanding all my pains
> Still there, a craving void remains.
>
> *Answer to a Lady*, [**144**, vol. xxxvi, pp. 180–182]

After leaving Frederick's court, Voltaire spent the next twenty-five years in exile at Ferney on the border between France and Switzerland. There he tended his garden, wrote, fostered a thriving local business community, welcomed visiting figures, and interceded on behalf of a number of families for whom state justice had gone awry. In the last year of his life, he received news that Paris was no longer off-limits to him, and the Comédie-Française invited him to stage a play if he would but write a new one. He complied, wrote *Irène*, coached the actors, saw the performance in a crowded, cheering theatre, and died shortly thereafter of prostate cancer complications [**101**, p. 375]. He was eighty-four. As a fitting last word, a line from *The Temple of Taste*:

> Farewell, my much loved friends, farewell.

CHAPTER **XII**

Riddle Resolutions

What is the answer to Voltaire's riddle?

To re-pose the riddle, recall that Micromégas, a philosopher in exile from another planetary system, visits earth in 1737, and finds a shipload of mathematicians, who have been testing an abstract theory, returning from taking measurements near the arctic circle. They engage in a lengthy dialogue about what they know and how they know it: the measurement of objects, the structure of the universe, their dreams, and their souls. As a parting gift, the giant gives man a book with the answers to all things. Later, when the philosophers open it, they find it empty, whereupon the story ends.

Why? What does an empty book mean?

We give several possible answers to this open-ended riddle, freely using the examples developed in previous chapters, as well as using new passages from Rabelais, Swift, Voltaire, and others.

Answer 1: The absent-minded professor. The simplest answer to Voltaire's riddle is that it was a comedy of errors.

Perhaps Micromégas, like many scientists, was absent-minded. A classic description of this trait appears in *Gulliver's Travels*: on the floating-island-in-the-sky of Laputa is a community of scientists all of whom have this syndrome. To help them function in society, each scientist employs a flapper, a servant who carries a ball or bladder attached to a short stick.

> The flapper is employed to attend his master in his walks, and upon occasion to give him a soft flap on his eyes, because he is always so wrapped up in cogitation, that he is in manifest danger of bouncing his head against every post, and in the streets, of justling others. [**139**, pp. 174–5]

As Micromégas and the secretary of the Scientific Society of Saturn are on a scientific expedition to earth, they bring along "a splendid assortment of mathematical instruments" as well as a retinue of servants, who in Scene IV of Chapter I, serve breakfast, consisting of two mountains. Micromégas' baggage may have been compact, and his books, writing material, and reducing equipment may have been in some disarray. A simple resolution to the riddle is that Micromégas, as an absent-minded scientist, simply picked up a blank book journal instead of the intended book.

Another stereotypical trait of the scientist—and more of the professor—is a propensity to live in an ivory tower pursuing theories having little to do with the real world. As a pertinent example of this, de Fontenelle, the secretary of the French Academy of Sciences once introduced Maupertuis to a social gathering, saying

> I have the honor of presenting M. de Maupertuis, who is a great mathematician and who nevertheless is not a fool. [**140**, p. 6]

Evidently, de Fontenelle had met many mathematicians who were otherwise. In the satirical poem, *The Temple of Taste*, Voltaire describes a thinker Bardou, who perhaps has over-inflated the validity of his own observations and theories:

> Bardou then cried out, "The world's in an error, and will always continue so; there's no God of Taste, and I'll prove it thus." Then he laid down a proposition, divided and subdivided it; but nobody listened, and a greater multitude than ever crowded to the gate [of the Temple of Taste]. [**144**, vol. xxxvi, p. 50]

That is, at least for Bardou, the public dismissed his endless theories as the foolishness that it was, and went on.

Thus, another resolution to Voltaire's riddle is that Micromégas was merely fooling himself: his book may have been filled with meaningless explanations that Micromégas merely thought were correct. So when the secretary of the academy, who just happened to be de Fontenelle at the time of the polar expedition, looked at Micromégas's book, he immediately perceived it to be the work of a fool—as he had already seen many fools in his life. Therefore he judged the book to be blank.

The story is often told of Ramanujan who taught himself mathematics as a young man in India and who subsequently sent some of his voluminous calculations to English mathematicians for them to consider. More than one mathematician scanned his assortment of formulas, and dismissed them as useless scribblings—until G. H. Hardy took a careful look at them, and discovered a genius of number theory. That is, another riddle resolution is that the secretary of the French Academy may have been an absent-minded savant who passed judgment on Micromégas's script too quickly.

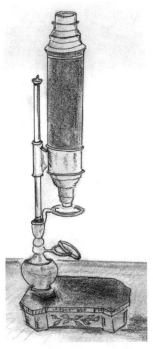

Figure 1. Sketch of an Italian compound microscope, 1726, the Petrus Patronus

The standard interpretation of the ending of the *Micromégas* story is that it is a caricature of de Fontenelle, secretary of the Academy from 1697–1739. In 1737, de Fontenelle was eighty years old. It could be that Voltaire was merely making fun of an old yet powerful curmudgeon with poor eyesight and poor judgment. The keepers of the inner courts to institutions are oftentimes well past their prime. But the irony goes deeper than that. Voltaire had published his *Elements of Newton's Philosophy* in 1738. He had hoped for a good review from the distinguished secretary, someone he counted as a friend. But de Fontenelle had probably made up his mind long before he opened Voltaire's book on Newton. When de Fontenelle actually looked at it, as we saw in Vignette IV, he thereafter concluded, "Ah, just as I thought, there is really nothing of significance herein." It is with this very scene and image that Voltaire concludes his story: a universal philosopher of gargantuan insight gives a mite of a secretary a copy of his life's observations who in turn can see nothing at all.

Although this dismissal of Voltaire by de Fontenelle could have been part of the initial inspiration for the final scene of *Micromégas*, Voltaire acknowledged the validity of de Fontenelle's sentiments. As we saw in Vignette IV, after publishing *The Elements* in 1738, Voltaire gave up trying to do mathematics, and focused on what he did do well. And thus as any good poet and storyteller, he packs into the final scene of *Micromégas* images ripe for alternative interpretations, some of which we delineate in the ensuing possible answers.

Answer 2: A scaling threshold. The book that Micromégas gave the philosophers was to be written "in very small script just for them," as one translation reads [**101**, p. 221]. It could very well be that in scaling his writing for human consumption, the script vanished.

Had Micromégas stood upright on earth, his head would be far above earth's breathable atmosphere. Voltaire made his hero much too tall for earth. Perhaps that was his intention, as Micromégas is a Frederick the Great figure, and no one human should be allowed to be that powerful. Nevertheless, let us assume that a 23 mile high Micromégas man is possible; how small can he scale script so that he is yet able to verify the result as being legible to humans?

Since Micromégas is about 20,000 times taller than the average man, it is reasonable to assume that the length and width of a page in one of his books was about 20,000 times the length and width of a page in a typical book in any library on earth. So in order for Micromégas to make the book for the French philosophers, he needed to reduce his print by a factor of about twenty thousand.

Can Micromégas' eyes, together with the technology of the eighteenth-century, resolve objects this small?

In his student days on Sirius, "Micromégas dissected small insects measuring nearly 100 feet in diameter, a size invisible to ordinary Sirian microscopes." However, even when looking through diamond magnifiers from his necklace, Micromégas could only see humans imperfectly. In the technology of Voltaire's day, with the use of microscopes pioneered by van Leeuwenhoek, Swammerdam, Hooke, and Hartsoeker, images could be magnified up to at most 270 times, less than two percent of what Micromégas required. Figure 1 shows a microscope from Voltaire's era. Thus, Micromégas probably could not read the book that he gave to the philosophers. He had no way of checking the quality of his reduction of the script. Therefore another simplistic answer to Voltaire's riddle is that the reduction process used in producing the book may have been fundamentally flawed, perhaps rendering illegible results despite the best of intentions—so accounting for a blank book.

Although earth's artisans have written legible sentences on grains of rice, and although earth's technicians have reduced text to microfilm so that long passages can be written within the dot at the end of this sentence, a limit exists as to how script can be scaled downwards before it vanishes. Reducing a text of script having area dimension one square foot by a factor of twenty thousand is probably possible in our era. Reducing it by a factor of one trillion, say, is impossible. The problem is the discrete structure of our universe. If we expect to find sameness at every positive scaling level, we will be sorely disappointed. Paper and ink when reduced repeatedly ultimately become individual molecules, individual atoms, individual subatomic particles, and then nothing at all.

At rock bottom, the universe appears to be empty. If we look too closely at a pebble grasped in our hands, it disappears before our eyes. As discussed in Chapter II, the universe fails to be a continuum of self-similarity, just as Lucretius described so long ago.

Here's a related question: what would the philosophers see if they looked at Micromégas's full-sized Sirian pages? Would they see the same thing as Micromégas?

Not necessarily.

For example, Figure 2 is a double-minded fractal. The figure appears to say "Yes" at the topmost level, then "No" at the secondary level, and "Yes" at the tertiary level, and so on, all the way down. Figure 6 of Exercise 1c shows a close-up view of the fractal at its third and fourth level. The fractal's meaning, significance, interpretation is dependent upon the viewing scale.

Does this fractal idea help resolve the riddle?

When Micromégas looks at his copy of the book, perhaps he can see all the answers. But when we, the French philosophers, look at that book from another level, all we see is negation and emptiness. It could be, in a strange way, that we are looking at the pages in the wrong way.

Let me give a more concrete example of the difference in using various scaled perspectives. Rabelais uses the word *cod* frequently. Now "cod" can mean many things, ranging from a piece of dead fish, to the male sex organ, to a fool. The image of any noun can be sharpened by adjectives. Usually we use at most a few adjectives to shape the tone for each noun in normal

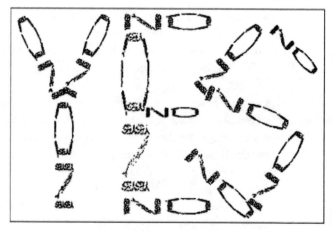

Figure 2. A Yes-No fractal

speech and text. But Rabelais surpasses us all. In Book III, Chapter 26 [**110**, pp. 306–308], he uses no less than 272 adjectives for cod, starting with

> dainty, mellow, lead-coloured, knurled, suborned, desired, stuffed, speckled, finely mettled, arabian-like, trussed-up-greyhound-like, ... harcabuzing cod.

The list reads like Lewis Carroll's *Jabberwocky*. I do not know what half the words above may mean in the context of modifying cod, and I would rather not be called any of them. Two chapters later, Rabelais gives several hundred more adjectives for cod as if he is adding them as an afterthought. In our day, and furnished with a word processor, he probably would have merged the lists. As it is, his lists are gargantuan, and exhibit scale. With respect to one's point of view, Rabelais has either given us a rich universe, fully fleshed with what is meant by cod, or he has given us an empty universe, heaping up words to no purpose, as if coloring a page completely with just one color, so that ultimately it says nothing at all.

Answer 3: A practical joke. Micromégas laughed uncontrollably when a French philosopher said that the universe was created solely for man. Perhaps he thought that such a world view warranted a practical joke.

A pertinent, classic practical joke is told by Hans Christian Andersen in *The Emperor's New Clothes*. In that story, a tailor convinces an emperor of his need for superior clothes. The emperor eagerly consents, whereupon the tailor makes a great show of weaving the new clothes, when in fact the clothes are purely imaginary. The emperor boldly wears the invisible clothes before a crowd of courtiers and admiring subjects as depicted in Figure 3, whereupon a little boy who knows no better says, "Why, he's naked."

Presumably, Micromégas gave his book to the French philosophers after their shipwreck. Surely one of them, if not all of them, opened the book to see what could be seen before they arrived in Paris. For once the ship was repaired, it took the team more than a month to reach Paris. What they saw of the book was enough for them to present it to the secretary of the French Academy of Sciences, who, like the little boy in Andersen's story, was honest enough to see that the book had no content.

Rabelais has a similar story, except instead of invisible clothes, it features an inaudible debate.

Thomas More was a contemporary of Rabelais. He was an erstwhile advisor to Henry VIII of England, a Catholic martyr, and writer of *Utopia* of 1516, a name that means "No Where," and the place of an ideal society which More describes in great detail. Throughout his work, Rabelais talks of Utopia and what can be done to get there. In book ii, Rabelais stages a debate between Pantagruel whom some scholars say represents the crown prince who will be Henry II of France, and "a certain learned man, named Thaumast, a great scholar of England," who most likely is Thomas More [**154**, p. 89]. At any rate, we shall refer to Thaumast as Thomas. As Thomas wishes to "prove whether Pantagruel's knowledge was so great as was reported," he proposes a format for debate unlike any other debate. It was not to be one with arguments pro and con, nor declamations, nor mere statistics. It will be by signs alone, without speaking, "for the matters are so abstruse, hard, and arduous, that words, proceeding from the mouth of man, will never be sufficient for unfolding of them to my liking" [**110**, p. 176]. Thomas sounds like the tailor persuading the emperor to adopt new clothes.

Although Pantagruel accepts the challenge, as honor gives him no other option, the giant enlists the help of his sidekick Panurge—who represents John Calvin (1509–1564) according to some scholars [**46**, p. 265]—to argue for him. At the agreed upon hour, the debaters begin. For hours the two trade signs, while the crowd listens—that is, watches. Rabelais describes the debate of gesticulations without flagging for three pages. Here is a snippet [**110**, p. 179]:

> Panurge suddenly lifted up in the air his right hand, and put the thumb thereof into the nostril of the same side, holding his four fingers straight out, and closed orderly in a parallel line to the point of his noise, shutting the left eye wholly, and making the other wink with a profound depression of the eyebrows and eye-lids.

Figure 3. The emperor's new clothes [7], watercolor by Edmund Dulac (1911)

Riddle Resolutions

Meanwhile the crowd understands nothing of most of the signs. They cheer nonetheless. And Thomas realizes that he has met a bigger practical joker than himself, and "begins to sweat great drops," whereupon Thomas soon concedes, announcing to the crowd,

> Out of the heart of England, [I have come] to confer with Pantagruel about the insoluble problems, both in magic alchymy, the cabala, geomancy, astrology, and philosophy. Yea he [and his student] hath to me discovered the very true well, fountain, and abyss of the encyclopedia of learning. In fine I will reduce into writing that which we have said and concluded, and will hereafter cause them to be printed, that every one may learn as I have done [**110**, p. 182].

That is, Thomas has promised a printed book with the answers to many things, if not all things. But unless it was *Utopia*, such a book was never written. It is like the blank book of Micromégas.

Answer 4: Invisible ink.

Another way to resolve the riddle is that the book could have been written with invisible ink. Perhaps some simple way of treating its pages could be applied so as to reveal its secrets.

Rabelais details such an occurrence with his giant Pantagruel. Whereas Voltaire's Micromégas gives a blank document, Rabelais's Pantagruel receives one. Pantagruel's document is a letter from a Parisian lady. Upon opening it, and seeing it blank, he thought

> [perhaps] the leaf of paper was written upon, but with such cunning and artifice that no man could see the writing at first sight. Therefore, to find it out, he set it by the fire, to see if it was made with sal ammoniac, soaked in water. Then put he it into the water, to see if the letter was written with the juice of tithymalle. After that, he held it up against the candle, to see if it was written with the juice of white onions.

He went on to use seven more tests. "But after all experiments, he could find out nothing" [**110**, pp. 189–190].

Pantagruel's letter was indeed blank. But how about Micromégas's book? Suppose none of Rabelais's tricks works on Micromégas's pages. What then?

People have five senses: sight, sound, touch, smell, and taste. Saturn's race had several hundred, while Micromégas's race had a thousand.

Such richness of alien sensory perception may well intimidate an earthling. Mr. A. Square from Flatland of Chapter III was certainly intimidated by the

Figure 4. A message?

sphere who penetrated his universe. The sphere kindly gave A. Square an enlightening tour of three-space. But afterwards, how can A. Square explain to his fellow citizens what he has seen and deduced? He cannot, and so writes his memoirs in prison.

If Micromégas has a thousand senses, books in his library are probably quite different than a book in eighteenth-century Paris. For example, if we gave the eighteenth-century French academicians a compact disk, a long-playing record, or a magnetic tape reel, they would be hard pressed to see immediately that the medium had any content.

It could be that the French savants' senses are simply fooled by the book and they fail to recognize the form of a message altogether. A simple, graphic example of a message that might be hidden from our senses, at least for a small while, is shown in Figure 4. At first glance it might appear as splotches of geometrical shapes—meaning nothing at all. But if we focus on the background—and look at the page upside down—and from a distance, the one-word message pops out as background becomes foreground, and we see an intelligible word.

One element of Voltaire's writing style is his ability to interpose positive and negative space, of foreground and background, which is the trick behind the graphic of Figure 4. In many passages, Voltaire builds a word foreground, which in and of itself forms no statement worthy of being said, appearing somewhat like vapid musak played on elevators or on call-waiting devices. But if we focus on what is implicitly unsaid, all at once the message pops out as a powerful statement—a verbal time bomb.

For example, in editing scene iii of *Micromégas* for this book, I was tempted to cut a passage where Voltaire talks about reading a "manuscript in the library of an illustrious archbishop," and says that he "cannot praise him enough" and will write "a long article on him and his sons." For months, the passage sounded like musak to me. And then, when thinking of something altogether different, I understood—*Oh, an archibishop is celibate; one living according to priestly vows should have no sons*. Voltaire was poking fun at a cleric, in this case, Archbishop Tencin for having censored a particular work [**148**, p. 26]. That which had appeared as pointless praise, now, with a slight change in perception, becomes pointed ridicule.

Thus, as one riddle resolution: the secretary of the French Academy when looking at Micromégas's pages simply did not yet get it.

Answer 5: The philosopher king. Long ago, Socrates, through the pen of Plato, asked a question [**110**, p. 177]:

> Inasmuch as philosophers only are able to grasp the eternal and unchangeable, and those who wander in the region of the many and variable are not philosophers, I must ask you which of the two classes should be the rulers of our State?

The answer most of us select, as we are only given two options, is the philosopher king. How will we recognize such a person? Here is part of the answer.

> He whose desires are drawn toward knowledge in every form will be absorbed in the pleasures of the soul, and will hardly feel bodily pleasure—I mean, if he be a true philosopher and not a sham one. [**105**, book vi, p. 216]

Throughout history, people have yearned for heroes, kings, messiahs, someone to lead toward a perfect society. We think of the wisdom of Solomon, the stoic Roman emperor Marcus Aurelius, the round table of Arthur, the elusive Prester John, Marco Polo's Kublai Khan, or the court of Suleiman the Magnificent.

When Frederick first wrote to Voltaire in 1736, he did so as heir apparent to the throne and as one who hungered to rule in an enlightened manner [**144**, vol. xxxviii, *Letter from the king to Voltaire*, p. 164]:

> Your poetry comprises a complete course of morality, by which men are taught both to act and to think, in a manner that whoever has read your works is fired with an ambition to tread in your footsteps.

Voltaire was impressed with the lad. In response, as discussed in Vignette II, Voltaire wrote an early version of *Micromégas* for the prince in 1739. In the story, the future king appears as a giant, larger than life, astride a comet, perhaps on the great comet that Halley predicted would soon reappear, coming in splendid glory to assume an earthly throne. He comes not with the quest for great power but with the heart of a philosopher bent upon inquiry into the nature of all things. Under him, society could experience a golden age. Frederick, in the mind of Voltaire and in the minds of many others, could become a genuine philosopher king.

After years of urging on Frederick's part, Voltaire became poet in residence at Frederick's court. But for Voltaire, with respect to kings, the old proverb, "familiarity breeds contempt," was valid. After a year, Voltaire saw manipulation, absorption in military matters, and trappings of the usual despot. For Voltaire, the idea of Frederick as philosopher king grew hollow, just as the book of Micromégas, the legacy of Frederick the Great given to all men, according to the poet, was equally hollow. He was a sham philosopher after all.

Even as Voltaire's disillusionment with Frederick grew, so too grew a cynicism about human nature [**144**, vol. viii, *Cromwell*, p. 29]:

> In the great game of human life, men begin with being dupes, and end in becoming knaves. An ardent novice at twenty often becomes an accomplished rogue at forty.

Of his veritable idol who chose to spend the last part of his life searching out Scripture, Voltaire wrote [**144**, vol. xii, *Plato*, p. 215],

> How could Newton, the greatest of men, comment upon the Apocalypse [*The Revelation of St. John the Divine*]? I fancy I see eagles, who after darting into a cloud go to rest on a dunghill.

Let me tell a personal story.

A few years ago, my wife, my two grade-school children, and I were returning from a year in Botswana, Africa, where I had been teaching at the university in Gaborone. We decided to stop in France. A few days in Paris would be a pleasant, postponed anniversary gift to my wife. *Where did we go?* Besides the well-known sites, we visited the catacombs. We entered at a small building along a nondescript street. A hole spiraled down worn stone steps. The blocks in the walls were rough stone. Bare light bulbs cast shadows along narrow low corridors. The passageways were made in the days of the Roman Empire of two thousand years ago. Along the walls were scrawled phrases in French.

> *Toute chair n'est que poussière*—All flesh is dust.
>
> *La vie est une mince volute de fumèe*—Life is a wisp of smoke.

Further along were piles of—*sticks?* Closer inspection revealed them to be bones: femurs, tibias, ribs. In the catacombs, millions of skeletons are piled from floor to ceiling in orderly fashion, the loose bones arranged in geometrical patterns—lines, crosses, diamonds, floral arrangements. Skulls as paperweights on table tops of whitened bone heaps returned our gaze through empty eye sockets. For a mile we hiked through this macabre graveyard beneath the city of romance.

We are all human. No man is a god, including the Fredericks who display such promise.

Answer 6: Illusion. Many have asked the question,

Why is there anything?

So too did Voltaire [**144**, vol. ix, *Force*, p. 92]. But unlike most of us, he enters into poetic dialogue with nature, and she responds [**144**, vol. xii, *Nature*, p. 74],

I know nothing about the matter. Pray go and inquire of Him who made me.

In a despondent tone, Voltaire writes [**144**, vol. xii, *Occult qualities*, pp. 51–52],

The more I read, the more I meditate, and the more I acquire, the more am I enabled to affirm that I know nothing.

What means this darkness?

Voltaire understood many things. He admired Newton, and his explanation of the laws of motion. He made a point of telling the public about this amazing intellectual achievement. The list of scientists and their recent accomplishments appearing in *Micromégas* is impressive. Voltaire wrote constantly and at great length and with magical prose. He was intrigued with the quest to know. But complete understanding, as he perceived it, eluded him. Knowledge of most any one thing is partial, and compared to what is unknown about that thing, the partial knowledge is almost nothing. He seemed to abhor the idea that matter was all there was in the universe, reluctantly concluding [**144**, vol. xiii, *Somnambulists and Dreamers*, p. 259],

What am I, therefore, if not a machine?

No answers seemed to exist for where ideas originate, what is the mind, and what is this thing called spirit? He both relished and ridiculed debate. He argued metaphysics with the best of his peers. But for what value?

> When men have disputed well and long on matter and spirit, they always end in understanding neither one another nor themselves. [**144**, vol. xiii, *Soul*, p. 272]

There are no easy, satisfying answers to life and its significance. Voltaire was genuinely irritated with the established ways of thinking that proclaimed and demanded otherwise. Come, he says, let us agree together.

> Let each of us boldly and honestly say, "How little it is that I really know." [**144**, vol. xii, *Monsters*, p. 18]

How could a pioneer of the Enlightenment utter such blasphemy?
The answer is, experience.

For example, in the excitement of understanding some of what Newton had achieved in the *Principia*, Voltaire and Emily experimented with fire in 1737 trying to understand its nature, as we saw in Vignette IV. When they started their experiments, perhaps they imagined that they would be able to measure many things about fire. As the experiments went on, they reduced their expectations to that of being able to measure some things, and at the end, to that of being able to measure a few things. That is, experimentation can easily lead to humility, and, especially for those of us who are clumsy in the laboratory, to despondency.

This basic problem of not knowing everything cannot be exorcized. For example, think of forecasting the weather. Today's weather predictions are good for about three days, fair for a week, and useless at a month [**52**, pp. 18–21]. No matter how much information we gather, the bounds for accurate weather forecasting are unlikely to improve significantly. Predicting trends in the stock market is probably just as difficult.

As another example, think of Newton's law of gravity. As we saw in Chapter III through VII, the idea has far-reaching implications; we can deduce that a planet orbits its sun in an ellipse, that planets flatten when spun, that this flattening of the planet produces a precession of its polar axis, and that free-falling particles through these orbs follow hypocyloid-type paths. But when many bodies interact, what then? For example, what is the orbit of Halley's Comet? As we saw in Chapter IV, Clairaut showed that Jupiter and Saturn delayed its predicted return by two years. In general, with more than two sizeable heavenly bodies in a system, we can but approximate a body's path. Human knowledge includes many theorems, and each theorem has its own, usually narrow, set of hypotheses which must be met before conclusions can be drawn. But when these hypotheses are not met, as happens more oftentimes than not, what then? We use approximation techniques, heuristic guesswork, and, in the words of an eminent computer scientist, "a point where faith takes over" [**69**, p. 192].

As a last example, think of intelligence. Micromégas and the Saturnian dwarf speculated whether the mite-like humans they discovered had souls; ultimately they decided that no test can answer that question. In our day, we speculate whether the machines we build might one day have intelligence. Alan Turing (1912–1954) proposed a test. In two remote rooms, place a machine M in one and a human H in another. If a judge, by asking questions of M and H and appraising their answers, cannot decide which is which, then, by default, Turing says that the machine is intelligent [**69**, pp. 595–599]. Such a way of knowing, while quite clever, is anything but objective. Subjectivity, by its very nature, includes an element of unknowability, an element with which we must be satisfied in our decision-making procedures; otherwise we will end up making little progress at all in the grand scheme of our lives and the enhancement of our communities.

Amidst the general state of our understanding of all things, one can almost hear Voltaire saying today, *How little it is that we will ever really know.*

Martin Gardner, a popular expositor of mathematics, most well-known for the long tenure of his column in *Scientific American*, echoes some of these thoughts in a critique of the explanation that self-consciousness and free-will are merely wisps of fancy left behind by the intertwining dynamics of strange loops.

> Why is our universe mathematically structured? Why does it, as [Stephen] Hawking recently put it, bother to exist? Why is there something rather than nothing? How do the butterflies in our brain—or should I say bats in our belfry—manage to produce the strange loops of consciousness? There may be advanced life forms in Andromeda who know the answers. I sure don't. And neither do you. [**50**, p. 854]

That is, we delude ourselves if we say we know when we really do not know. As we have seen, Voltaire suspected that there were no absolute answers to the questions that he most desired to know, and as a result, the answers with which we amuse ourselves and pride ourselves in reaching are probably mostly illusions as well—which accounts for the blankness of Micromégas's book. The quest to measure all things and to know all things is forever doomed to be incomplete.

Yet maybe we have read too much into Micromégas's promises about his book. His specific words: in his book, man "would see the answer and end of all things." So maybe, rather than promising all the answers, Micromégas promised only one answer, the answer to what will occur when time is no more. The second law of thermodynamics says that entropy in the universe ever increases with time—which means that eventually all the stars in the sky will grow cold, life will cease, and order will be no more. That scenario is emptiness indeed. H. G. Wells captured this idea in *The Time Machine*, when his traveler goes to the end of time on earth. All that remains of earth's great civilizations is some crabs scuttling sideways on a beach—like a blank page from Micromégas's book.

Answer 7: Tabula Rasa. The one man that Voltaire probably admired more than Newton was John Locke, of whom he said [**144**, vol. xxxix, *Locke*, pp. 33–34],

> There surely never was a more solid and more methodical understanding, nor a more acute and accurate logician, than Locke.

Locke describes man at birth as a blank sheet, a *tabula rasa*. As we grow and experience the world, we write on that book, creating a personal identity and an understanding of the world. The idea is an old one, for in *On the Soul (De Anima)*, Aristotle had first suggested that man was an *unscribed tablet*, and St. Thomas Aquinas, who rediscovered Aristotle toward the end of the Middle Ages, coined the phrase *tabula rasa*.

So, a last resolution to Voltaire's riddle is that Micromégas's book was blank because it was a tabula rasa. We as a people, with our collective inquisitiveness, will write in that ideal book, filling it with discoveries, understanding, further questions, and renewed hope. The book is written as we learn. In that sense, Micromégas's book indeed contains all the answers to all the questions about the universe. However, the book is never complete at any specific time.

In *The Temple of Taste*, Voltaire imagines an ideal library containing the best of the world's literature. Since some of that literature has yet to be written, the library is not yet full. Indeed, from an optimistic perspective, it is probably mostly empty, as we hope that many good books are yet to be written. In mathematics, as we noted in Vignette V, there is a similar ideal library. Actually, the library is a single ideal volume, called *The Book*. Its pages contain the best of the proofs of all the theorems that can be imagined. If we were allowed a glimpse of *The Book* and were only permitted to see the best proofs that have currently been discovered by man, the book would probably appear mostly blank—because there is so much more to be discovered.

Riddle Resolutions

Figure 5. A deep Hubble look into blank space, courtesy of NASA

More simply, think of an integer analogy. Imagine a time-keeper enumerating the positive integers $\mathcal{N}$, counting them one at a time in order from one to infinity. Imagine that he writes them down in a book as he counts them. We will allow him a finite amount of time to write any integer. Of course, the time-keeper will never count all of $\mathcal{N}$. But as long as the office of time-keeper is handed down from one generation to the next, for any particular integer n, there will be a time when n is counted and appears in the book. However, if the book was designed so as to contain all of $\mathcal{N}$, at any time the book will appear mostly blank. Now identify $\mathcal{N}$ with $\mathcal{Q}$, the set of the questions about the universe that will ever be asked by man. $\mathcal{Q}$ should be a *countable* set—a set that can be written as a sequence. For up to any given time, there are a finite number of people who have ever lived, each of whom have had at most a countable number of cogent questions, most of which have already been asked by others. Imagine that the book of Micromégas contains the elements of $\mathcal{Q}$ as man finds corresponding, suitable answers. Thus, for any specific question, there could be a time when a good answer along with the question might appear in the book. Of course, as a result of that answer, there may be many more questions. However, Micromégas's book is a big one, big enough to contain all of the questions and answers about the universe. So there will always be plenty of pages left to fill—and the book always appears mostly blank.

Bertrand Russell and Alfred North Whitehead actually attempted to write such a book, producing their monumental treatise of applying a production-of-statements algorithm to an axiom system, which would generate the set of all conceivable statements, each of which could be tested for theoremhood. If such a plan worked, ultimately every theorem could be proved, at least in theory. However, in 1932 Kurt Gödel, like a Voltaire figure, showed that such a scheme was impossible, that there will always be truths that are unreachable by any series of legitimate steps within a given consistent axiom system that includes the basic properties of the integers.

Gödel's result suggests that the universe of ideas—and by extension—the universe itself is perhaps unfathomable, even as Stephen Hawking said in a 2003 lecture [64]:

> What is the relation between Gödel's theorem and whether we can formulate the theory of the universe? If there are mathematical results that cannot be proved, there are physical problems that cannot be predicted.

That is, we will never understand the universe completely. More, in the language of Chapter II, it could be that the set of possible theorems and established facts about the universe may constitute a set of measure zero in a sea of truths and facts that we may never know. For such a model, there will always be more to discover. The apparent complexity of the mind—*How do we think?*—and the apparent complexity of matter—*What is gravity, anyway?*—unveil the richness of the universe in which we live. That is good news, says Hawking [64],

> I am glad that our search for understanding will never come to an end, and that we will always have the challenge of new discovery. Without it, we would stagnate. Gödel's theorem ensured there would always be a job for mathematicians.

Of course, many do not need Gödel to reach the same conclusion. Here is de Fontenelle in 1686,

> For goodness sake, let us admit that there will still be something left for future centuries to do. [32, p. 34]

Another riddle-resolving element suggested by the *tabula rasa* idea is the very nature of the *tabula rasa* itself. For a tabula rasa is a human being: mind, spirit, and body. Each person is part of nature—which of course we try to analyze and understand. Many are the times when we ask of ourselves,

> Why am I here?

or of nature in general about some phenomenon,

> How do you do it?

And nature is mute, or gives a Voltaireian response,

> I do not know. You figure it out!

Mathematicians have a long history of trying to understand the pattern of the prime integers. And the best that we can show for three thousand years of work on the question is that the pattern of the primes is the pattern of the primes. So too, it could be that philosophers will ultimately conclude that the pattern of life is the pattern of life. With such a perspective, when the secretary of the French Academy looked into the book of Micromégas, all he saw was the book of nature. There was no analysis and understanding in the book. It just was. And it is up to us to make sense of it.

To summarize—parts of the book of nature we do understand; the other parts are like blank pages in as far as writing down what we can say about it. However, if we focus on a particular part that might seem blank—perhaps devoting years of our lives to understanding a particular phenomenon—some progress is made. This, of course, is what scientific investigation and proving theorems is all about—insight and hard work. Galileo says it well [**48**, pp. 237–238]:

> Philosophy is written in this great book—I mean the universe—which stands continually open to our gaze, but it cannot be understood unless one first studies the language and the characters in which it is written. It is written in the language of mathematics, and its characters are triangles, circles, and other geometrical figures, without which it is humanly impossible to understand a single word of it.

We end this book with a heavenly example of apparent blankness not being blank: to an observer in a secluded place—perhaps far out in a desert, or on a ship in the south seas—the Milky Way splashes itself against the night sky. To say the least, the sight is awesome. However, parts of this canopy look blank. When the Hubble Telescope is trained upon any apparently dull patch of the universe, its deep-space cameras unveil a bewildering array of stars and galaxies such as is shown in Figure 5. If a page of nature looks blank, a closer look almost invariably shows otherwise.

Such might be the nature of Micromégas's book. Such *is* the nature of our universe.

Exercises

1. (a) Generate a double-minded fractal that starts with "No" on the outermost level and "Yes" on the secondary level, and "No" on the tertiary level, and so on forever.

   (b) Generate a triple-minded fractal that cycles through the alternatives "Yes-Maybe-No."

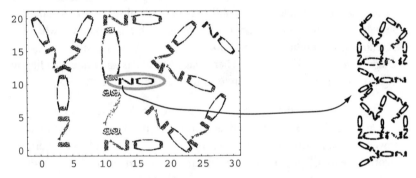

Figure 6. Zooming in on a fractal

   (c) Adapt the CAS code of Chapter II so as to zoom in on any portion of the fractal. For example, Figure 6 zooms in on the left-hand leg of the "N" in "No" in the middle arm of the "E" of "Yes."

2. (a) Douglas Adams (1952–2001) in *The Hitchhiker's Guide to the Galaxy*, contained in the collection [2], has hero Arthur Dent stumble across the answer to the ultimate question in the universe. What is this succinct answer? Contrast Voltaire's story of *Micromégas* with Adams's.

   (b) Isaac Asimov (1920–1992) in *The Last Question* [11] poses the ultimate question to a Multivac, *Can the second law of thermodynamics be reversed?* What is the computer's final answer?

3. (a) Read Lewis Carroll's *Jabberwocky* and contrast it with Rabelais's list of adjectival modifiers of cod.

(b) Roger Penrose discusses the differences between a human mind and a machine in *The Emperor's New Mind* [**102**]. How does he use Hans Christian Andersen's story as described on p. 286 in his argument?

4. (a) In Voltaire's quote on p. 292, he also includes Locke as being an eagle landing on a dunghill. What did Locke do to warrant Voltaire's dismay?

   (b) Besides Frederick the Great, find a candidate in history who comes close to being a genuine philosopher king. Justify your answer.

5. What does Voltaire mean by the last line of his poem, *Answer to a Lady*, on p. 279?

6. Try this experiment on your copy machine: Take an image such as the microscope of Figure 1. Reduce it by half onto a sheet of paper. Take that sheet, and reduce its image by half onto another sheet. Repeat this last step a total of five times. Then reverse the process, that is, magnify the last sheet by a factor of two, and repeat five times. Compare the last sheet with the original.

7. Suppose someone says, "I know," with respect to some theorem, issue, or result. List a handful of ways that their statement can be true. For example, here is one way:

   I remember working through the proof once. Now, given enough time, I should be able to convince myself that it is so.

8. Statistical reasoning as used in weather prediction and stock market investment is a combination of objective and subjective evaluation. Give other examples of this way of knowing.

9. Why does mathematics seem to be the key to understanding the universe as Galileo suggests on p. 299?

10. Can you think of other rationales besides the seven given in this chapter as to resolving Voltaire's riddle of the blank book?

Appendix

1. Average value of a function f over an interval

Let $a = x_0 < x_1 < x_2 < \cdots < x_n = b$ be a partition of $[a, b]$ with $x_{i+1} = x_i + \Delta x$ where $\Delta x = (b - a)/n$, for integers i with $0 < i < n$. The average of f at the n uniformly spaced x values x_i, $1 \leq i \leq n$, is $\sum_{i=1}^{n} f(x_i)/n$, which is the same as $\sum_{i=1}^{n} f(x_i)\Delta x/(b - a)$, whose limit as $n \to \infty$ is defined to be the average value of f over $[a, b]$:

$$\frac{\int_a^b f(x)\,dx}{b - a}.$$

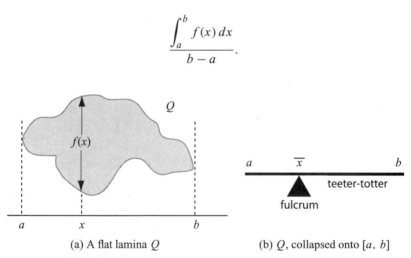

(a) A flat lamina Q (b) Q, collapsed onto $[a, b]$

Figure 1. Finding the x-coordinate of the center of mass, $\bar{x}$

2. Centroid of a flat lamina

A flat lamina Q is an object that lies in the xy-plane and has uniform density. Suppose Q lies between $x = a$ and $x = b$, with $a < b$, and has constant density δ, as shown in Figure 1.a. Suppose $f(x)$ is a function giving the cross-sectional vertical width of Q along the vertical line at x for each x with $a \leq x \leq b$. Let $\bar{x}$ be the x-coordinate of the center of mass of Q. Consider this mind-experiment:

> For each x between a and b, assume that all of the mass of Q coalesces along the vertical line at x to a single point at $(x, 0)$, namely, the mass $\delta f(x)\, dx$. The mass-moment of this mass about the point $(\bar{x}, 0)$ is the mass times its distance from $\bar{x}$, namely, $\delta f(x)(x - \bar{x})\, dx$. Just as with a teeter-totter on the playground, as in Figure 1(b), the line segment $[a, b]$ of all of these point-masses balances at the fulcrum $(\bar{x}, 0)$ when the sum of all of the mass-moments is zero.

That is, $\bar{x}$ satisfies the integral equation

$$\int_a^b \delta f(x)(x - \bar{x})\, dx = 0, \tag{1}$$

which means that

$$\bar{x} = \frac{\int_a^b x f(x)\, dx}{\int_a^b f(x)\, dx} = \frac{\int_a^b x f(x)\, dx}{\text{Area of } Q}. \tag{2}$$

3. Pappus's theorem

If region Q of Figure 1(a) is spun about the vertical line $x = c$, where $c \leq a$, then the volume of the resultant solid of revolution is given by the integral

$$\int_a^b 2\pi(x - c) f(x)\, dx, \tag{3}$$

where $f(x)$ is the vertical cross section height of Q at x. This technique of determining volumes is called the shell method, as the area of a cylindrical shell generated by spinning the cross-sectional height of Q at x about the axis is the circumference of the shell's base times its height.

Let Q be a region in the plane, as in Figure 1. Let C be the centroid of Q. Let $c \leq a$. Pappus said that the volume of P, the solid of revolution obtained by spinning Q about the axis $x = c$, is $2\pi r A$, where r is the distance of Q's centroid from the axis and where A is the area of Q.

To see that this is true, from (3) and (2), the volume of P is

$$\int_a^b 2\pi(x - c) f(x)\, dx = 2\pi \left(\frac{\int_a^b x f(x)\, dx}{A} A - cA \right) = 2\pi(\bar{x} - c) A.$$

4. Rotational inertia

Consider a simple slingshot as a point mass of m attached to a mass-less string of length r. The rotational inertia I of the sling with respect to m being whirled in a circle with radius r is the product $r^2 m$. Thus as a simple toy, if r is small, the slung mass (with small I) won't damage one's surroundings, but if r is large, the toy (with large I) soon becomes a wrecking ball.

This nomenclature is meant to preserve analogous ideas between straight-line motion and rotational motion. That is, the kinetic energy of a point-mass m moving with speed v is $\frac{1}{2}mv^2$. If a simple sling is being whirled at ω radians/sec, then the kinetic energy of the point-mass is $\frac{1}{2}m(r\omega)^2$, where $r\omega$ is the rotational speed of the mass along its circuit. That is, the kinetic energy is $\frac{1}{2}I\omega^2$, so that I in a rotational motion context corresponds to m in a straight-line motion context.

More generally, the rotational inertia of a solid S with respect to an axis $\mathcal{L}$ of rotation is the sum of all of its mass-moments. That is, let us say that at the point (x, y, z) in S, its mass is $\delta(x, y, z)\,dx\,dy\,dz$ and its distance is $r(x, y, z)$ from $\mathcal{L}$, where $\delta(x, y, z)$ is the density function of S. Thus the rotational inertia I of S about $\mathcal{L}$ is the triple integral

$$I = \int\int\int_S \delta(x, y, z)\, r^2(x, y, z)\, dx\, dy\, dz.$$

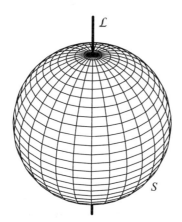

Figure 2. Sphere rotating about its pole

Example Let S be a homogeneous sphere of mass M and radius R and let $\mathcal{L}$ be the z-axis. Then $\delta(x, y, z) = 3M/(4\pi R^3)$, $r(x, y, z) = \sqrt{x^2 + y^2}$, and its rotational inertia about $\mathcal{L}$ is

$$I = \frac{3M}{4\pi R^3} \int\int\int_S x^2 + y^2 \, dx \, dy \, dz.$$

In cylindrical coordinates, this becomes

$$I = \frac{3M}{4\pi R^3} \int_{-R}^{R} \int_0^{\sqrt{R^2-z^2}} \int_0^{2\pi} r^3 \, d\theta \, dr \, dz = \frac{2}{5} MR^2.$$

5. Exponential growth

Exponential growth is perhaps the most fundamental and simple rate of change exhibited in nature. Its differential equation is $y' = ay$ where $y(t)$ is some real-valued function in terms of the parameter t and a is some constant. Given an initial value $y(0)$, its solution is $y = e^{at} y(0)$. Sometimes we encounter systems of intertwined exponential growth such as

$$x' = ax + by \text{ and } y' = cx + dy$$

where a, b, c, and d are constants and both x and y are real-valued functions. If we let $z(t) = (x, y)$ and

$$A = \begin{bmatrix} a & b \\ c & d \end{bmatrix},$$

then the system can be written as

$$z'(t) = Az(t). \tag{4}$$

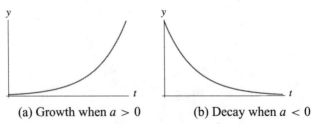

(a) Growth when $a > 0$ (b) Decay when $a < 0$

Figure 3. Exponential behavior

Even though (4) is a matrix equation, the form of its solution follows the one-dimensional case. That is, the solution to (4), given an initial value $z(0)$,

Appendix

is

$$z(t) = e^{At}z(0). \tag{5}$$

Since the series for e^{at} is

$$e^{at} = 1 + at + \frac{a^2 t^2}{2!} + \frac{a^3 t^3}{3!} + \cdots = \sum_{n=0}^{\infty} \frac{a^n t^n}{n!},$$

we define

$$e^{At} = I + At + \frac{A^2 t^2}{2!} + \frac{A^3 t^3}{3!} + \cdots = \sum_{n=0}^{\infty} \frac{A^n t^n}{n!}, \tag{6}$$

where I is the identity matrix. The derivative of the right-hand side of (6) is

$$\begin{aligned}
(e^{At})' &= A + A^2 t + \frac{A^3 t^2}{2!} + \frac{A^4 t^3}{3!} + \cdots = A(I + At + \frac{A^2 t^2}{2!} + \frac{A^3 t^3}{3!} + \cdots) \\
&= A e^{At} \\
&= A + A^2 t + \frac{A^3 t^2}{2!} + \frac{A^4 t^3}{3!} + \cdots = (I + At + \frac{A^2 t^2}{2!} + \frac{A^3 t^3}{3!} + \cdots)A \\
&= e^{At} A,
\end{aligned}$$

which means that (5) indeed solves (4). Furthermore, A and e^{At} commute multiplicatively.

Example 1. Solve the differential equation $x'' + x = 0$.

To do this, let $y = x'$, and let $z = (x, y)$, so that the original equation becomes $y' = -x$, or, as a system, $z' = Mz$, where

$$M = \begin{bmatrix} 0 & 1 \\ -1 & 0 \end{bmatrix}.$$

The powers of M produce a cyclical sequence: $M^{4n} = I$, $M^{4n+1} = M$, $M^{4n+2} = -I$, and $M^{4n+3} = -M$, where n is a non-negative integer. Thus by (6),

$$\begin{aligned}
e^{Mt} &= I\left(1 - \frac{t^2}{2!} + \frac{t^4}{4!} - \cdots\right) + M\left(t - \frac{t^3}{3!} + \frac{t^5}{5!} - \cdots\right) \\
&= I \cos t + M \sin t = \begin{bmatrix} \cos t & \sin t \\ -\sin t & \cos t \end{bmatrix}.
\end{aligned} \tag{7}$$

By (5), the solution is $z = e^{Mt}z(0)$, which means that $x(t) = a\cos t + b\sin t$ where a and b are general constants. (In particular, $a = x(0)$ and $b = x'(0)$.) To achieve a more compact form, let $c = \sqrt{a^2 + b^2}$, so that

$$x(t) = c\left(\frac{a}{c}\cos t - \left(-\frac{b}{c}\right)\sin t\right) = c(\cos\phi\cos t - \sin\phi\sin t) = c\cos(t+\phi)$$

where ϕ is the *phase angle* for which $\cos\phi = a/c$ and $\sin\phi = -b/c$.

Example 2. Solve the differential equation $x'' + x = k$, where k is some constant.

As in Example 1, let $y = z'$. The differential equation can now be written as $z' = Mz + \mathbf{b}$, where $\mathbf{b} = (0, k)$. Rewrite the matrix equation as $z' - Mz = \mathbf{b}$ and multiply through by the *integrating factor* e^{-Mt} to obtain

$$\left(e^{-Mt}z\right)' = e^{-Mt}z' + e^{-Mt}(-M)z = e^{-Mt}\mathbf{b}.$$

Integrating gives

$$e^{-Mt}z = \int e^{-Mt}\mathbf{b}\,dt = -\int e^{-Mt}(-M)M^{-1}\mathbf{b}\,dt = -e^{-Mt}M^{-1}\mathbf{b} + \mathbf{c},$$

where $\mathbf{c} = (\alpha, \beta)$ is some constant vector. Multiply through by e^{Mt},

$$z = -M^{-1}\mathbf{b} + e^{Mt}\mathbf{c}.$$

By (7), $x = \alpha\cos t + \beta\sin t + k$, or

$$x = \gamma\cos(t+\phi) + k,$$

where $\gamma = \sqrt{\alpha^2 + \beta^2}$ and ϕ is a phase angle.

Appendix Item 9 gives an algorithm to find a closed form for e^{At} that works well for various $n \times n$ matrices A, where n is a positive integer.

6. The gamma function

$\Gamma(p)$ was discovered by Leonhard Euler when he was 22 in 1729, in response to a question from Christian Goldbach about what the values of numbers like 5.5 factorial should be [29]. The function was named by Adrien-Marie Legendre who in 1825 generated an extensive table of its values. For any positive number p,

$$\Gamma(p) = \int_0^\infty t^{p-1}e^{-t}\,dt. \tag{8}$$

Appendix

Using integration by parts, we see that

$$\Gamma(p+1) = p\Gamma(p), \tag{9}$$

for any positive number p. Since $\Gamma(1) = 1$, then by (9), $\Gamma(n+1) = n!$ for any positive integer n.

A surprising and useful gamma value is

$$\Gamma\left(\frac{1}{2}\right) = \sqrt{\pi}. \tag{10}$$

To see that (10) is true, the substitution $x^2 = t$ gives $\Gamma(1/2) = \int_0^\infty 2e^{-x^2} dx$ and, using polar coordinates,

$$\Gamma^2\left(\frac{1}{2}\right) = \int_0^\infty \int_0^\infty 4e^{-x^2}e^{-y^2} dx\, dy = \int_0^\infty \int_0^{\pi/2} 4re^{-r^2} d\theta\, dr = \pi.$$

7. The Laplace transform

The Laplace transform was another discovery of Leonhard Euler, and is named for Pierre-Simon Laplace, who used it in studying probability. For any function $f(t)$, its Laplace transform $F(s)$ is

$$F(s) = \mathcal{L}[f(t)](s) = \int_0^\infty e^{-st} f(t)\, dt, \tag{11}$$

where s is the variable of the transform. For example, the Laplace transform of the constant function 1 is $1/s$. By integration by parts,

$$\mathcal{L}[t^n](s) = \frac{n!}{s^{n+1}}, \tag{12}$$

where n is any positive integer. As perhaps is no surprise, the Laplace transform of $1/\sqrt{t}$ involves $\Gamma(1/2)$, for, with $u = st$,

$$\mathcal{L}\left[\frac{1}{\sqrt{t}}\right](s) = \int_0^\infty \frac{e^{-st}}{\sqrt{t}} dt = \frac{1}{\sqrt{s}} \int_0^\infty \frac{e^{-u}}{\sqrt{u}} du = \frac{\Gamma(1/2)}{\sqrt{s}} = \sqrt{\frac{\pi}{s}}. \tag{13}$$

Thus, by (12), (13), and integration by parts,

$$\mathcal{L}[t^{n/2}](s) = \frac{\Gamma(1+n/2)}{s^{1+n/2}}, \tag{14}$$

where n is any positive integer.

Here are some more useful identities where k is a constant and $s > 0$:

- $\mathcal{L}[f'(t)](s) = s\,F(s) - f(0)$,
- $\mathcal{L}[\cos(k\,t)](s) = \dfrac{s}{s^2 + k^2}$,
- $\mathcal{L}[\sin(k\,t)](s) = \dfrac{k}{s^2 + k^2}$.

8. Volume of a ball in n-space

For each positive integer n, let $V_n(r)$ be the set of all points in n-space that are no more than distance r from the origin. We imagine that $V_n(r) = k_n r^n$ where k_n is some constant, as each dimension, by symmetry, should contribute the same factor to volume in $\mathcal{R}^n$. To find k_n, we follow Lasserre [77] and compute $V_n(r)$ by using the Laplace transform (14). The key to his argument is the exponential property, $e^{p+q} = e^p e^q$ for any numbers p and q. To use this idea, we compute $V_n(\sqrt{t})$ for the volume of a ball of radius $\sqrt{t} \geq 0$ whose equation is $t = \sum_{i=1}^{n} x_i^2$. We know that $V_n(\sqrt{t}) = k_n t^{n/2}$, whose Laplace transform by (14) is

$$\mathcal{L}[V_n(\sqrt{t})](s) = \int_0^\infty e^{-st} k_n t^{n/2}\,dt = k_n \frac{\Gamma(1 + n/2)}{s^{1+n/2}}, \qquad (15)$$

where k_n is a constant. We also know that

$$V_n(\sqrt{t}) = \int_{B_n(\sqrt{t})} 1\,dx = \int_{\sum_{i=1}^n x_i^2 \leq t} 1\,dx,$$

an iteration of n integrals with variables $x = (x_1, x_2, \ldots, x_n)$, whose Laplace transform is

$$\mathcal{L}[V_n(\sqrt{t})](s) = \int_0^\infty e^{-st} \left(\int_{\sum_{i=1}^n x_i^2 \leq t} 1\,dx \right) dt. \qquad (16)$$

The integration in (16) is over the solid paraboloid $t \geq \sum_{i=1}^{n} x_i^2$ in $\mathcal{R}^{n+1}$. When the order of integration is $dx\,dt$, the outer integral ranges over $0 \leq t < \infty$; for the inner integral, for a fixed t value, the integration ranges over the cross-section of the paraboloid at level t, a disk of radius $\sqrt{t}$ in $\mathcal{R}^{n+1}$ as indicated in Figure 4(a). With the order of integration $dt\,dx$, the integration for the first n integrals ranges over all of $\mathcal{R}^n$; for the innermost integral, the range of t values in the paraboloid for a fixed point x in $\mathcal{R}^n$ is the interval

Appendix

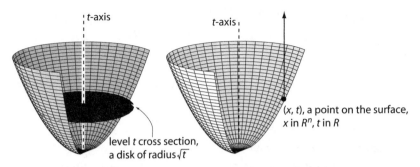

(a) Integration order $dx\, dz$
(b) Integration order $dz\, dx$

Figure 4. The surface of the solid $t \geq \sum_{i=1}^{n} x_i^2$ in $\mathcal{R}^{n+1}$

$\sum_{i=1}^{n} x_i^2 \leq t < \infty$, as indicated in Figure 4(b). With the order of integration reversed, (16) becomes

$$\mathcal{L}[V_n(\sqrt{t})](s) = \int_{\mathcal{R}^n} \left(\int_{\sum_{i=1}^{n} x_i^2 \leq t} e^{-st}\, dt \right) dx$$

$$= \frac{1}{s} \int_{\mathcal{R}^n} e^{\left(-s \sum_{i=1}^{n} x_i^2\right)} dx = \frac{1}{s} \prod_{i=1}^{n} \int_{-\infty}^{\infty} e^{-s x_i^2}\, dx_i. \qquad (17)$$

From (10), $\Gamma(1/2) = \sqrt{\pi}$, so that for each i,

$$\int_{-\infty}^{\infty} e^{-s x_i^2}\, dx_i = 2 \int_{0}^{\infty} e^{-s x_i^2}\, dx_i$$

$$= (2/\sqrt{s}) \int_{0}^{\infty} e^{-x_i^2}\, dx_i = \Gamma(1/2)/\sqrt{s} = \sqrt{\pi/s}.$$

Thus (17) becomes

$$\mathcal{L}[V_n(\sqrt{t})](s) = \frac{(\pi/s)^{n/2}}{s} = \frac{\pi^{n/2}}{s^{1+n/2}}. \qquad (18)$$

By (15) and (18), $k_n = \pi^{n/2}/\Gamma(1 + n/2)$. Therefore

$$V_n(r) = \frac{\pi^{n/2} r^n}{\Gamma(1 + n/2)}. \qquad (19)$$

9. A representation of e^{Mt} for some square matrices M

We compute e^{Mt} where

$$M = \begin{bmatrix} 0 & -b & 0 & b \\ b & 0 & -b & 0 \\ 0 & -a & 0 & a \\ a & 0 & -a & 0 \end{bmatrix}, \tag{20}$$

for any numbers a and b with $a \neq b$.

Let the Laplace transforms of the functions $x(t)$, $y(t)$, $u(t)$, and $v(t)$ be $X(s)$, $Y(s)$, $U(s)$, and $V(s)$. Let $z(t) = (x, y, u, v)$, so that $Z(s) = (X, Y, U, V)$. The solution to the equation $z'(t) = Mz(t)$ is $z(t) = e^{Mt}z(0)$ by (5). For the moment, let $z(0) = (1, 0, 0, 0)$. With this value for $z(0)$ then $z(t)$ is the first column of the matrix e^{Mt}, and the (1, 1) entry of e^{Mt} is $x(t)$.

The system $z'(t) = Mz(t)$ and its Laplace transform can be written as two systems of four equations each:

$$\begin{cases} x'(t) = b(-y+v), \\ y'(t) = b(x-u), \\ u'(t) = a(-y+v), \\ v'(t) = a(x-u), \end{cases} \qquad \begin{cases} sX - 1 = b(-Y+V), \\ sY = b(X-U), \\ sU = a(-Y+V), \\ sV = a(X-U). \end{cases}$$

We rewrite the second of these systems as $NZ = \begin{bmatrix} 1 & 0 & 0 & 0 \end{bmatrix}^T$, where

$$N = \begin{bmatrix} s & b & 0 & -b \\ -b & s & b & 0 \\ 0 & a & s & -a \\ -a & 0 & a & s \end{bmatrix}.$$

Thus $Z = N^{-1} \begin{bmatrix} 1 & 0 & 0 & 0 \end{bmatrix}^T$, which means that the (1, 1) entry of N^{-1} is X. With a CAS, we compute the (1, 1) entry of N^{-1} as

$$X = \frac{a^2 - ab + s^2}{s(a^2 - 2ab + b^2 + s^2)} = \frac{1}{a-b}\left(\frac{a}{s} - \frac{bs}{s^2 + (a-b)^2}\right).$$

Therefore the (1, 1) entry of e^{Mt} is

$$x = \frac{1}{a-b}(a - b\cos((a-b)t)).$$

In like fashion, we calculate the remaining entries of the first column of e^{Mt}.

Appendix

To find the second column of e^{Mt} we repeat the steps, but this time with the initial value of z as $z(0) = (0, 1, 0, 0)$. Similarly, we calculate the remaining two columns of e^{Mt}, to finally get

$$e^{Mt} = \frac{1}{a-b} \begin{bmatrix} a - b\cos\omega & -b\sin\omega & -b + b\cos\omega & b\sin\omega \\ b\sin\omega & a - b\cos\omega & -b\sin\omega & -b + b\cos\omega \\ a - a\cos\omega & -a\sin\omega & -b + a\cos\omega & a\sin\omega \\ a\sin\omega & a - a\cos\omega & -a\sin\omega & -b + a\cos\omega \end{bmatrix}, \tag{21}$$

where $\omega = (a-b)t$.

10. Newton's path through the earth: finding r in terms of θ and t

Both of the nonlinear differential equations of (VIII.1) are of the form

$$\frac{d^2 w}{du^2} + \alpha^2 w = \frac{\beta^2}{w^3}, \tag{22}$$

where α and β are constants.

Since we wish to solve the two equations for the initial conditions, $r(0) = R = 1/z(0)$, $\frac{dr}{dt}|_{t=0} = 0$ and $\frac{dz}{d\theta}|_{t=0} = 0$, the initial conditions for (22) are $w(0) = \delta$ for some $\delta > 0$ and $\frac{dw}{du}|_{t=0} = 0$. To solve this equation, let $\rho = \frac{dw}{du}$, which means that $\frac{d^2w}{du^2} = \frac{d\rho}{du} = \frac{d\rho}{dw}\frac{dw}{du} = \rho\frac{d\rho}{dw}$. Hence (22) becomes

$$\rho\, d\rho = \left(\frac{\beta^2}{w^3} - \alpha^2 w\right) dw, \tag{23}$$

where $\rho = 0$ when $w = \delta$. Integrating (23), using the boundary conditions, and solving for ρ gives

$$\frac{dw}{du} = \rho = \pm\sqrt{\beta^2\left(\frac{1}{\delta^2} - \frac{1}{w^2}\right) - \alpha^2(w^2 - \delta^2)},$$

which, using the substitution $q = w^2 - \delta^2$, becomes

$$\frac{\delta}{2\sqrt{\beta^2 q - \alpha^2\delta^2 q(q + \delta^2)}}\, dq = \pm du,$$

where $q = 0$ when $u = 0$. Integrating, using the initial condition, and rewriting in terms of w and u gives

$$\sin^{-1}\left(\frac{2\alpha^2\delta^2}{\beta^2 - \alpha^2\delta^4}\left(w^2 - \frac{\beta^2 + \alpha^2\delta^4}{2\alpha^2\delta^2}\right)\right) = \pm 2\alpha u - \frac{\pi}{2},$$

which simplifies to

$$w^2 = \frac{\beta^2 + \alpha^2\delta^4}{2\alpha^2\delta^2} - \frac{\beta^2 - \alpha^2\delta^4}{2\alpha^2\delta^2}\cos(2\alpha u).$$

Using the double angle formula for cosine, this expression simplifies to

$$w = \sqrt{\delta^2\cos^2(\alpha u) + \left(\frac{\beta}{\alpha\delta}\right)^2\sin^2(\alpha u)}. \tag{24}$$

We choose the positive square root in (24) because $w > 0$ near $u = 0$.

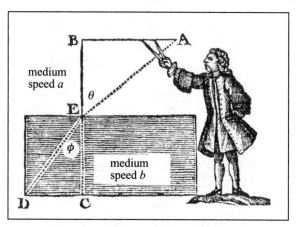

Figure 5. Snell's law diagram from Voltaire's 1738 *Elements of Newton* [**142**, p. 73]

11. Snell's law

Snell's law, discovered by Willebrord Snell in 1621, says

$$\frac{\sin\theta}{a} = \frac{\sin\phi}{b}, \tag{25}$$

where θ and ϕ are the angles of inclination of the wave or small particle away from a normal to the boundary between two mediums, and where a

Appendix

is the relative speed of a wave in the first medium, such as air, and b is the relative speed in the second medium, such as a rectangular crystal, as shown in Figure 5.

To derive (25), we need a model for a wave or a small particle moving along a path. René Descartes suggested an intuitive model in 1657. He likened a wave crossing the boundary between two mediums to a ball rolling across a floor from wood onto a rug. Since an ideal ball has just a single contact point with the floor, whereas a wave should have some breadth contact with its medium, we modify Descartes' ball into the next most simple model, and use a two-point contact with the floor. We use a barbell: two disks $\hat{A}$ and $\hat{B}$ connected by a bar of length d as shown in Figure 6. Each disk makes contact with the floor at a single point, and each traces a path as the barbell rolls along. As the barbell rolls, we stipulate that the disks never slip. What this condition means is that the plane of each disk meets the floor along the tangent line to the path made by the disk.

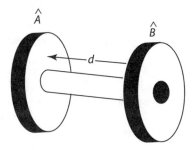

Figure 6. Descartes' barbell

As the barbell approaches a straight-line boundary between two mediums, as shown in Figure 7, once disk $\hat{A}$ crosses the boundary, it rolls at a different speed than $\hat{B}$. This means that the barbell starts turning, and continues turning until $\hat{B}$ also crosses the boundary. If $a > b$, then $\hat{B}$ moves further than $\hat{A}$ during the turn so after the turn, the new angle of inclination of the barbell with respect to the boundary is less than θ.

Except for a change in scale, the bending of the barbell's path as it crosses the boundary is invariant with respect to d. That is, for each d, we choose a unit length equal to the length of the barbell, and a unit time so that the relative speeds of a and b are the speeds of the barbell in the two mediums. Thus we may take d as 1, and interpret it as any distance, including the case when the two disks are arbitrarily close together, which gives a way of approaching the behavior of Descartes's rolling ball model.

Let $A(t) = (x(t), y(t))$ and $B(t) = (u(t), v(t))$ be the positions of $\hat{A}$ and $\hat{B}$ at time t. Let $A(0) = (0, 0)$ and $B(0)$ be above the x-axis.

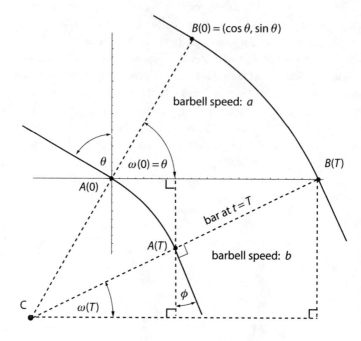

Figure 7. Barbell crossing boundary, $\theta = \pi/3$, $a = 2$, $b = 1$, $T \approx 0.6$

The unit vector from $A(t)$ to $B(t)$ is $(-y', x')/b$ and the unit vector from $B(t)$ to $A(t)$ is $(v', -u')/a$, as long as $A(t)$ is on or below the x-axis and $B(t)$ is on or above the x-axis. Since $A(t)$ and $B(t)$ are connected by a bar of unit length 1, then

$$A(t) = B(t) + \frac{1}{a}(v'(t), -u'(t)) \quad \text{and} \quad B(t) = A(t) + \frac{1}{b}(-y'(t), x'(t))$$

for $0 \leq t \leq T$, since $\|(-y'(t), x'(t))\| = b$ and $\|(v'(t), -u'(t))\| = a$. This system can be rewritten as

$$\begin{aligned} x'(t) &= b(-y + v), & y'(t) &= b(x - u), \\ u'(t) &= a(-y + v), & \text{and} \quad v'(t) &= a(x - u). \end{aligned} \quad (26)$$

Appendix

Let $z(t)$ be the column vector $[\ x\quad y\quad u\quad v\ ]^T$ and let M be the 4×4 matrix as given by (20). In terms of z and M, (26) becomes

$$z'(t) = Mz(t) \quad \text{with} \quad z(0) = [\ 0\quad 0\quad \cos\theta\quad \sin\theta\ ]^T,$$

whose solution is $z(t) = e^{Mt}z(0)$ where e^{Mt} is given by (21).

To find T, the time it takes for the entire barbell to cross the boundary, as in Figure 7, we set the last component of $z(t)$ to zero,

$$-a\cos\theta\sin((a-b)t) + (-b + a\cos((a-b)t))\sin\theta = 0,$$

which simplifies to

$$\sin((a-b)T - \theta) = -\frac{b}{a}\sin\theta, \tag{27}$$

where T has replaced t. The unit vector from $A(T)$ to $B(T)$ is $(p, q) = B(T) - A(T) = (u(T) - x(T), -y(T))$. After the barbell passes the boundary, both $A(t)$ and $B(t)$ trace out lines in the direction $(q, -p)$. Let ϕ be the angle of inclination of $(q, -p)$ away from the normal to the boundary. Thus $\sin\phi = q$, (since $(q, -p)$ is a unit vector). Since $y(T)$ is the second entry of $z(t) = e^{MT}z(0)$, then

$$\frac{\sin\phi}{b} = -\frac{y(T)}{b} = \frac{1}{b(a-b)}(b\sin((a-b)T)\cos\theta$$
$$-b\cos((a-b)T)\sin\theta + b\sin\theta),$$

which by (27) simplifies to

$$\frac{\sin\phi}{b} = \frac{1}{a-b}(\sin((a-b)T - \theta) + \sin\theta)$$
$$= \frac{1}{a-b}\left(-\frac{b}{a}\sin\theta + \sin\theta\right) = \frac{\sin\theta}{a},$$

which is Snell's law 25.

For a contrasting (non-elementary) derivation of Snell's law involving Maxwell's equations, see Born [**19**, pp. 36–38].

12. The radial version of Snell's law

We adapt Snell's Law to concentric layers. Figure 8 depicts a circle of radius r_1 within a circle of radius r_2, both having center O, and with A, B, and C as

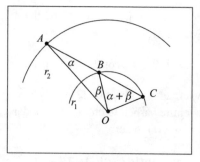 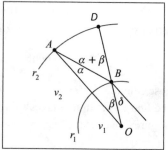

(a) A little geometry (b) Through the boundary at B

Figure 8. Adapting Snell's law to the circle

collinear points on the circles as indicated. Let $\angle OAB = \alpha$ and $\angle AOB = \beta$. Therefore

$$\angle OBC = \alpha + \beta, \tag{28}$$

so $\angle BCO = \alpha + \beta$, since $\triangle OBC$ is isosceles. By the law of sines,

$$\frac{\sin \alpha}{r_1} = \frac{\sin(\alpha + \beta)}{r_2}, \tag{29}$$

so

$$r_2 \sin \alpha = r_1 \sin(\alpha + \beta). \tag{30}$$

Now modify Figure 8a into Figure 8b. The inner circle has radius r_1 and the outer circle has radius r_2. Suppose that a wave or a small particle travels with speed v_1 within the inner circle and with speed v_2 between the circles. Let points O, B, and D be collinear. Let the angles α, β, and δ be as indicated. By (28), $\angle ABD = \alpha + \beta$. By Snell's law (25)

$$\frac{r_1 \sin(\alpha + \beta)}{v_2} = \frac{r_1 \sin \delta}{v_1}, \tag{31}$$

which by (30) gives

$$\frac{r_2 \sin \alpha}{v_2} = \frac{r_1 \sin(\alpha + \beta)}{v_2}. \tag{32}$$

By (31) and (32), we have

$$\frac{r_2 \sin \alpha}{v_2} = \frac{r_1 \sin \delta}{v_1}. \tag{33}$$

Appendix

For a wave passing through a series of concentric layers, (33) holds between succesive layers. Thus (33) holds for all layers, including the case when the layers form a continuum, which in turn means that

$$\frac{r \sin \alpha}{v} = c \tag{34}$$

at each point P along the path of a wave through a continuum of radial layers, where c is an constant, v is the medium speed of the wave at radial distance r, and α is the angle between a tangent to the wave's path at P and a ray from the center O of the concentric layers.

Cast of Characters

Edwin A. Abbott (1838–1926) is the author of *Flatland* (1884). He became headmaster of the City of London School at age 26.

Alcuin of York (c. 735–804) was a scholar from a learning center at York, England, famed for its focus on the seven liberal arts. Charlemagne (c. 747–814) invited him to France to tutor the young nobles, including himself. Due in part to Alcuin's arguments for tolerance, Charlemagne abolished the death penalty for paganism in 797.

Hans Christian Andersen (1805–1875) was a Danish story-teller. His first job was as apprentice to a tailor. One of his most memorable stories is that of the emperor's new clothes.

Archimedes's (c. 287–c. 212 BC) most famous saying is about the lever, "Give me a place to stand, and I will move the earth." According to legend, he was slain during the Roman siege and sack of Syracuse. Apparently he was in the midst of proving a theorem and was drawing circles in the sand. His last words were to a Roman soldier: "Do not disturb my circles."

Aristotle (384–322 BC) was a student of Plato. He wrote extensively on everything from physics to poetry. He was also the tutor of Alexander the Great (356–323 BC).

Cyrano Hercule Savinien de Bergerac (1619–1655) was a French soldier and playwright. He studied under Pierre Gassendi (1592–1655) who is sometimes referred to as the Galileo of France. Assimilating some of Gassendi's ideas on the heavens, de Bergerac wrote romances about life on the moon and the sun. The first person that de Bergerac's astronaut meets upon landing on the moon is the Old Testament prophet Elijah, who had traveled there in an iron chariot by continually throwing a magnet toward the moon.

George Berkeley (1685–1753) is the Father of Idealism. He championed the idea that we can merely perceive matter, rather than knowing it in an absolute sense. For example, place your right hand in hot water and your left hand in cold. Then plunge both into lukewarm water. The right hand perceives it to be cold water and the left hand perceives it to be hot. Berkeley goes on to suggest that we are but ideas in the mind of God—Platonism taken to the extreme.

Johann Bernoulli (1667–1748) and his older brother Jacob Bernoulli (1654–1705) were Swiss mathematicians. Johann taught both Leonhard Euler and Pierre Maupertuis. The elder's name is immortalized in the Bernoulli trials of statistics. The younger Bernoulli is most famous for the brachistochrone riddle as discussed in Chapters VII and VIII. Both Bernoullis were adamant Leibnizians in the great Leibniz-Newton debate about who discovered calculus.

Pierre Bouguer [pronounced "Boo-gair"] (1698–1758) is called the Father of Naval Architecture because of his expertise in ship design. He was a member of the equatorial team that measured three degrees of arc along a line of longitude at the equator in the 1730's. Once Maupertuis abdicated his seat in the French Academy in the 1740's, Bouguer took his place.

Lewis Carroll was the pen-name of Charles Lutwidge Dodgson (1832–1898) who was a mathematician and author, most notably of the Alice stories. My favorite quote from him is, "Of what use is a book without pictures?"

Giovanni Domenico Cassini (1625–1712) and his son Jacques Cassini (1677–1756) were French astronomers. Giovanni was the first to implement Galileo's idea of using the moons of Jupiter to determine longitude on earth. Jacques was the first to publish useable tables on Jupiter's moons so that anyone with both table and telescope could determine longitude, provided the weather was good and Jupiter was in the night sky. Both astronomers championed a lemon-shaped earth.

William Caxton (c. 1415–1492) introduced the printing press to England. In his encyclopedia, *Mirrour of the World* (1481), as well as describing Hesiod's problem of a body falling down a hole to the center of the earth as presented herein in Vignette VIII, he includes the reciprocal one. However, rather than agreeing with Hesiod that it would take nine days for an anvil to fall from heaven to earth, the *Mirrour* says that it would take a great stone one hundred years to fall that distance [23, p. 72].

Anders Celsius (1701–1744) was a Swedish astronomer who joined Maupertuis's polar expedition. He invented the Celsius temperature scale, except that boiling water was 0° and freezing water was 100°. It was Carolus Linnaeus (1707–1778) in 1745 who reversed the measure.

Princess Charlotte's (1745–1808) more complete name is Friederika Charlotte Leopoldine von Brandenburg Schwedt, a niece of Frederick the Great. As a teen-ager she was tutored by Leonhard Euler, who wrote philosophical-mathematical letters to her from 1760–63. In *Letter L,* Euler reminds her how Voltaire used to laugh about Maupertuis's idea of a hole going to the center of the earth. She would have been about six years old at the time of the laughter.

Alexis Claude de Clairaut [pronounced "Kluh-roh"] (1713–1765) was home schooled. He became a member of the French Academy of Sciences at age eighteen due to his demonstrated mastery of mathematics. He was one of Voltaire's mathematics tutors, and he served as Maupertuis's lieutenant on the polar expedition of 1736–37. He also accounted for the long delay in Halley's Comet expected return in 1757, due to the gravitational influence of Jupiter and Saturn.

Charles Marie de La Condamine (1701–1774) was a member of the equatorial team that measured arclength along a line of longitude at the equator in the 1730s. On his return trip to Paris from Quito, La Condamine began by going down the Amazon. Along the way he collected samples of chinchona bark and seedlings, which reputedly contained an antidote to malaria. Unfortunately, while boarding a homebound ship on the Brazilian coast, his small boat was swamped, washing away his seedlings. Fortunately, he salvaged some bark samples, which Carolus Linnaeus carefully catalogued—and which ultimately helped lead to the development of quinine. Voltaire patterned much of *Candide*'s hero after La Condamine.

Nicolaus Copernicus (1473–1543) advocated a heliocentric solar system, and circulated his ideas in manuscript form in 1514. Like Newton, he worked on monetary reform, except in Prussia rather than England. He formulated the principle now known as Gresham's Law: *bad money drives good money out of circulation.*

Dante Alighieri (1265–1321) wrote *The Divine Comedy* in about 1300 while in exile from his beloved Florence. The word *comedy* is used in the sense of a medieval belief in an ordered universe, which is also the sense in which the French named their theatre the Comédie-Française, a theatre that primarily staged tragedies.

René Descartes (1596–1650) is known as the Father of Modern Philosophy. The cartesian coordinate system bears his name. He introduced exponential notation. His most famous quote is, "I think, therefore I am." He championed the notion of celestial vortices as the mechanism behind gravity. Like Voltaire, he also was a hypochondriac; and he preferred not to rise before noon.

Albrecht Dürer (1471–1528) was an artist from northern Europe during the Renaissance. He preferred printing from woodblocks to painting on canvases because prints were so much easier to market than paintings.

Emily, more formally, Gabrielle Émilie Le Tonnelier de Breteuil, Marquise du Châtelet (1706–1749) was Voltaire's companion in newtonizing—as she called it—for fifteen years until her death during child-birth.

Eratosthenes (276–194 BC) is most famous for calculating the earth's circumference. He is also credited with devising the system of latitude and longitude. According to legend, once he became blind, he starved himself to death.

Paul Erdős (1913–1996) was an itinerant mathematician who published nearly 1500 papers, most of them with co-authors. He is second only to Euler in the sheer lifetime volume of mathematics produced as measured by page-count.

Leonhard Euler (1707–1783) was a student of Johann Bernoulli, and became, among other things, the mathematician in residence for Frederick the Great for twenty-five years. Of his skill and insight, Laplace says, "Read Euler, he is the master of us all."

Bernard le Bovier de Fontenelle (1657–1757) held the office of secretary of the French Academy of Sciences in perpetuity—actually for forty-two years. He was also a member of the French Academy, seat number 27. Here is his characterization of what a mathematician does [**32**, p. 64]:

> If you grant a mathematician the least principle, he'll draw a conclusion from it that you must grant him too, and from that conclusion

another, and in spite of yourself he'll lead you so far you'll have trouble believing it.

Frederick the Great (1712–1786) transformed Prussia into a European power. Besides playing the flute and being a master military tactician, Frederick enjoyed raising pet greyhounds.

Galileo (1564–1642) is sometimes called the father of modern science. After determining the periods of the four largest moons of Jupiter, he realized that they could be used as a clock for anyone with a telescope, and thus could be used to determine longitude on the high seas, a method refined by the Cassini's and utilized until the invention of reliable clocks and, more recently, global-positioning-satellites.

Gargantua was a fictional giant in French folklore whom Rabelais introduced to the world, and from whom the word *gargantuan* is derived.

Kurt Gödel (1906–1978) was a logician most famous for his incompleteness theorems. Like his friend Albert Einstein (1879–1955), he took refuge in the United States from Nazi Germany in the 1930's. Once his wife died, like Eratosthenes in his blindness, he starved himself to death.

Edmond Halley (1656–1742) was an all-around savant. Besides encouraging Newton to publish the *Principia* and getting a great comet named after himself and doing many other things, Halley pioneered the diving bell. In the 1690s he made dives in the River Thames to sixty feet, lasting hours at a time, replenishing his air with a system of weighted barrels filled with fresh air.

Godfrey Harold Hardy (1877–1947) was an English mathematician who prided himself on doing only pure mathematics. From his autobiography he describes his work, "I have never done anything 'useful.' No discovery of mine has made, or is likely to make, directly or indirectly, for good or ill, the least difference to the amenity of the world." Of course, he was wrong, as cryptologists and nuclear physicists have subsequently utilized his ideas extensively. He mentored Srinivasa Ramanujan (1887–1920) starting in 1913, and together they did great work in number theory.

Stephen Hawking (b. 1942) nearly failed to finish graduate studies in physics, as he was diagnosed with Lou Gehrig's disease when he was 21, and he saw little point in expending energy on a degree when he would soon die. Fortunately, he persevered. His most famous work deals with black holes. He was the Lucasian Professor of Mathematics at Cambridge from 1980 to 2009.

Henry II (1519–1559) became King of France in 1547. Henry was an avid sportsman, and enjoyed participating in jousts. He died when a lance pierced the visor of his helmet during a joust in celebration of his daughter's marriage. Some commentators on Rabelais contend that the Pantagruel character is a Henry II figure when he was the crown prince. Henry IV is a ninth cousin of Henry II.

Henry IV (1553–1610) became King of France in 1594. Raised as a Huguenot, Henry was invited by the powers that be to be married in Paris in 1572; many Protestants gathered for the ceremony, viewing the festivities as a hopeful sign from the religious wars of the time; but a few days after the wedding, thousands of Protestants were slaughtered in Paris during the infamous St. Bartholomew's Day Massacre. Nevertheless, to promote the well-being of the state, twenty-two years later Henry converted to Catholicism upon being crowned king. Voltaire retold the king's story in his epic poem *The Henriade*.

Hipparchus (190–129 BC) is credited with discovering the precession of the poles. He also compiled the first star charts as well as the first trigonometry tables. He gathered together all of what the world then knew about astronomy.

Christiaan Huygens (1629–1695) was a Dutch mathematician who loved to tinker. Among his inventions is the pocket watch. Encouraged by his friend Blaise Pascal, he wrote the first text on probability theory in 1657.

Thomas Jefferson (1743–1826) was a student of the Enlightenment and the third president of the United States. Of Jefferson's insight and learning, President John F. Kennedy quipped to a dinner party of forty-nine Nobel Prize winners in 1962, "This is the most extraordinary collection of talent, of human knowledge, that has ever been gathered together at the White House, with the possible exception of when Thomas Jefferson dined alone" [**129**, p. 347].

Johnannes Kepler (1571–1630) published his first two laws of planetary motion in *Astronomica Nova* (1609). Interestingly, the results were ignored by his contemporary giant colleagues, Galileo and Descartes.

Athanasius Kircher (1602–1680) is sometimes referred to as the last Renaissance man, that is, one who knew most of what there was to know at the time. Among his many speculations, he imagined that plagues were spread by microorganisms and that large fossil bones were the remains of giant men. In his inimical style, Voltaire characterizes this Jesuit scholar [**146**, p. 59, fragment 181]:

> An engraver executed certain characters at random, and Kircher explained them.

Joseph Louis Lagrange (1736–1813) succeeded Leonhard Euler as mathematician in residence at Frederick the Great's Berlin Academy, his tenure lasting twenty years. Among other things, he is credited with conceiving the calculus of variations.

Pierre-Simon Laplace (1749–1827) is sometimes called the Newton of France. According to legend, Napoleon observed that Laplace had not once mentioned the Author of the Universe in his treatise *Méchanique Céleste*, whereupon Laplace averred, "Sire, I had no need for that hypothesis," whereafter Lagrange said that it was a pity, as that hypothesis "explains so many things" [**118**, p. 305].

Henri Lebesgue (1875–1941) extended the notion of length in the hopes of being able to measure the length of any set of real numbers, and thereby extended the notions of area, volume, and anything involving measure.

Georges-Louis Leclerc, Comte de Buffon (1707–1788), a mathematician and naturalist, was a friend to both Maupertuis and Voltaire. Charles Darwin (1721–1764) credits him with being among the first to advance the idea of natural selection. Many calculus students are familiar with Buffon's riddle of the falling needle: Given a floor of parallel strips of wood, each of the same width w, allow a needle of length L to fall onto the floor, with $L \leq w$. What is the probability that the needle lies across a boundary line between two strips? The answer is $2L/(w\pi)$.

Gottfried Wilhelm Leibniz (1646–1716), along with Newton, was the co-discoverer of calculus. His over-arching philosophy has often been summarized as, "We live in the best of all possible worlds." Some scholars feel that Voltaire's Dr. Pangloss character in *Candide* is a caricature of Leibniz, while others feel that Pangloss caricatures Maupertuis.

John Locke (1632–1704) became in 1667 the personal physician of the politician Lord Ashley, Earl of Shaftesbury. From extended physician-patient dialogue, Locke speculated upon the social contract between the governed and the governors. His idea of the natural rights of man became the cornerstone of Thomas Jefferson's *Declaration of Independence*.

Louis XV (1710–1774) commissioned two expeditions, sending one to the equator and the other to the arctic circle, to test Newton's claim that the earth was flattened at the poles. Louis XV has the dubious honor of having had the most famous of all royal mistresses: the Marquise de Pompadour (1721–1764), a woman of legendary charm and grace. Voltaire was a favorite under her tenure. Mainly upon her recommendation, Louis appointed Voltaire as the Royal Historiographer as well as allowing Voltaire a seat in the French Academy.

Lucretius (c. 99–c. 55 BC) is a Roman poet whose major work explores a model for the structure of the universe. It is said that he went mad upon taking a love potion, and thereafter committed suicide.

Pierre-Louis Moreau de Maupertuis [pronounced "moh-pair-twee"] (1698–1759) was the leader of the arctic expedition of 1736–37 which measured a degree of arc along a line of longitude in Lapland, so verifying Newton's prediction about a flattened earth. As some of today's calculus students might agree, Maupertuis once categorized infinite series as "the most disagreeable thing in mathematics" [**140**, p. 43].

Julien Offray de La Mettrie (1709–1751) was the personal physician to Frederick the Great. He championed the pleasures of the senses, and concluded that man was a machine. It is said that in treating himself of severe indigestion, he mistakenly bled himself to death. Of his stature, Frederick the Great wrote, La Mettrie "was merry, a good devil, a good doctor, and a very bad author. By not reading his books, one can be very content" [**21**, pp. 74–75].

Micromégas is a giant from Sirius and a Frederick the Great figure in Voltaire's story *Micromégas*.

Napoleon (1769–1821) became First Consul in 1799, taking control of France after the disorder of the French Revolution. One year earlier, he led a military campaign to Egypt along with a team of 167 scientists and mathematicians. One of their discoveries was the Rosetta Stone.

Isaac Newton's (1643–1727) first major mathematical discovery was the generalized binomial theorem in 1665. In 1705, he was knighted by Queen Anne for his work as Warden of the Mint, not for his achievements in physics and mathematics. Newton was an alchemist and his experiments with chemicals may have resulted in mercury poisoning, which could account for a mental eccentricity late in life.

Pantagruel, the son of Gargantua, is a fictional giant introduced by Rabelais in 1532.

Paracelsus (1493–1541), originally named Phillip von Hohenheim, also called himself Philippus Theophrastus Aureolus Bombastus von Hohenheim. Paracelsus was a physician who focused on developing chemical medicines. He is known as the Father of Toxicology, and formulated the principle, "All things are poison; only the dose permits something not to be poisonous."

Blaise Pascal (1623–1662) was a French philosopher and mathematician. As a young man of nineteen, Pascal invented a mechanical calculator to add and subtract so as to help his father, who was tax assessor for the state, with his endless sums. Pascal may be the most quoted of all mathematicians. My favorite among his pensées is, "The heart has reasons of which the mind knows nothing."

Plato (c. 427–c. 347 BC) founded the Academy in Athens. He was a student of Socrates (c. 470–399 BC) and the teacher of Aristotle. He taught that the world around us is a shadow of reality, and that inquiry is the road to knowledge.

Pliny the Elder (23–79) loved to collect curious facts from the books which he had read to him at all hours of the day. He is perhaps the kindest literary critic the world has known, for he said [**106**, vol. i, p. xi],

No book is so bad but that some part of it has value.

He died from poisonous gases while investigating the volcanic eruption at Vesuvius that buried Pompeii.

Rabelais (1494–1553) was a Franciscan friar, then a Benedictine monk and a medical doctor. His writings involve the giants Gargantua and

Pantagruel, and their roguish, wine-drinking friends, who discuss ideas of utopian life.

Jean-Baptiste Rousseau (1671–1741) was a French poet whose libelous verses earned him permanent exile from Paris in 1712. The more famous Rousseau, Jean-Jacques (1712–1778), was a poet and philosopher. His *The Social Contract* (1762) begins, "Man is born free, and everywhere he is in chains."

Bertrand Arthur William Russell (1872–1970) teamed with Alfred North Whitehead (1861–1947) to write *Principia Mathematica* (1913) whose underlying premise was demolished by Gödel's work twenty years later. In 1950, Russell won the Nobel prize in literature.

Tobias Smollett (1721–1771) edited a collection of Voltaire's work, translated to English, in 38 volumes. He began his career as a naval surgeon, but turned to literature, writing satire in the tradition of Rabelais and Swift. One of his novels, *The History and Adventures of an Atom* (1768), is the somewhat coarse journal of an omniscient atom who has resided in the minds of various statesmen of the past, and was inspired by his readings of *Micromégas*. Instead of making his character the largest man, he made him the smallest.

Willebrord Snell (1580–1626) was a Dutch astronomer and mathematician. In 1617, he published his results, using triangulation, of measuring the distance of one degree along a line of longitude connecting the villages of Alkmaar and Bergen op Zoom. His measure in kilometers was 107.4 km versus the actual measure of 111 km. His technique was basically the same as used by the equatorial and the polar expeditions of the 1730s.

Solomon (c. 1000–c. 928 BC) was king of Israel, and is reputed to have been the wisest of all men. Here is is one of the many proverbs attributed to him (Proverbs 3:27):

> Withhold not good from them to whom it is due, when it is in the power of your hand to do so.

A. Square is the hero of *Flatland*, written by Edwin Abbott in 1884. A. Square is literally a square.

Jonathan Swift (1667–1745) was an Irish priest and the foremost of satirists in the English tongue. He often used the pen-name Lemuel Gulliver.

Voltaire (1694–1778) is the pen-name of François-Marie Arouet. He was a word wizard of the eighteenth-century. In a quote that typifies his style and philosophy [**144**, vol. v, *Angels*, p. 222]:

> There are persons who have resolved all questions; which once occasioned a man of sense and wit to say of a grave doctor, "That man must be very ignorant, for he answers every question that is asked him."

Comments on Selected Exercises

Chapter I

1. (a) In the passage referred to in footnote 8, Micromégas's nose should be 1/21 of his height, which gives 5714 feet, about 619 feet less than Voltaire's length for the nose.

   (c) In scene I, Voltaire uses the ratio 5 feet is to 9000 leagues as 120,000 feet is to x where x is the circumference of the Sirian's world. Solving for x gives 216,000,000 leagues.

   (d) We assume that the supergiant's foot is 8000 miles, the diameter of the earth. Since a man in Voltaire's day was about five feet tall, then G is about 40,000 miles tall. We take an ordinary man to be about 0.001 miles tall. Let x be the radius of G's world. Solving $0.001/4000 = 40000/x$ gives $x \approx 1600$ a.u. where 1 a.u. is about 100 million miles. No, the largest supergiant stars yet found have radii up to about seven a.u.

3. (c) Vignette II discusses some of these ideas.

4. (a) The distances of the earth, Mars, Jupiter, and Saturn from the sun are about 1, 1.524, 5.203, and 9.54 astronomical units respectively. To write these measurements in miles, multiply by 93 million. Remember that a league is about three miles.

   (b) See the case study in Chapter IV. Figure IV.6 shows the relative sizes of the planets' orbits along with an orbit similar to Halley's Comet, except moved into the ecliptic plane.

   (c) Since the periods of the earth, Mars, and Jupiter are 1, 1.88, and 11.86 years respectively, let $E(t) = 2\pi t$, $M(t) = 2\pi t/1.88$, and $J(t) = 2\pi t/11.86$ be the polar angle change of earth, Mars, and Jupiter since the travelers landed on Jupiter. A function that gives the sum of the magnitudes of the angle differences between the planet pairs earth and Mars, earth and Jupiter, and Mars and Jupiter is

$$f(t) = \mathrm{mod}(|E(t) - M(t)|, 2\pi) + \mathrm{mod}(|E(t) - J(t)|, 2\pi)$$
$$+ \mathrm{mod}(|M(t) - J(t)|, 2\pi).$$

We wish to find a positive time t in years when $f(t)$ is small. Graphing f over a 200 period year period gives Figure 9.

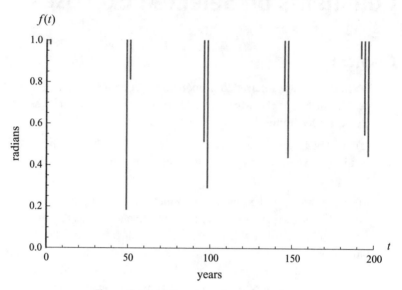

Figure 9. Propitious times for a comet fly-by

From the graph of Figure 9, it appears as if a good arrangement occurs once every 50 years. Zooming in on the graph gives a better value of 49.15 years. At 49.15 years, the alignments between any two of the three planets are off by about two degrees from the alignments at time 0, and the relative positions of the three planets are about 54° counterclockwise about the sun from what they were at time 0.

5. (a) The solution of $x^5 = 100$ is about 2.52.

(b) Pliny the Elder gives an early explanation [**106**, book ii, section lxiv, pp. 295–297]:

> The ceaseless revolution of the cosmos around the earth forces her immense globe into the shape of a sphere.

Descartes's theory involves a little more squeezing.

(c) The last section of Chapter II discusses some approaches to this problem.

Comments on Selected Exercises

6. (a) My favorite rendition of Aquinas's proofs: Love is so beautiful an idea that it must necessarily exist. But what is love? Ah, God is love.

7. (a) The main idea is that nature exhibits sameness at many different levels of being or perception. The section entitled *small, medium, and large* of Chapter II addresses this idea.

10. Chapter XII offers several answers.

(a) Below neck (Jacques de Molay)   (b) On shoulder (Odin sculpture)

Figure 10. Position of the mantle's buckle

Chapter II

1. (c) Figure 10 shows two possible placements for the mantle's buckle. Thirty spans plus 9.5 feet for the head is about 32 feet. Taking this value as half of Nimrod's height makes him about 64 feet tall.

   (d) By the hint in canto xxxiv, since Dante is one tenth the height of Nimrod, then Nimrod was less than a tenth the length of Satan's arm, which means that Satan's arm was over 500 feet long. Since a body is about three arm lengths long, then Dante's estimate for Satan's height is about 1500 feet.

2. (a) (2) Any union of open intervals that contains B also contains A.

   (4) Let $\epsilon > 0$. Let $\mathcal{O}$ be a union of open intervals containing $A \setminus B$ for which $m^*(\mathcal{O}) < m^*(A \setminus B) + \epsilon/2$. Let $\mathcal{U}$ be a union of open intervals containing B for which $m^*(\mathcal{U}) < \epsilon/2$. Then

$$m^*(A) \leq m(\mathcal{O} \cup \mathcal{U}) \leq m^*(\mathcal{O}) + m^*(\mathcal{U}) < m^*(A \setminus B) + \epsilon$$

Figure 11. Another forest fern

for every $\epsilon > 0$, which means that $m^*(A) \leq m^*(A \backslash B)$. The reverse inequality follows by (2).

(5) Let C be a set of real numbers and D be a set of outer measure 0. By (4) $m^*(C) = m^*(C \backslash D) = m^*(D' \cap C)$. By (2) $m^*(D \cap C) = 0$. Thus D is a measurable set, and so $m(D) = 0$.

(b) Let $L = (-\infty, a)$ be an interval where a is any real number. We show that L is a measurable set, a first step in showing that any interval is a measurable set. Let C be a set of real numbers with finite outer measure. Let ϵ be some small positive number. There exists a family of open intervals I_α whose union contains C for which

$$\sum \ell(I_\alpha) < m^*(C) + \epsilon.$$

Let $J_\alpha = I_\alpha \cap L$ and $K_\alpha = I_\alpha \cap (a, \infty)$. Then

$$\begin{aligned} m^*(C) &\leq m^*(C \cap L) + m^*(C \cap L') \\ &\leq m^*(C \cap L) + m^*(\{a\}) + m^*(C \cap (a, \infty)) \\ &\leq \sum \ell(J_\alpha) + 0 + \sum \ell(K_\alpha) \\ &= \sum \ell(I_\alpha) < m^*(C) + \epsilon, \end{aligned}$$

which means that $m^*(C) = m^*(C \cap L) + m^*(C \cap L')$, making L measurable.

Comments on Selected Exercises

3. Let F be the field of all rational functions $p(x)/q(x)$ with coefficients of p and q being integers evaluated at π, where q is not the zero function. Since the set of all polynomials with integer coefficients is *countable*, that is, the set can be made into a sequence, then so is the set of all rational functions with integer coefficients, which means that F is countable. And any countable set of real numbers has measure 0.

4. See reference [119].

5. See reference [120].

6. (b) $T_1(x, y) = 0.45(0, y)$, $T_2(x, y) = 0.6R(\pi/4).(x, y) + (0, 1)$, $T_3(x, y) = 0.6R(-\pi/4).(x, y) + (0, 1)$

7. (b) The new fern is shown in Figure 11.

9. (a) $r_m = \dfrac{7927}{2\cos\theta} \approx 243{,}300$ miles.

(b) $88°22'$ (assuming a diameter of 7927 miles).

(c) From part (a) we take the distance between the centers of the earth and moon as $(243300 + 7927/2)$. Therefore Hipparchus' guess for the diameter of the moon is $d_m = (243300 + 7927/2)\sin(.25\pi/180) \approx 2158$ miles. If we take Hipparchus' guesses for r_m as 280,000 miles and d_m as 2666 miles, we can find Hipparchus' guess for d_e.

10. (a) The best running time for a marathon is a little under 2 hours and 4 minutes. Running three consecutive marathons in nine hours is most impressive.

(b) At latitude $35°$ the radial distance from the earth's axis is about $R = 3300$ miles, which means that the local time differential between the endpoints of Philonides's run is about $80/(\omega R) \approx 6$ minutes, where $\omega = 2\pi/\text{day}$. That is, the apparent time to run west should be about 8 hours 54 minutes, and the apparent time running east should be about 9 hours 6 minutes. Perhaps Philonides took a nap on the way home.

(c) About $88.5°$ N.

Chapter III

1. We start this exercise by looking at specific dimensions for Flatland. Imagine that Flatland's baseline is 100 yards long and is aligned with the x-axis so that one endpoint is at O, the origin, and the other endpoint is at $(100, 0)$. By (III.13), $a(50, s) = -2\tan^{-1}(50/s)$ and the direction of

this acceleration is $(0, -1)$. In particular, the acceleration due to gravity in this model at $(50, 10)$ is $-2\tan^{-1}(5) \approx -2.7468$ yds/sec^2. What point on the extreme left side of Flatland has this same gravitational acceleration? To find this point $P = (0, s)$ for some $s > 0$, by the same reasoning as was used in deriving (III.13), we know that the downward acceleration a_1 on point P is

$$a_1 = -\int_0^{100} \frac{s}{x^2 + s^2}\, dx = -\tan^{-1}\frac{100}{s},$$

while the rightward acceleration A_2 on point P is

$$a_2 = \int_0^{100} \frac{x}{x^2 + s^2} = \ln\sqrt{\frac{100^2 + s^2}{s^2}}.$$

Solving $a_1^2 + a_2^2 = (\tan^{-1}(5))^2$ for s gives $s \approx 9.89$. That is, at $P \approx (0, 9.89)$ the acceleration of gravity is $(2.31, -1.47)$, the same magnitude as at $(0, 10)$, yet instead of straight down it is directed 32.5° south of east. Now repeat these calculations for a variety of points between 0 and 50, and so obtain a gravitherm. Figure 12 gives a few of these curves.

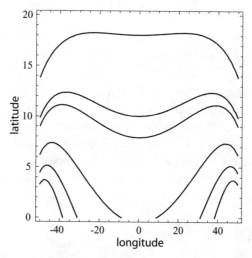

Figure 12. Some Flatland gravitherms

3. (a) Since distance can be defined so that the radius of a planet is unit length, then the acceleration $a(s)$ due to the planet's gravity on a point

Comments on Selected Exercises

$s > 1$ units above the center of the planet is

$$a(s) = 2 \int_0^1 \int_{-\sqrt{1-x^2}}^{\sqrt{1-x^2}} \left(\frac{y-s}{x^2 + (y-s)^2} \right) dy \, dx.$$

After the inner integration is performed,

$$a(s) = -\int_0^1 \ln(s^2 + 1 - 2s\sqrt{1-x^2}) - \ln(s^2 + 1 + 2s\sqrt{1-x^2}) \, dx.$$

Use integration by parts with $f(x) = \ln(s^2 + 1 - 2s\sqrt{1-x^2}) - \ln(s^2 + 1 + 2s\sqrt{1-x^2})$ and $g'(x) = 1$. Since $xf(x)]_0^1 = 0$, then

$$a(s) = -\int_0^1 \frac{2sx^2}{\sqrt{1-x^2}(s^2 + 1 - 2s\sqrt{1-x^2})}$$
$$+ \frac{2sx^2}{\sqrt{1-x^2}(s^2 + 1 + 2s\sqrt{1-x^2})} dx.$$

With the change of variable $\sin\theta = x$,

$$a(s) = -\int_0^{\frac{\pi}{2}} \frac{2s \sin^2 \theta}{s^2 + 1 - 2s\cos\theta} + \frac{2s \sin^2 \theta}{s^2 + 1 + 2s\cos\theta} d\theta.$$

With the change of variable $z = \sin(\theta/2)$, an old substitution trick more fully explained in Exercise IV.7,

$$a(s) = -\int_0^1 \frac{16sz^2}{(1+z^2)^2((s-1)^2 + (s+1)^2 z^2)}$$
$$+ \frac{16sz^2}{(1+z^2)^2((s+1)^2 + (s-1)^2 z^2)} dz.$$

After breaking the inner expression into partial fractions and simplifying, $a(s)$ is

$$-\frac{2(s^2+1)}{s} \int_0^1 \frac{1}{1+z^2} dz + \frac{(s+1)^2}{s} \int_0^1 \frac{1}{1 + \left(\frac{s+1}{s-1}z\right)^2} dz$$
$$+ \frac{(s-1)^2}{s} \int_0^1 \frac{1}{1 + \left(\frac{s-1}{s+1}z\right)^2} dz,$$

which is

$$a(s) = -\frac{1}{s}\left(2(s^2+1)\tan^{-1}(1) - (s^2-1)\right.$$
$$\left.\left(\tan^{-1}\left(\frac{s+1}{s-1}\right) + \tan^{-1}\left(\frac{s-1}{s+1}\right)\right)\right),$$

or

$$a(s) = -\frac{2}{s}\left(2(s^2+1)\frac{\pi}{4} - (s^2-1)\frac{\pi}{2}\right) = -\frac{2\pi}{s}.$$

That is, in Flatland, the gravitational acceleration induced by a planet on a particle above its surface varies inversely with the distance of the particle from the planet's center.

(b) Use (III.3): $v^2(s) = 2\int_h^s a\,dy + v_0^2$, where $a(y) = -k/y$ with k as some positive constant, h as the initial distance of the thrown ball, and v_0 as the initial upward speed of the ball. As s increases, $\int_h^s a\,dy$ decreases without bound. Therefore there must be a distance s where $v(s) = 0$, and thereafter the ball falls as it rose.

4. (a) See the section on Kepler's laws of planetary motion in Chapter IV.
(b) In four dimensions, (III.19) becomes

$$\frac{d^2z}{d\theta^2} + z = \frac{kz}{h^2},$$

which can be written as

$$\frac{d^2z}{d\theta^2} = \left(\frac{k}{h^2} - 1\right)z = \lambda z,$$

where $\lambda = k/h^2 - 1$. If $\lambda = 0$, then $z = A\theta + b$ for some constants A and b, which means that $r = 1/z$ is either a circle or its distance from the origin goes to zero with time. If $\lambda < 0$, we recognize the differential equation as the characterization of simple harmonic motion. In this case $z(\theta) = A\cos(\sqrt{-\lambda}\theta + \phi)$, for some constants A and ϕ, which means that $r = 1/z$ will go off to infinity. If $\lambda > 0$, the solution is $z = A\cosh(\sqrt{\lambda}\theta + \phi)$. Since the hyperbolic cosine increases without bound as $\theta \to \infty$, then $r = 1/z$ approaches zero with increasing time.

5. Solving $z = 1$ for a solution near $\theta = 4$ gives $\theta_c \approx 4.29503$. Thus $\lambda = \theta_c/(2\pi) \approx 0.683576$. The polar plot of $r(\lambda\theta)$ where $r = 1/z$,

with z being the solution of the differential equation of this exercise, is called the regularized orbit of planet P. Figure 13 shows the regularized orbit corresponding to Figure III.14; the dotted curve is the polar ellipse $r = 0.4182527/(1 - 0.5817472\cos\theta)$; the point O is the focus of the polar ellipse.

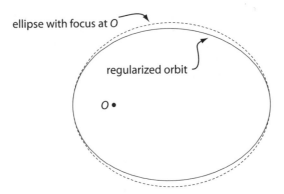

Figure 13. A polar ellipse versus the normalized orbit

6. Let $r(\theta)$ be the solution to the differential equation. To find the angle θ_y for which the polar plot of $r(\lambda\theta)$ is a maximum, let $y(\theta) = r(\lambda\theta)\sin\theta$ and solve for $y' = 0$, obtaining $\theta_y \approx 0.929855$. The semi-minor axial length is $b = y(\theta_y) \approx 0.476296$, and the semi-major axial length is $a = (1 + r(\lambda\pi))/2 \approx 0.632212$. The polar plot of the ellipse ($a\cos\theta + a - r(\lambda\pi)$, $b\sin\theta$) as a dashed curve is shown superimposed on the regularized orbit of Exercise 6 in Figure 14. They are close, but are not the same curve. Their differences are more noticeable when $k = 10$, for example.

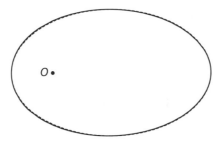

Figure 14. A translated ellipse versus the normalized orbit

7. The ray from the origin in the direction $(\cos\theta, \sin\theta)$ is $t(\cos\theta, \sin\theta)$, where t is a positive parameter. It intersects the ellipse at a unique point $(a\cos\phi + p, b\sin\phi)$ for some angle ϕ. Hence $(t\cos\theta - p)/a = \cos\phi$ and $(t\sin\theta)/b = \sin\phi$. Square both equations, then add them, and solve for t using the quadratic formula, obtaining the desired parametrization.

8. The polar plot of the ρ-precession of the ellipse is the dotted curve in Figure 15. Not a bad fit!

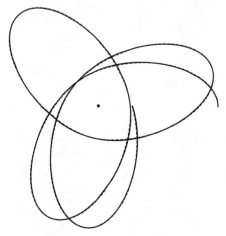

Figure 15. The preccessed approximation versus a Flatland orbit

9. As can be seen in the graph of Figure 16, the function $Z(\theta)$ is trying to solve the differential equation $z''(\theta) + z = 5/z$. The expression $ZZ'' + Z^2$ is tantalizingly close to a constant value of 5.

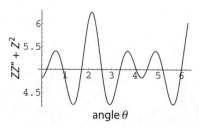

Figure 16. Trying to be 5

10. (a) For $n \neq 4$, try $a^{4-n} = k/h^2$. For $n = 4$, z being constant means that $k = h^2$.

(b) Suppose $z = a$ corresponds to a stable trajectory for a planet X. Imagine giving X a small nudge, so that its new trajectory corresponds to $z = a + w$ where w is some small variable change in z, which means that w^k is negligible for any integer $k > 1$. When $n = 4$, substituting $a + w$ into (III.19) in place of z, with $k = h^2$, gives $\frac{d^2w}{d\theta^2} = 0$, which in turn means that $w \approx c\theta$ (since w is assumed to be nonconstant), which means that w becomes large as θ increases, contradicting the assumption that w remains small. For $n > 4$, substituting $a + w$ into (III.19) in place of z gives, by part (a),

$$\frac{d^2w}{d\theta^2} + (a+w) = \frac{k}{h^2}(a+w)^{n-3} \approx a^{4-n}(a^{n-3} + (n-3)a^{n-4}w),$$

which simplifies to

$$\frac{d^2w}{d\theta^2} \approx (n-4)w.$$

When $n \geq 5$, this differential equation is hyperbolic, and again w becomes large as θ increases.

Chapter IV

1. Solve $T(\pi/2) = 150$ using (IV.13), and obtain $e \approx -0.532$.

2. Jupiter's orbit has eccentricty $e = 0.048$ and its period is 11.86 years. Change the function $T(\theta)$ in (IV.13), replacing 365 with 11.86. Then compute $T(5\pi/4) - T(3\pi/4)$ to get the maximum possible time lapse, namely 3.23 years.

3. (a) One astronomical unit is 1.496×10^8 km, and the sun's radius is 6.96×10^5 km. So the comet came within 235 thousand kilometers from the surface of the sun. To put this in context, Mercury is about 58 million kilometers from the sun's surface.

   (b) 889 AU.

4. (a) Solving $a - c = 0.586$ and $a + c = 35.1$ gives $a = 17.843$ AU and $c = 17.257$ AU, which means that $b = 4.535$ AU and $e \approx -0.967$. We assume that Jupiter's orbit is a circle with radius 5.203 AU. Thus

Galley's orbit is

$$r = \frac{b^2/a}{1 + e\cos\theta} \approx \frac{1.153}{1 - 0.967\cos\theta}.$$

Solving $r = 5.203$ gives $\theta \approx 0.635$ radians. With (IV.13) suitably modified (replace 365 days with 75.3 years), $T(0.635)$ gives $Q \approx 36.57$ years.

(b) One way to solve this open-ended problem is to set up a system of differential equations involving the sun, Jupiter, and the comet. For simplicity, assume that the orbit of Jupiter is a circle with center at the sun, and assume that Jupiter's influence on the comet extends only to the time when the comet is within 100 million miles of its orbit. At time $t = 0$ assume that the comet is 100 million miles out beyond Jupiter's orbit. From the information given you will be able to determine the comet's velocity at $t = 0$. You'll need to experiment with your system of equations so that at the time that the comet crosses Jupiter's path, Jupiter is ten million miles west of the comet. Once you have established that information, you will be able to find the time and velocity at which the comet is 100 million miles inside Jupiter's orbit, and you will be well on your way to solving this problem. For an example of the kind of system of differential equation you must set up, see [**122**, pp. 62–65].

5. With the model of the comet and the earth of Figure 17, with $\theta = 60°$ and D in millions of miles as the distance between the comet and the earth where the triangle, whose vertices are the earth, the comet head, and the tail end, is an isosceles triangle, we have $\cot(\theta/2) = D/(55/4)$ which means that the comet came within 24 million miles of earth.

6. Using Figure 17 with $\theta = 90°$, as in Exercise 5, gives a tail length of about 6.4 million miles as the comet passes earth. Comparing this value versus the result of Exercise 5 suggests that rather than assuming in Exercise 5 that the tail near earth should be half of the tail length at aphelion, it should be about one ninth.

7. (c) Since $z = \tan(\theta/2)$, then $\cos(\theta/2) = 1/\sqrt{1+z^2}$ and
$$\cos\theta = 2\cos^2(\theta/2) - 1 = 2/(1+z^2) - 1,$$
which in turn gives the third identity in part (c).

(d) To obtain the second identity of part (d), observe that $\sin\omega = \alpha z/\sqrt{1+\alpha^2 z^2}$ and $\cos\omega = 1/\sqrt{1+\alpha^2 z^2}$.

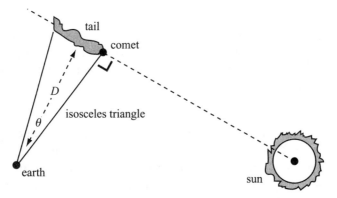

Figure 17. A comet's tail near aphelion

8. (a) To prove this identity let $\tan\alpha = c\tan\beta$ and $\chi = \alpha - \beta$. By the addition identity for the tangent function,

$$\frac{\tan\beta + \tan\chi}{1 - \tan\beta \tan\chi} = \tan(\beta + \chi) = c\tan\beta.$$

Solving for $\tan\chi$ gives

$$\tan\chi = \frac{(c-1)\sin\beta\cos\beta}{1 + (c-1)\sin^2\beta},$$

which gives (IV.17) since $\alpha = \beta + \chi$.

9. (a) With these perihelion and aphelion values, $a = 20.15$, $b = 3.4641$, $c = 19.85$, and $e = 0.985112$, which means that Halley's Comet trajectory can be taken as

$$r(\theta) = \frac{b^2/a}{1 + e\cos\theta} \approx \frac{0.5955}{1 + 0.9851\cos\theta}.$$

The angle at which this orbit crosses the orbits of Mars and Jupiter at $M = 1.524$ AU and $J = 5.203$ AU, respectively, are $\theta_m \approx 2.23755$ and $\theta_j \approx 2.6881$ radians. Even though $J - M \approx 3.679$ AU, the distance along the elliptical arc between the orbits of Jupiter and Mars is

$$\int_{\theta_m}^{\theta_j} \sqrt{r^2(w) + (r'(w))^2}\, d\theta \approx 3.87 \text{ AU}.$$

Chapter V

1. The readings should be about the same. However, as is pointed out in geophysics texts such as [81, pp. 49–50], gravity is 0.05186 m/sec² less

at the equator than at the poles. So Emily's weight at the equator is about half of one percent less than what she weighs at the poles; that is, she weighs 2.6 newtons less at the equator than at the north pole. The reasons for this anomaly are that the earth is not a homogeneous body and that the gravitational acceleration due to mass alone is not a radially symmetric function.

3. (a) Let the three vertices of a triangle be at the coordinate-pairs given by A, B, C. Show that the point $(A + B + C)/3$ is on each median. In particular, show that this point is on the line segment from A to $(B + C)/2$, the midpoint of the side opposite of A.

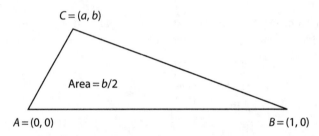

Figure 18. A triangle in a convenient general position

(b) Any non-trivial triangle can be rotated and translated so that an edge lies along the x-axis and the top-most vertex is between (and above) the other two vertices, as shown in Figure 18. Furthermore, we position the left-most vertex at the origin A, and adopt a length unit so that the right-most vertex is at $B = (1, 0)$. Let $C = (a, b)$ be the coordinates of the remaining vertex, which means that $0 \le a \le 1$ and $b > 0$. We analyze the case when $0 < a < 1$. By the center of mass entry from the appendix,

$$\bar{x} = \frac{\int_0^a x(bx/a)\,dx + \int_a^1 x(b(x-1)/(a-1))\,dx}{b/2} = \frac{a+1}{3},$$

which is the x-coordinate of $(A + B + C)/3$ as given in part (a).

(c) From part (a), the distance from A to $(A + B + C)/3$ is twice the distance from $(B+C)/2$ to $(A+B+C)/3$, because $(A+B+C)/3 - A$ is twice $(B + C)/2 - (A + B + C)/3$.

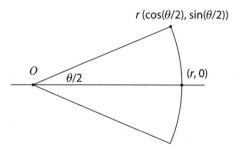

Figure 19. A sector of radial length r, central angle θ

(d) From the center of mass entry from the appendix, and with the aid of Figure 19,

$$\bar{x} = \frac{\int_0^{r\cos(\theta/2)} x(2\tan(\theta/2)\, x)\, dx + \int_{r\cos(\theta/2)}^{r} x(2\sqrt{r^2 - x^2})\, dx}{r^2 \theta/2},$$

which when simplified is

$$\bar{x} = \frac{2\sin(\theta/2)(r + r\cos\theta + 2r\sin^2(\theta/2))}{3\theta},$$

which approaches $2r/3$ as $\theta \to 0$.

5. When $a = b$, the result is immediate. So assume that $b > a$, in which case the integral can be written as

$$\int \frac{\cos\phi}{(a + (b-a)\sin^2\phi)^{3/2}}\, d\phi.$$

With $z = \sin\theta$, the integral can be rewritten as

$$a^{-3/2} \int \frac{1}{\left(1 + \left(\sqrt{\frac{b-a}{a}}\, z\right)^2\right)^{3/2}}\, dz.$$

With $\tan\omega = \sqrt{(b-a)/a}\, z$, the integral can be rewritten as

$$\frac{1}{a\sqrt{b-a}} \int \cos\omega\, d\omega = \frac{\sin\phi}{a\sqrt{a\cos^2\phi + b\sin^2\phi}}.$$

6. (a) Let $x = r(\theta)\cos\theta$ and $y = r(\theta)\sin\theta$. Then $\dfrac{x^2}{a^2} + \dfrac{y^2}{b^2}$ simplifies to 1, which is the standard equation of an ellipse.

(b) The area of a quarter ellipse is $\dfrac{\pi ab}{4}$. Since

$$\int_0^a bx\sqrt{1 - \dfrac{x^2}{a^2}}\, dx = \dfrac{a^2 b}{3},$$

then the x-coordinate of the centroid is $\dfrac{4a}{3\pi}$, which by symmetry means that the y-coordinate of the centroid is $\dfrac{4b}{3\pi}$.

7. Newton gives ratios of the pole diameter to the equatorial diameter of Jupiter as ranging from 11 to 12 and from 13.5 to 14.5. The polar radius of Jupiter is 66,854 km and the equatorial radius is 71,492 km.

10. (a) Consider this thought experiment: Extend the radius of the earth to r and slow earth's rotation so that its period matches T and so that the satellite is lightly resting on earth's surface at its equator. As such the gravitational acceleration exactly balances centripetal acceleration. That is,

$$\dfrac{gR^2}{r^2} = r\left(\dfrac{2\pi}{T}\right)^2,$$

which gives

$$T^2 = \dfrac{4\pi^2}{gR^2} r^3,$$

which is (IV.8).

(b) Assume that the moon's period is two weeks. Then Kepler's third law gives 246,000 km.

(c) 42,340 km

Chapter VI

1. The volume V of the earth is $V \approx 1.0834 \times 10^{21}$ m^3. So $\alpha = (V - 4\pi R^3/3)/V$.

Comments on Selected Exercises   349

2. The plane of the sun's path has equation $z = y\tan\theta$ and the cone obtained by rotating the ray OS about the z-axis has equation $\tan\phi = z/\sqrt{x^2+y^2}$. Arbitrarily take the y-coordinate of S as $y = 1$, so its z-coordinate is $z = \tan\theta$ and its x-coordinate satisfies $r = \sqrt{x^2+1} = \tan\theta/\tan\phi$, where r is the length of $(x, y, 0)$. Since $\sin\delta = y/r$, then $\sin\delta = 1/r = \tan\phi/\tan\theta$, giving the desired result.

3. Using the identity $2\sin^2\alpha = 1 - \cos(2\alpha)$ and the substitution $\theta = 2\alpha$ gives

$$\int_0^{2\pi} \frac{\sin^2\alpha}{1+a^2\sin^2\alpha}\,d\alpha = 4\int_0^{\pi/2} \frac{\sin^2\alpha}{1+a^2\sin^2\alpha}\,d\alpha$$
$$= 2\int_0^{\pi} \frac{1-\cos\theta}{a^2+2-a^2\cos\theta}\,d\theta.$$

Using the substitution $z = \tan(\theta/2)$, a trick explained in Exercise IV.7.a, transforms this integral into

$$4\int_0^{\infty} \frac{z^2}{(1+(1+a^2)z^2)(1+z^2)}\,dz$$
$$= \frac{4}{a^2}\int_0^{\infty} \frac{1}{1+z^2} - \frac{1}{1+(1+a^2)z^2}\,dz,$$

which equals $\dfrac{2\pi(1+a^2-\sqrt{1+a^2})}{a^2(1+a^2)}$.

4. Since $\tan\phi = \tan\theta\sin\delta$, then $\sin\phi = \tan\theta\sin\delta/\sqrt{\tan^2\theta\sin^2\delta+1}$, as illustrated in Figure 20. Thus the average value of $\sin^2\phi$ over the interval $[0, 2\pi]$ is

$$\frac{1}{2\pi}\int_0^{2\pi}\sin^2\phi\,d\delta = \frac{\tan^2\theta}{2\pi}\int_0^{2\pi}\frac{\sin^2\delta}{\tan^2\theta\sin^2\delta+1}\,d\delta$$
$$= \frac{1+\tan^2\theta-\sqrt{1+\tan^2\theta}}{1+\tan^2\theta},$$

which simplifies to $1-\cos\theta$.

5. (a) See the example in Item 4 in the appendix on rotational inertia.

6. (b) The change would be minor. Both the moon's motion and the sun's motion with respect to the earth are minor factors for the period of earth's pole; the major factor is the earth's spin about its axis.

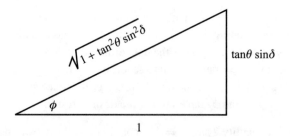

Figure 20. A convenient right triangle.

7. To solve the problem, with $e = 0.5$, update the functions t and T. Solve $T(\theta) = -31$, getting -1.44 radians, which means that 23 years ago perihelion and the winter solstice coincided. Next, define $\Omega(\tau) = -2\pi(\tau+23)/100$, and compute $T(\Omega(25))$, getting day -164 or July 20 as the winter solstice. Spring is $T(\Omega(25) + \pi/2) \approx -31$. That is, the spring equinox occurs on about December 1 in year 25.

Chapter VII

1. (a) $\theta = 2\sin^{-1}\left(\dfrac{\sin\left(\frac{\pi}{1-\lambda}\right) + \frac{1}{\lambda}\sin\left(\frac{\pi}{1-\lambda}\right)}{1 - 1/\lambda}\right)$.

2. (a) $m + n$,

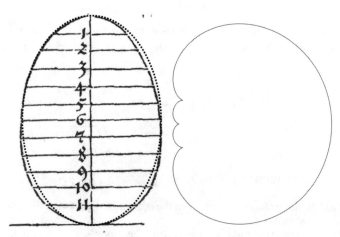

(a) An egg versus an ellipse (dotted curve)   (b) A simple Dürer curve

Figure 21. A pair of Dürer curves

(b) One arch of the hypocycloid has length $\dfrac{8}{1+\lambda}$, whereas one arc of the epicycloid has length $\dfrac{8}{|1-\lambda|}$.

3. See Figure 21(a).

4. (b) Figure 21(b) shows the graph of $4(\cos\theta, \sin\theta) + 3(\cos(2\theta), \sin(2\theta)) + 2(\cos(3\theta), \sin(3\theta)) + (\cos(4\theta), \sin(4\theta))$.

6. See Figure 22(a).

8. The match is surprisingly close for the upper loop of the curve as shown in Figure 22(b).

(a) A limaçon on Dürer's (b) A 4-cusped hypocycloid on Caxton's

Figure 22. Curves superimposed on woodcuts

10. (a) $\beta(t - \sin t,\ 1 - \cos t)$.

Chapter VIII

1. (a) If $\rho = dw/du$, then

$$\frac{d^2 w}{du^2} = \frac{d\rho}{du} = \frac{d\rho}{dw}\frac{dw}{du} = \rho\,\frac{d\rho}{dw}.$$

2. (b) Since $r^2 \dfrac{d\theta}{dt} = h$, the change in area ΔA per change in time Δt is given by

$$\Delta A = \int_t^{t+\Delta t} \frac{r^2}{2}\frac{d\theta}{d\tau}\,d\tau = \int_t^{t+\Delta t} \frac{h}{2}\,d\tau = \frac{h}{2}\Delta t. \tag{35}$$

(c) With $x' = \frac{dx}{dt}$ and $y' = \frac{dy}{dt}$, $\frac{s^2}{x^2}\frac{dv}{dt} = \sec^2 v \frac{dv}{dt} = \frac{xy' - yx'}{x^2}$.
Therefore

$$s^2 \, dv = (xy' - yx') \, dt$$
$$= \left(R\cos(\sqrt{k}\,t) \frac{h}{R\sqrt{k}} \sqrt{k}\cos(\sqrt{k}\,t) - \frac{h}{R\sqrt{k}} \sin(\sqrt{k}\,t) \right.$$
$$\left. (-R\sqrt{k})\sin(\sqrt{k}\,t) \right) dt$$
$$= h \, dt.$$

(d) $\Delta \mathcal{A} = \frac{1}{2}\int_t^{t+\Delta t} s^2 \frac{dv}{d\tau}\, d\tau = \frac{1}{2}\int_t^{t+\Delta t} h \, d\tau = \frac{h}{2}\Delta t$.

(e) By (b) and (d), from time 0 to t, the curves of each parametrization both sweep out area $ht/2$ and since both parametrizations start at the initial condition $(R, 0)$ and proceed in the counterclockwise direction, the arclengths generated by them over this time period are the same.

3. (a) Since $\Theta(t)$ is the area under the always positive, periodic function ρ over the interval from 0 to t, $\Theta(t)$ increases without bound.

(b) From Figure 23a, for any t with $\pi/(2\omega) < t < \pi/\omega$, the area under the curve from $\pi/(2\omega)$ to t is equal to the area from $\pi/\omega - t$ to $\pi/(2\omega)$. Therefore the area from 0 to t is the area from 0 to π/ω minus the area from 0 to $\pi/\omega - t$, which gives the desired result.

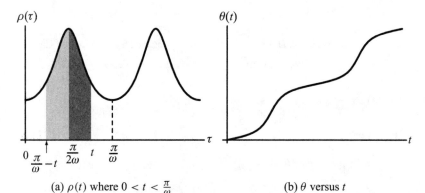

(a) $\rho(t)$ where $0 < t < \frac{\pi}{\omega}$ (b) θ versus t

Figure 23. The relation between θ and t

(e) Let $\chi = \alpha - \beta$ where $\tan \alpha = c \tan \beta$. By the addition identity for the tangent function,

$$\frac{\tan \beta + \tan \chi}{1 - \tan \beta \tan \chi} = \tan(\beta + \chi) = c \tan \beta.$$

Solving for tan χ gives

$$\tan \chi = \frac{(c-1)\sin\beta\cos\beta}{1+(c-1)\sin^2\beta},$$

which means that

$$\alpha = \beta + \tan^{-1}\left(\frac{(c-1)\sin\beta\cos\beta}{1+(c-1)\sin^2\beta}\right).$$

(f) The inverse tangent term of (e) is bounded and periodic because its argument is periodic, continuous and has positive denominator. The graph of θ versus time is shown in Figure 23(b).

5. (a) $\mathcal{H}(1/3, 0, t)$.

 (b) Try $\alpha = 1/b$ when $b > 0$.

 (c) $3\mathcal{T}(1, 2, 2/3, 1/3, 2t/3)$
 $= (2\cos(2t/3) + \cos(4t/3), \, 2\sin(2t/3) + \sin(4t/3))$,
 which parametrizes the same curve as
 $(2\cos(2t) + \cos(4t), \, 2\sin(2t) + \sin(4t))$
 $= (2\cos(3t-t) + \cos(t+3t), \, 2\sin(3t-t) + \sin(t+3t))$
 $= (3\cos t \cos(3t) + \sin t \sin(3t), \, -3\sin t \cos(3t) + \sin(3t)\cos t)$,
 which parametrizes the same curve as
 $3(\cos(t/3)\cos t + \frac{1}{3}\sin(t/3)\sin t, \, -\sin(t/3)\cos t + \frac{1}{3}\cos(t/3)\sin(t))$
 $= 3\mathcal{H}(1/3, 1/3, t)$.

6. (a) Show that the first two components of Γ are of the form $a\mathcal{H}(p, p, q\,t)$ for some numbers a, p, and q.

10. Here are a few items: organ regeneration/farming, artificial intelligence, manned exploration of Mars, genome projects, subatomic structure with the CERN particle accelerator.

Chapter IX

1. Assume that the time lapse between strides for both dog and hare are the same. Let x be the number of strides taken by each beast. Solving $9x = 150 + 7x$ gives $x = 75$ strides for the capture to occur. Thus the dog runs 675 feet before overtaking the hare.

2. $y = \dfrac{x^2}{4c} - \dfrac{c\ln x}{2}$.

3. (a) $6\frac{2}{3}$ seconds; (b) 0.657 seconds.

6. Modify procedure swan so that $M(t) = \cos(t + \pi/6, \sin(t + \pi/6)$. Set $R = 1$ and $r = 0.01$ (where r represents the maximum distance between the spider and the fly for which it can be said that the spider has captured the fly). Experiment with various values of c so that the spider and fly rendezvous at $(-1, 0)$, getting $c \approx 1.156$, which means that the spider moves 1.156 as fast as the fly. Figure 24 shows the corresponding pursuit curve.

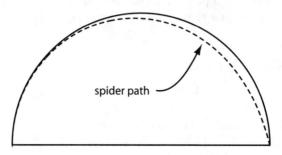

Figure 24. Spider-fly rendezvous at the left-hand point

10. Let the first snail (out of n snails) start at $(1, 0)$. By symmetry, the trail of the ith snail will be a copy of the trail left by the first snail, rotated counterclockwise by $\omega = 2\pi i/n$. The matrix which when multiplied to a 2-dimensional vector rotates the vector by ω radians counterclockwise is

$$M(\omega) = \begin{bmatrix} \cos\omega & -\sin\omega \\ \sin\omega & \cos\omega \end{bmatrix}.$$

We adapt procedure swan to draw the pursuit curves of the n snails, calling it procedure snail. We set $r = 0.01$, so that the procedure halts when the snails get within r units of each other.

```
snail[n_, ε_]:=Block[{x, X, δ, ω, m, i, bag, sack, flag, purse},
    x = {1, 0}; bag = {x}; flag = True; ω = 2π/n; m = 0;
    purse = {};
    While[flag, X = M[ω].x - x; δ = √X.X;
        If[(δ <r) || (m> 300), flag = False]; m = m + 1;
```

Comments on Selected Exercises

Figure 25. Snail-trails with $n = 7$

```
x = x + X ε/δ; bag = Append[bag, x]];
Do[sack[i]=ListPlot[Transpose[M[(i-1)ω].Transpose[bag]],
  PlotJoined → True, AspectRatio → Automatic,
  DisplayFunction → Identity];
  purse = Append[purse, sack[i]], {i, 0, n-1}];
Show[purse, DisplayFunction → $DisplayFunction, Axes → None]];
```

Figure 25 is the result of executing snail[7, 0.01].

Chapter X

2. (b) 1800 and 1916 where $q(t) = -9.32987 + 0.0134359t - 3.61518 \times 10^{-6}t^2$.

3. Define a kilo-cubit as one thousand cubits, abbreviated as kc. The area difference is $4 - \pi$ kc^2, which is over 200,000 square meters—that's a lot of vegetables.

4. Yes, it is a metric. We prove the triangle property. Let $Z = (z_1, z_2)$. We assume that $|a - b| \le |a - c| + |c - b|$ for any three real numbers $a, b,$ and c. Then

$$D(X, Y) \le \max\{|x_1 - z_1| + |z_1 - y_1|, |x_2 - z_2| + |z_2 - y_2|\}.$$

Since $|x_i - z_i| + |z_i - y_i| \leq D(X, Z) + D(Z, Y)$ for $i = 1$ and $i = 2$, then $D(X, Y) \leq D(X, Z) + D(Z, Y)$ for any Z.

8. $p \approx 3.65$.

Chapter XI

2. (a) Use (II.3) and (II.4) to obtain an estimate using the simple model of a stationary earth. The velocity of the coconut at distance s from the earth is

$$v(s) = \sqrt{-2gR^2 \int_R^s \frac{1}{y^2}\,dy + v_0^2} = \sqrt{2gR^2\left(\frac{1}{s} - \frac{1}{R}\right) + v_0^2}.$$

Experiment with different guesses for v_0 so that $\int_R^Q 1/v\,ds = 2$ days. Thus v_0 should be close to 11.195 km/sec. To obtain a better estimate, use a model involving a system of differential equations such as the case study of Jules Verne's shot to the moon in [**122**, pp. 61–65].

(b) The volume of rope (uncompressed) is 0.00375 km^3. Rather than using the density of 1 gram per meter, we assume that the density of Pantagruel's rope is half that of water, or about 0.5 grams per cubic centimeter. In this case the rope's total mass is about 1.875 billion kilograms or a weight on earth of about two million tons. For comparison, the weight of a fully loaded Saturn V moon rocket was 3500 tons. It is unlikely that Pantagruel's harpoon throw can lift all of this rope into space. On the other hand, if Pantagruel is fully 2 km tall, then the volume of his body is about 0.2 km^3, so perhaps he can heave the harpoon and its rope as portrayed in the story.

3. Here is one way to get close to Pliny's result. Assimilate the ideas of Pythagoras and Aristarchus to guess that the distance from the sun to the zodiac is $D + 19D = 20D$ rather than $D + 2D = 3D$. Then $R/(40D) \approx 1/93.3$. Can you get closer to 1/96?

4. (a) We assume the mountain is a right circular cone with radius of the base $r = 7.5$ km and height $h = 5$ km. We wind the rope around the mountain so that its coils form a coat of thickness s on the surface of the mountain, somewhat like snow on its slopes. The change in the radius of the cone Δr and the change in the height Δh of the cone when wearing its rope coat is $\Delta r \approx 1.8s$ and $\Delta h \approx 1.2s$, and the change in

the mountain's volume is $\Delta V \approx 0.00375$ km^3 by Exercise 1. Since the volume of a cone is $V = \frac{1}{3}\pi r^2 h$ and

$$\Delta V \approx \frac{\partial V}{\partial r}\Delta r + \frac{\partial V}{\partial h}\Delta h = \frac{2\pi r h}{3}\Delta r + \frac{\pi r^2}{3}\Delta h \approx (1.2\pi r h + 0.4\pi r^2)s,$$

then $s \approx 1.6$ cm, which means that the rope coat about Spindle Mountain is no more than one coil thick.

(b) The human scalp has about 100,000 hairs. Taking shoulder-length hair of about one foot gives a length of 100,000 feet, or almost 20 miles. Assuming Pantagruel's height as one mile, about 1000 times the height of a man, gives the total length of a head of his hair of 20,000 miles, less than 10% of the distance to the moon.

5. (b) The phenomenon of the harpoon following the sun was almost immediate. In a vortex model, the vortex effect near the earth will be marginal; if a vortex model was accurate, the phenomenon of the harpoon following the sun would have been gradual.

7. The lack of air. The rope could break in multiple places. Pantagruel could lose his grip on the rope.

8. Let us suppose that we can see a bright red disk of radius 1 cm at a distance of 100 meters. Similarly then, a bright red disk of radius 25 miles could be seen on the moon at a distance of 250,000 miles.

9. (a) For simplicity we assume that the speed of the harpoon when it struck Oahu Island is less than the tangential speed of the earth at Oahu. Since Oahu is at about latitude 21.5° N, then its distance from the polar axis is $r \approx 5955$ km. Assuming that the rope wraps around this line of latitude means that the chest at the end of 375,000 km of rope lands 836 km west of Oahu, which is about 8 degrees west of the island.

(b) For simplicity we assume that the rope is coiled about the line of latitude 25° N, 5800 km away from the earth's pole. The crew finds 10 coils of rope near the Canary Islands. Reasoning as we did in Exercise 9 means that the harpoon struck the earth at longitude 64° W near the Virgin Islands, and the treasure chest landed at longitude 168° W, just east of the Midway Islands.

10. (b) In each of the stories, the protagonists more or less get lost and encounter strange peoples as they work their way home.

Chapter XII

2. (a) 42.

   (b) Let there be light.

3. (a) Below is the first stanza of Carroll's poem in the original English.

   > 'Twas brillig, and the slithy toves
   > Did gyre and gimble in the wabe:
   > All mimsy were the borogoves,
   > And the mome raths outgrabe.

5. See footnote 41 of Chapter I.

References

1. Edwin A. Abbott, *The Annotated Flatland: a romance of many dimensions*, annotated by Ian Stewart, Perseus, Cambridge, MA, 2002.
2. Douglas Adams, *The Ultimate Hitchhiker's Guide*, Random House, 1996.
3. C. L. Adler and J. Tanton, π is the minimum value for pi, *College Math. J.*, 31 (2000) 102–106.
4. Martin Aigner and Günter M. Ziegler, *Proofs from the Book*, Springer, 1998.
5. Robert N. Andersen, Justin Stumpf, and Julie Tiller, Let π be 3, *Math. Mag.*, 76 (2003) 225–231.
6. Ken Alder, *The Measure of All Things*, The Free Press, New York, 2002.
7. Hans Christian Andersen, *The Snow Queen and other Stories from Hans Andersen*, illustrated by Edmund Dulac, Hodder & Stoughton, London, 1911.
8. Anonymous, *Reynard the Fox,* Dana Estes, 1901.
9. R. C. Archibald and H. P. Manning, Remarks and historical notes [on the pursuit problem], *Amer. Math. Monthly*, 28 (1921) 91–93.
10. Aristotle, *On the Heavens*, translated by W.K.C. Guthrie, Harvard University Press, Cambridge, 1958.
11. Isaac Asimov, The last question, *Science Fiction Quarterly*, November (1956) 7–15.
12. Clara Sue Ball, The early history of the compound microscope, *Bios*, 37 (1966) 51–60.
13. Michael Barnsley, *Fractals Everywhere*, Academic Press, 1988.
14. G. A. Barton, *Archaeology and the Bible*, American Sunday-School Union, 1913.
15. P. Beckmann, *A History of π*, Golem, 1971.
16. David Beeson, *Maupertuis: an intellectual biography*, The Voltaire Foundation, Oxford, 1992.
17. E. T. Bell, *Men of Mathematics*, Simon & Schuster, New York, 1965.
18. Bernadette Brady, Four Galilean horoscopes: an analysis of Galileo's astrological techniques, *Culture and Cosmos*, 7 (1997) 113–144.

19. M. Born and E. Wolf, *Principles of Optics*, Pergamon Press, Oxford, 1965.
20. Dionys Burger, *Sphereland*, Apollo Editions, New York, 1965.
21. Thomas Carlyle, *History of Friedrich II of Prussia*, vol. xvi, Project Gutenberg EBook, 2008.
22. D. Castellanos, The ubiquitous π, part I, *Math. Mag.*, 61 (1988) 67–98.
23. William Caxton, *Mirrour of the World*, edited by Oliver H. Prior, Oxford University Press, 1913.
24. S. Chandrasekhar, *Ellipsoidal Figures of Equilibrium*, Yale University Press, 1969.
25. R. Chasse, How to make π equal to three, (the sequel), *Amer. Math. Monthly*, 99 (1992) 751.
26. Peter Colwell, *Solving Kepler's Equation over Three Centuries*, Willmann-Bell, 1993.
27. James A. Connor, *Kepler's Witch*, Harper, 2004.
28. Dante, *The Divine Comedy I: Hell*, translated by Dorothy Sayers, Penguin, 1949.
29. Philip J. Davis, Leonhard Euler's integral: a historical profile of the gamma function, *Amer. Math. Monthly*, 66 (1959) 849–869.
30. Michael A. B. Deakin and Hans Lausch, The Bible and π, *Math. Gazette*, 82 (1998) 162–166.
31. Cyrano de Bergerac, *Voyages to the Moon and the Sun*, translated by R. Aldington, Orion Press, 1962.
32. Bernard le Bovier de Fontenelle, *Conversations on the Plurality of Worlds*, translated by H. A. Hargreaves, University of California Press, 1990.
33. Antoine de Saint-Exupéry, *The Little Prince*, translated by Katherine Woods, Harcourt, Brace & World, New York, 1943.
34. A.K. Dewdney, *The Planiverse: computer contact with a two-dimensional world*, Copernicus, 2001.
35. Underwood Dudley, π_t: 1832–1879, *Math. Mag.*, 35 (1962) 153–154.
36. William Dunham, *Euler: The Master of Us All*, Mathematical Association of America, 1999.
37. ———, personal e-mail communication, November 28, 2008.
38. Albrecht Dürer, *The Complete WoodCuts of Albrecht Dürer*, edited by Willi Kurth, Crown Publishers, New York, 1941.
39. ———, *Underwesung der Messung (The Art of Measurement with Compass and Straightedge)*, Nuremberg, 1538; an English translation: Octovo Press, 2003.
40. Gerald A. Edgar, *Measure, Topology, and Fractal Geometry*, Springer-Verlag, 1990.

41. Espenak, Fred. 2006. Twelve year planetary ephemeris: planetary phenomenon. *NASA Reference Publication 1349*. Greenbelt, Maryland: Goddard Space Flight Center. http://sunearth.gsfc.nasa.gov/eclipse

42. Leonhard Euler, *Letters to a German Princess*, vol. i, Harper, 1837.

43. Russell Euler and Jawad Sadek, The π's go full circle, *Math. Mag.*, 72 (1999) 59–63.

44. W. M. Feldman, *Rabbinical Mathematics and Astronomy*, Hermon Press 1978.

45. Paula Findlen, editor, *Athanasius Kircher: the last man who knew everything*, Routledge, 2004.

46. Albert Frey, *Sobriquets and Nicknames*, Ticknor Press, 1888.

47. Galileo Galilei, *Dialogue Concerning the Two Chief World Systems—Ptolemaic & Copernican*, translated by Stillman Drake, University of California Press, 1967.

48. ———, *The Assayer* in *Discoveries and Opinions of Galileo*, edited by Stillman Drake, Anchor, 1957.

49. ———, *Le Opere di Galileo Galilei*, Nazionale Edizione, vol. ix, Barbèra, 1968.

50. Martin Gardner, Do Loops Explain Consciousness? Review of *I am a Strange Loop*, *Notices of the American Mathematical Society*, 54 (2007) 852–854.

51. Owen Gingerich, *The Eye of Heaven: Ptolemy, Copernicus, Kepler*, The American Institute of Physics, 1993.

52. James Gleick, *Chaos: Making a New Science*, Penguin, 1988.

53. Ludwig Goldscheider, *Leonardo da Vinci: the Artist*, Phaidon Press, 1948.

54. G. P. Gooch, *Madame du Châtelet and her lovers*, *The Contemporary Review*, 20 (1961) 648–653.

55. S. H. Gould, Gulliver and the moons of Mars, *Journal of the History of Ideas*, 6 (1945) 91–101.

56. A. Gray, *A treatise on gyrostatics and rotational motion*, Dover, 1959, a reprint of the 1918 edition.

57. John L. Greenberg, *The Problem of the Earth's Shape from Newton to Clairaut*, Cambridge University Press, 1995.

58. R. C. Gupta, On the values of π from the Bible, *Ganita-Bharati, Bulletin of the Indian Society for the History of Math.*, 10 (1988) 51–58.

59. J. B. S. Haldane, *Possible Worlds and Other Essays*, Chatto and Windus, 1930.

60. Thomas L. Hankins, *Jean D'Alembert: Science and the Enlightenment*, Taylor & Francis, 1990.

61. Edward R. Harrington, *Masks of the Universe: Changing Ideas on the Nature of the Cosmos*, Cambridge University Press, 2003.

62. A. S. Hathaway, Solution of problem 2801 by the proposer, *Amer. Math. Monthly*, 28 (1921) 93–97.

63. Stephen Hawking, *A Brief History of Time*, Bantam, New York, 1990.

64. ———, Gödel and the end of physics, Cambridge-MIT Institute Distinguished Lecture series given on 23 January 2003 at the Judge Institute, Cambridge University.

65. Roger Herz-Fischler, Dürer's paradox or why an ellipse is not egg-shaped, *Math. Mag.*, 63 (1990) 75–85.

66. F. Herzog, On the biblical value of π, *Centennial Review*, 18 (1974) 176–179.

67. Andreas K. and Alice K. Heyne, *Leonhard Euler: a man to be reckoned with*, translated by Alice K. Heyne and Tahu Matheson, Birkhäuser, 2007.

68. Michael Rand Hoare, *The Quest for the True Figure of the Earth*, Ashgate, 2005.

69. Douglas Hofstadter, *Gödel, Escher, Bach: an Eternal Golden Braid*, Basic Books, 1979.

70. Christiaan Huygens, *Cosmotheoros*, Timothy Childe, 1698.

71. Washington Irving, *Knickerbocker's History of New York*, (book i, chapter v), in *The Works of Washington Irving*, vol. iv, New York, Peter Fenelon Collier, 1897.

72. A. S. Kaufman, Determining the length of the medium cubit, *Palestine Exploration Quarterly*, 116 (1984) 120–132.

73. Anthony Kenny, *The Five Ways: Saint Thomas Aquinas' Proofs of God's Existence*, Routledge Press, 1991.

74. Morris Kline, *Mathematics: a cultural approach*, Addison-Wesley, 1962.

75. Gustave Lanson, *Voltaire*, translated by R. A. Wagoner, John Wiley, 1966.

76. C. S. Lewis, *On Stories and Other Essays on Literature*, edited by Walter Hooper, Harcourt Brace Jovanovich, 1982.

77. Jean B. Lasserre, A quick proof for the volume of n-balls, *Amer. Math. Monthly*, 108 (2001) 768–769.

78. D. Anthonii Lewenhoeck, Observationes de natis è semine genitali animalculis, *Phil. Trans.*, 12 (1677–78) 1040–1046.

79. ———, Part of a letter, dated June 9, 1699, concerning the animalcula in semine humano, *Phil. Trans.*, 21 (1699) 301–308.

80. Tung-Po Lin and R. C. Lyness, Fast tunnels through the earth, problem 5614, *Amer. Math. Monthly*, 76 (1969) 708–709.

81. William Lowrie, *Fundamentals of Geophysics*, Cambridge University Press, 1997.

82. Lucretius, *The Way Things Are*, translated by Rolfe Humphries, Indiana University Press, 1968.

83. Martin Luther, *Luther's Works*, 55 volumes, edited and translated by John W. Doberstein, Muhlenberg Press, Philadelphia, 1959.

84. Pierre de Maupertuis, *Vénus physique suivi de Lettre sur le progrès des sciences*, edited by Patrick Tort, Aubier-Montaigne, Paris, 1980.

85. L. A. McFadden, P. R. Weissman, R. V. Johnson, *Encyclopedia of the Solar System*, Academic Press, 2008.

86. J. Meeus, π and the Bible, *J. of Recreational Math.*, 13 (1980–81) 203.

87. Nancy Mitford, *Voltaire in Love*, Harper, New York, 1957.

88. Antonio Montes, et al., The perimeter of an oval: problem 10227, *Amer. Math. Monthly*, 101 (1994) 688–689.

89. F. V. Morley, A curve of pursuit, *Amer. Math. Monthly*, 28 (1921) 54–61.

90. John Morley, *Voltaire*, McMillan, London, 1913.

91. Shane Morrison and Jared Newton, *Floating Islands, High Voltage, and... Physics*, Senior honors paper, King College, 2007.

92. Paul J. Nahin, *When Least is Best*, Princeton University Press, 2004.

93. ———, *Chases and Escapes*, Princeton University Press, 2007.

94. Reviel Netz and William Noel, *The Archimedes Codex*, Da Capo Press, 2007.

95. Isaac Newton, *The Principia, Mathematical Principles of Natural Philosophy*, translated by I. Bernard Cohen and Anne Whitman, University of California Press, 1999.

96. Paul T. Nicholson, *Brilliant Things for Akhenaten: the production of glass, vitreous materials and pottery at Amarna Site 0.45.1*, Egypt Exploration Society, 2007.

97. R. Norwood, How to make π equal to three, *Amer. Math. Monthly*, 99 (1992) 111.

98. ———, More on π, *Amer. Math. Monthly*, 100 (1993) 577.

99. Réginal Outhier, *Journal d'un vorage au Nord, En 1736 & 1737*, Paris, Piget & Durand Press, 1744.

100. J. B. Payne, The validity of the numbers in *Chronicles*, part I, *Bibliotheca Sacra*, 136 (1979) 109–128.

101. Roger Pearson, *Voltaire Almighty*, Bloomsbury, 2005.

102. Roger Penrose, *The Emperor's New Mind*, Oxford University Press, 1999.

103. F. K. Pizor and T. A. Comp, eds., *The Man in the Moone: an Anthology of Antique Science Fiction and Fantasy*, Sidgwick and Jackson, 1971.

104. Henry Platanakis, e-mail communication to Joseph Fitsinakis, forwarded to Andrew Simoson, November 2, 2008.

105. Plato, *The Republic*, translated by H. Jowett, Random House, no date.

106. Pliny, *Natural History*, Harvard University Press, 1949.

107. Plutarch, *Plutarch's Moralia*, vol. xii, translated by H. Cherniss and W. C. Helmbold, Harvard University Press, Cambridge, MA, 1957.

108. J. S. Pickering, *1001 Answers to Questions about Astronomy*, New York, Grosset & Dunlap, 1958.

109. A. S. Posamentier and N. Gordon, An astounding revelation on the history of π, *Math. Teacher*, 77 (1984) 47, 52.

110. Rabelais, *The Works of Rabelais*, illustrated by Gustave Doré, privately printed, no date.

111. C. B. Read, Did the Hebrews use 3 as a value for π? *School Science and Math.*, 64 (1964) 765–766.

112. James Reston, *Galileo*, Beard Books, Washington, D. C., 2000.

113. Rudy Rucker, The gravitational force of large 3D and 4D slabs, *Mathematica* notebook of October 24, 2000, e-mail communication, March 3, 2005.

114. Giorgio de Santillana, *The Crime of Galileo*, University of Chicago Press, 1978.

115. David Shavin, A review of [**140**], *21st Century*, (Spring 2004) 48–51.

116. William Shea, Galileo then and now, (conference paper), Center for Philosophy of Science, University of Pittsburgh, October 11–14, 2007.

117. G. F. Simmons, *Differential Equations*, McGraw-Hill, 1972.

118. G. F. Simmons and S. G. Krantz, *Differential Equations*, McGraw Hill, 2007.

119. A. J. Simoson, An *Archimedean* paradox, *Amer. Math. Monthly*, 89 (1984) 114–116, 125.

120. ———, On two halves being two wholes, *Amer. Math. Monthly*, 91 (1984) 190–193.

121. ———, The trochoid as a tack in a bungee cord, *Math Mag.*, 73 (2000) 171–184.

122. ———, *Hesiod's Anvil: falling and spinning through heaven and earth*, Mathematical Association of America, 2007.

123. ———, Finding spring on planet X, *Primus*, 17 (2007) 228–238.

124. ———, Pursuit curves for the Man in the Moone, *College Math. J.*, 38 (2007) 330–338.

125. ———, Albrecht Dürer's trochoidal woodcuts, *Primus*, 18 (2008) 489–499.

126. ———, Solomon's sea and π, *College Math. J.*, 40 (2009) 22–32.
127. ———, A double-minded fractal, *Primus*, 19 (2009) 381–387.
128. ———, Black holes through the *Mirrour*, *Math. Mag.*, 82 (2009) 372–381.
129. James B. Simpson, *Simpson's Contemporary Quotations*, Houghton-Mifflin, 1988.
130. David Singmaster, Problems to sharpen the young: an annotated translation of *Propositiones ad acuendos juvenes*, the oldest mathematical problem collection in Latin, attributed to Alcuin of York, *Math. Gazette*, 76 (1992) 102–126.
131. Allyn Sloan, A. Farquhar, *Dog and Man: The Story of a Friendship*, Ayer Press, 1971.
132. David Eugene Smith, On the origin of certain typical problems, *Amer. Math. Monthly*, 24 (1917) 64–71.
133. Frank D. Stacey, *Physics of the Earth*, John Wiley & Sons, New York, 1969.
134. Sally Stephens, Cosmic collisions, *The Universe in the Classroom*: 23, Astronomical Society of the Pacific, 1993, at http://www.astrosociety.org/education/publications/tnl/23/23.html
135. M. D. Stern, A remarkable approximation to π, *Math. Gazette*, 69 (1985) 218–219.
136. P. A. Steveson, More on the Hebrews' use of π, *School Science and Math*, 65 (1965) 454.
137. Ian Stewart, *Flatterland*, Perseus, Cambridge, MA, 2001.
138. N. M. Swerdlow, Galileo's Horoscopes, *Journal for the History of Astronomy*, 35 (2004) 135–141.
139. Jonathan Swift, *Gulliver's Travels*, Signet, 1960.
140. Mary Terrall, *The Man who Flattened the Earth: Maupertuis and the Sciences in the Enlightenment*, University of Chicago, 2002.
141. Voltaire, *Candide*, translated by Theo Cuffe, Penguin Books, 2005.
142. ———, *The Elements of Sir Isaac Newton's Philosophy*, translated by John Hanna, Frank Cass & Co., London, 1967.
143. ———, *Eléments de la philosophie de Newton*, edited by Robert L. Walters and W. H. Barber, vol. 15 in *Les Œuvres Completes de Voltaire*, The Voltaire Foundation, Oxford, 1992.
144. ———, *Works of Voltaire*, 42 vols., Werner, 1904.
145. ———, *Micromégas, Zadig, Candide*, edited by Réne Pomeau, F. F. Flammarion, 1994.
146. ———, *Thoughts, Remarks, and Observations*, London, 1802.
147. ———, *Memoirs of the Life of Voltaire*, Printed for G. Robinson, 1784.

148. Ira O. Wade, *Voltaire's Micromégas: a study in the fusion of science, myth, and art*, Princeton University Press, 1950.

149. Stan Wagon, A smoother flight to the moon, *College Math. J.*, 39 (2008) 48.

150. Jessica Miller Waterman, *Mechanics in Flatland*, Senior honors paper, King College, 2007.

151. Robert Whitaker, *The Mapmaker's Wife*, Basic Books, 2004.

152. Florence Donnell White, *Voltaire's Essay on Epic Poetry: a study and an edition*, Phaeton Press, New York, 1970.

153. E. A. Whitman, Some historical notes on the cycloid, *Amer. Math. Monthly*, 50 (1943) 309–315.

154. M. P. Wilcocks, *The Laughing Philosopher: being a life of François Rabelais*, George Allen and Unwin, 1950.

155. C. C. Wylie, On King Solomon's molten sea, *Biblical Archaeologist*, 12 (1949), 86–90.

156. A. Zuidhof, King Solomon's molten sea and π, *Biblical Archeologist*, 45 (1982) 179–184.

157. Judith P. Zinnser, *La Dame d'Esprit*, Viking, 2006.

Index

Abbott, Edwin, 75, 321
acceleration, 80
Achilles, 223
Adam and Eve, 243
Aeneid, 42
air, 33
alchemist, 264
Alcuin, 223, 321
Alexander the Great, 69, 321
algorithm, 298
Alpha, xiii, 60
Alps, 268
Amazon River, 126, 323
American Indian, 219
Andersen, Hans Christian, 286, 321
Andes Mountains, 2
André the Giant, 41
angular velocity, 147
animacule, 25
anomaly angles, 106
Antarctica, 269
anti-gravity, 222
aphelion, 102
apogee, 200
Apollo, 222
apparent trajectory, 232
apple, 95
Aquinas, Thomas, 36, 38, 296
Archimedes, xiii, 60, 269, 321
Archimedes's paradox, 54
arclength, 132
arctic circle, 21, 124
Argonaut, 142
argument, 15
Aristarchus, 62
Aristotle, 33, 34, 121, 296, 321, 329
ark, 141
  of the covenant, 245

armlength, 45, 65
Arouet, François-Marie, 71
astrologer, 142
astronomical unit, 8, 47, 108, 343
Aswan, 60
asymptotic solution, 229
Augustine, 14
aurora borealis, 194
axe, ice, 23
axiom system, 298
axis, 145

Babylonians, 243, 252, 253
Bach, C. P. E., 1
Baltic Sea, 20, 22
barycenter, 64
Bastille, 11, 71, 160
bath, 242
Beatrice, 221
Bedlam, xv
Beijing, 18, 110
Benedictine, 259
Beowulf, 41
Bergerac, Cyrano de, 222, 321
Berkeley's ghosts, 52
Berkeley, George, 51, 72, 321
Berlin Academy, 191
Bernoulli
  trial, 322
  Jacob, 112, 322
  Johann, xii, 123, 177, 322
Beta, xiii, 60
binary tree, 67
black
  hole, 55, 198
  plague, 95
Bonaparte, Napoleon, 220
Botswana, 292

367

bottleneck, 176
Bouguer, Pierre, 2, 124, 134, 225, 322
Boyer, Jean-François, 10
brachistochrone, 207, 322
Brave New World, 258
brightness, 80
Brobdingnagians, 44
Buffon's riddle, 327
Buffon, Comte de, 193, 327
bungee cord, 167
Bunyan, John, 44
butterfly effect, xv

Caesar, Julius, 31
calculator, 329
calculus, 51
  of variations, 181, 327
calendar, Gregorian, 141
Calvin, John, 287
cancer, prostate, 280
Candide, 71, 328
cardioid, 174
Carroll, Lewis, 286, 322, 358
Cartesian, 122
Cassini, 191
  Giovanni, 322
  Jacques, 8, 322
Castel, Charles, 19
Castor, 32
Catherine the Great, 18
Catholic, 124
Caxton, William, 186, 189, 322
celestial sphere, 139
Celsius, Anders, 33, 124, 323
censorship, 175
centroid, 303
chaos theory, xv
Charlemagne, 223, 321
Charles IX, 176
Christianity, 237
Church, Catholic, 176
circle
  squarers, 242
  osculating, 138

circumnavigate, 267, 269
Cirey, 159
city of refuge, 244, 254
Clairaut, Alexis, 2, 7, 96, 98, 111, 124, 159, 323
clockwise, 115
collar, 249
Columbus, 222, 267
comet, 101, 272, 291, 325
  Galley, 112
  Halley's, 38, 49, 108, 112, 123
  Kegler, 110
  Kirch, 112
  sun-grazing, 112
  Swift-Tuttle, 110
complexity, 57
compression, 58
Condamine, la, 96, 124, 323
conic, 249
conjunction, 114, 141
conservation
  of angular momentum, 88
  of energy, 178, 207
constant
  of angular momentum, 88
  meta-universal, 81
constellation, 139
continued fraction, 246, 255
contraction, 56
conundrum, 223
Copernicus, 121, 174, 222, 225, 230, 263, 269, 277, 323
core sample, 267
correlation coefficient, 254
Cortile della Pigna, 43
Cosmotheoros, 217
countable, 337
  set, 297
counterclockwise, 115, 250
cross product, 146
crossbow, 265, 268
Crusades, 31
cryptology, 325
cubit, 41, 251, 252
cup, bottomless, 260
curvature, 138

curve,
    regular, 180
    simple, 180
    smooth, 180
cycloid, 174, 187
Cyclops, 42

da Gama, Vasco, 269
Daedalus, 218
Dante, 42, 221, 324
Darwin, Charles, 327
de Molay, Jacques, 335
debate, Leibniz-Newton, 322
Declaration of Independence, 328
Deism, 237
Delambre, Jean, 134
Derham, William, 12
Descartes, René, 2, 34, 38, 121, 315, 324
diamagnetic, 258
Diet of Europe, 19
dimension
    fourth, 90
    second, 75
    third, 75
Divine Comedy, The, 221, 324
DNA, 59
Dodgson, Charles, 322
dog, 139, 223, 225
    days of summer, 7
    star, 7
Domingo Gonsales, 222
Doré, Gustave, 46
Dr. Akakia, 194
Dr. Pangloss, 328
Dream, The, 221
Dürer, Albrecht, 43, 122, 139, 163, 185, 218, 324

earth
    model, homogeneous, 198
    flattened, 98
eccentricity, 101, 103, 111
eclipse
    lunar, 63
    solar, 61

ecliptic plane, 139
Eden, 273
Egypt, 7, 239
Egyptians, 243, 246
Einstein, Albert, 325
Elements of Newton's Philosophy, 98, 101
Elijah, 321
    of Vilnah, 245
ellipse, 101, 249
Elysian Fields, 273
Emily
    a figure of, 17
    and superstition, 238
    and Voltaire, 160
    different perspectives, xiii
    life summary, 324
    meets Voltaire, 96
emperor's new clothes, 286
energy, kinetic, 305
Enlightenment, 96, 258
Entebbe, 207
entropy, 296
epic poetry, 95
epicycloid, 164
epitrochoid, 164
equator, 2, 123
equinox, 103
    spring, 146
Eratosthenes, xiii, 60, 134, 324
Erdős, Paul, 119, 324
escape velocity, 78
Essential Nature, 10
Euclid, 10
Eudoxus, 62
Euler's equation, 153
Euler, Leonhard
    and calculus of variations, 180
    and Laplace transforms, 309
    and the apple, 96
    and the gamma function, 308
    as tutor, 189
    life summary, 324
    Sophia's story of, 1
Eusthenes, 263

experiment
  repeatable, 275
  thought, 348
exponential growth, 306
extension, 16

fathom, 28
Fermat, Pierre, xiii
fern, 336
Ferney, 118, 161, 280
field, 66, 337
fingerbreadth, 251
fixity, 77, 83
Flatland, xii, 75, 321
Florence, 324
Fontenelle, de, 13, 20, 21, 36, 98, 282, 283, 324
fractal, 30, 56, 285
Franciscan, 259
Frederick I (Sweden), 124
Frederick the Great
  a fan of Voltaire, 48
  and comets, 111
  and the Berlin Academy, 191
  and Voltaire's first visit, 17
  as Micromégas, 12
  dinner party, 1
  letter to Voltaire, 291
  life summary, 325
  militarization, 32
Frederick Wilhelm I, 24, 41, 48
French, 27
  Academy, 19, 36, 96, 123
  Revolution, 71, 134
Friar John, 270
frond, 57

Gagarin, Yuri, 12
Galileo, 65
  and Dante's riddle, 45
  and heliocentrism, 121
  and horoscopes, 139
  and simple harmonic motion, 197
  fall from the moon, 221
  inclines, 174
  life summary, 325

trial of, 11
gamma function, 308
Gangan, Baron, 7
Garden of Eden, 95
Gardner, Martin, 295
Gargantua, 45, 238, 260, 325
Gassendi, Pierre, 321
gematria, 245
Gemini, 32
geometry
  elliptical, 253
  non-Euclidean, xiv, 253
giant, 4, 41, 98
  big, 41
  bigger, 44
  biggest, 47
  height, 260
  super, 37
glass
  blown, 239
  rolled, 239
God, 16, 35, 246
Gödel, Kurt, 298, 325, 330
Godin, Louis, 124
Godwin, Francis, 221, 222
Goldbach, Christian, 308
Golden Altar, 252
Goliath, 41
gravimeter, 85
gravitherm, 90, 338
gravity, 12, 76, 122
  specific, 269
great circle, 207, 253
greatest lower bound, 53
Greenland, 21
Gregorian calendar, 154
Gregory, James, 254
Grendel, 41
Gulf of Bothnia, 23
Gulliver's Travels, 18, 29, 44, 258, 281

Haldane, J. B. S., 35, 44
Hall, Asaph, 18
Halley, Edmond, 101, 112, 325
handsbreadth, 243

Index 371

Hardy, G. H., 282, 325
Hartsoeker, Nicolas, 25
Hawking, Stephen, 198, 295, 325
Hebrew, 245
heliocentric, 323
Henriade, The, 13, 72, 326
Henry II, 287, 326
Henry IV, 13, 326
Henry VIII, 287
Hesiod, 189
Hexagon, A., 80
hinge, 148
Hipparchus, 62, 68, 142, 145, 326
historian, 23
historiographer, 10, 191
hole, 3, 189
holy, 251
Homer, 42, 71
homosexuality, 13
homotopy, 248
horoscope, 139
Hubble Telescope, 299
Huguenot, 160, 326
Huxley, Aldous, 258
Huygens, Christiaan,
 aliens, 217
 and Saturn's rings, 18
 life summary, 326
 size of aliens, 47
hypochondriac, 161
hypocycloid, 163, 164, 177, 207
hypotrochoid, 164

Icarus, 218
impact, earth-comet, 110
impenetrability, 16
inclination, 104
inclines, 174
incompleteness, 325
incompressible fluid, 126
India, 275
Indian Ocean, 61
inertia, 147
Inferno, 42, 64
infimum, 53
infinitesimal, 51

integer, 297
intelligence, artificial, 38
interval, 52
invisible ink, 289
Irving, Washington, 219
Islam, 237
ivory tower, 282

Jabberwocky, 286
Jansenist, 10
Janssen, Zacharias, 21
Jefferson, Thomas, 220, 326, 328
Jerusalem, 12
Jesuit, 10
Jesus, 237, 238
jet, 225
Josephus, 248
Jupiter, 18, 108, 111, 123, 343

Kegler, Ignatius, 18, 110
Kennedy, John F., 326
Kepler's laws, 99, 102
Kepler, Johannes, 101, 121, 139, 221, 326
kilo-cubit, 355
kinetic energy, 178, 204
King Arthur, 291
Kircher, Athanasius, 119, 327
Kirkherus, 119
Kittisvaara, 125
König, Samuel, 194
Kublai Khan, 291

Lagrange, Joseph, 180, 327
Lake Geneva, 263
lamina, 303
Laplace's equation, 153
Laplace, Pierre-Simon, 309, 327
Lapland, xi, 21, 33, 159
Laputa, 18, 258, 259, 281
latitude, 132, 324
Law, Gresham's, 323
law, inverse square, 101
league, 7
Lebesgue, Henri, 55, 327
Leclerc, Georges, 193

Leeuwenhoek, Anton van, 25
Legendre, Adrien-Marie, 308
Leibniz, Gottfried, 34, 35, 191, 327
lemon shape, 122, 322
length, 52
Lesbesgue, Henri, 53
Levites, 244
Lewis, C. S., 219, 258
life, shortness of, 15
light, 16, 80
  year, 32
lightning, 221
Lilliputians, 44
limaçon, 167, 172
line, 245, 247
  of best fit, 243
  of sight, 226
Line, Victoria, 76
Lineland, 54, 82
Linnaeus, Carolus, 323
lion, 178
Lisbon, 269
Little Prince, The, 223, 235
Locke, John, 34, 35, 296, 328
longitude, 322, 324
Lorentz-Fitzgerald contraction, 253
Louis XIV, 12
Louis XV, 124, 191, 328
Louvre, 34
love potion, 328
Lucian, 218, 221
Lucifer's Hammer, 110
Lucretius, 53, 217, 285, 328
Lulli, Jean-Baptiste, 12
lunar unit, 228
Lunatic, 220
Luther, Martin, 259
Lutheran, 124

Maasai, 72
Magellan, Ferdinand, 37, 269
magnet, nickel-cadmium, 258
Magog, xv
major axis, 249
Malebranche, Nicholas, 34, 35
Man in the Moon, 220

man-machine, 25
marathon, 337
Marco Polo, 291
Marcus Aurelius, 291
Mars, 18, 108
Marsden, Brian, 110
Masha'allah Ibn-Athari, 121
Masoretic text, 245
mass, 83
mathematician, 324
matrix, dynamic, 200
matter, 16, 34
Maupertuis, 282
  a deep hole, 323
  a pursuit problem, 225
  as Voltaire's tutor, 190
  charm of, 23
  friend of Emily, 17
  Lapland expedition, xi, 7, 124, 134
  life summary, 328
  list of proposals, 193
  member of the French Academy, 19
  president of Berlin academy, 2
  scandal, 159
Maxwell's equations, 317
measure, 51
  theory, 327
Mechain, Pierre, 134
median, 135
Mediterranean Sea, 20
Menaechmus, 250
Mercator projection, 234
meta-creature, 90
meter, 134
metric, 254
Mettrie, la, 2, 25, 328
Micromégas, 55, 98, 101, 111, 217, 291, 328
  a Frederick figure, 49
  and giants, 41
  and pacifism, 32
  and test to determine intelligence, 22
  and the scientific method, 26
  meaning of, 7

Index 373

on Saturn, 12
story summary, xi
the riddle, 281
microscope, 21, 25, 284
Midway Islands, 357
militarization, 32
Milky Way, 12
Milton, John, 95
mind, 16
experiment, 20
minor axis, 249
Mishnah, 243
mite, dust, 24
Mohammed, 237
Moldavia, xv
moon, 221, 222, 267
moon-men, 219
More, Thomas, 15, 260, 287
Mordaunt, Charles, 44
Moréri, Louis, 18
Moses, 41
Mountain, Spindle, 268
mufti, 10, 12, 18

Napoleon, 328
National Curve Bank, 227
National Science Foundation, 49
natural selection, 327
Nazareth, 12
Nazi Germany, 325
nebula, 12
Neptune, 108
Newton's method, 106
Newton, Isaac
a gravity model for Flatland, 77
a lemon or an orange?, 2
and Bernoulli's riddle, 178
and holes through the earth, 197
and orbits, 101
and the Apocalypse, 292
death, 95
earth as an orange, 122
life summary, 329
the apple, xi
Nicholas of Cusa, 174
Nimrod, 42, 64, 335

Niven, Larry, 110
north star, 108
Northwest Passage, 269
Notre Dame, 274
nova, 55
number
irrational, 54
rational, 54

obliquity, 21
Odin, 335
Oedipus, 71
ontology, 16
optimization, 191
orange, mandarin, 122
orbit, normalized, 341
Orwell, George, 258
outer measure, 53
Outhier, Reginald, 28, 124
oval, 250

Palestine, 31
Panama, 126
Pantagruel, xiii, 45, 260, 263, 287, 289, 329
Panurge, 263
Pappus, 130, 304
parabaloid, 310
parabola, 85
Paracelsus, 266, 329
Paradise Lost, 95
parallax, 193
Paris, 12, 23, 126, 267
parsec, 32
Pascal
Blaise, 10, 174, 326, 329
Jaqueline, 10
passage, northwest, 193
Pentagon, Benjamin, 79
penumbra, 63
perigee, 200
perihelion, 102, 142
perpetual motion, 194
Perseid showers, 110
Peter's Pine, 42
phase angle, 102, 229, 308

Philonides, 69
philosopher
   king, 1, 291
   sham, 291
Philosophical Letters, The, 96
Phoebe, 111
photon, 80
pine-cone, 42
Pisces, 139
plane, ecliptic, 104
Planström sisters, 23
Plato, xv, 35, 291, 329
Pliny the Elder, 14, 31, 61, 69, 275, 329, 334
Plutarch, 197, 217
Poe, Edgar Allan, 222
point-mass, 304
Pointland, 82
Polaris, 108
pole, 123
Pollux, 32
polygon, 76
Pomadour, Marquise de, 328
Pompeii, 329
potential energy, 178, 204
Potsdam, 1
   Giants, 24, 41, 48
Pournelle, Jerry, 110
precession, 93, 142, 326
Prester John, 291
Princess Charlotte, 1, 189, 323
Principia, 98, 101, 106, 122
Principia Mathematica, 330
probability, 326
proportion, 9
Protagoras, 36
Prudent, 124
pseudo-circle, 250, 253, 255
pseudo-metric, 253
Ptolemy, 21, 61
pursuit curve, 221, 227
pyramid, 194
Pyrenees, 267
pyrolithic graphite, xvi, 258
Pythagoras, 273
Pythagorean, 218

quadrant, 23
question, the ultimate, 300
Quito, 2, 123, 207

Rabbi Nehemiah, 247
Rabelais, xii, 29, 286
   and bigger giants, 45
   and Gargantua, 238
   life summary, 329
   the English, 259
Ramanujan, 282, 325
Rami bar Ezekiel, 248
Ramses the Great, 239
rational function, 337
Reamur, René, 30
reasoning, proportional, 98
reduction in order, 227
relativity, 253
retrograde, 123
   motion, 108
Reynard the Fox, 224
Richelieu, Cardinal de, 36
right-hand rule, 146
River Thames, 325
Rosetta Stone, 329
rotation, 58
   energy, 205
   inertia, 305
Rousseau, Jean-Baptiste, 119, 330
Rousseau, Jean-Jacques, 330
Royal Society, 178, 190, 221
Russell, Bertrand, 298, 330

Sagredo, 140
Sahara Desert, 269
Saint Amable, 237
Satan, Dante's, 45
Saturn, 12, 14, 108
Saturn V rocket, 356
scale, 284
Schiaparelli, Giovanni, 110
scholasticism, 34
scientific method, 27
scientism, 258
scientist, Swiftian, 258
Selenia, 273

self-similarity, 57
Septuagint, 242
sequence, 53
set
 countable, 297
 measurable, 55
 measure zero, 54
Shakespeare, William, 35
Sharp, Abraham, 254
shell method, 304
sidereal, 227
Sierpinski triangle, 56
simple harmonic motion, 197
singularity, 63
Sirius, xi, 7, 16
Smollett, Tobias, 195, 330
smooth, 52
Snell's law, 178, 314
Snell, Willebrord, 134, 314, 330
Soanen, Bishop, 18
social contract, 328
Social Contract, The, 330
Socrates, 291, 329
solar
 eclipse, 139
 system, 101, 108
solid of revolution, 304
Solomon, 238, 291, 330
Solomon's Sea, 241, 247
solstice, 103
 summer, 142
 winter, 142
Sorbonne, 260
South Pole, 269
span, 41, 65
spanning angle, 183
spermatozoa, 25
sphere, 77, 121
Sphereland, 80
spider, 168
spring, 101
Square, A., xii, 75, 85, 101, 330
St. Bartholomew's Day Massacre, 326
St. Peter's Basilica, 43
St. Petersburg, 1

standard
 deviation, 254
 double, 251
star
 charts, 326
 magnitude, 21, 32, 38
star, supergiant, 37, 333
Stöffler, Johannes, 141
stone
 kidney, 273
 philosopher, 194
Stuart, Ian, 76
Suleiman the Magnificent, 291
sun, midnight, 270
supergiant stars, 47
Swammerdam, Jan, 30
Swift
 Jonathan, xii, 29, 44, 258, 259
 life summary, 330
 Lewis, 110
Switzerland, 270
symmetry, four-fold, 250
synodic, 228
Syracuse, 321

tabula rasa, 35, 296
tack, 175
tailor, 286
Talmud, 242, 244, 247
tangent, 229
Tell, William, 263
temperature, 33, 323
Temple of Taste, The, 119, 121, 259, 282, 296
temple
 first, 252
 second, 251
Tencin, Archbishop, 18
test, Turing, 295
Thalemege, 269
The Book, 119
theism, 237
theorem, 298
Thor, 137, 201
threshold of hearing, 37
Time Machine, The, 296

Titan, 42
toise, 12, 17
top, 145
Torino, 209
Tornio, 20, 125
torque, 146
tractrix, 227
transform, Laplace, 309, 310
transit time, 183
Treaty of Belgrade, 31
triangulation, 330
trigonometry, 326
trochoid, 163
Tropic
   of Cancer, 142, 274
   of Capricorn, 142
Troy, 42, 146
True Cross, 266
True History, The, 218, 221
Turing, Alan, 295
Tuttle, Horace, 110

Ulysses, 42
unitary fractions, 246
United Nations, 19
Utopia, 15, 260, 272, 287

Valhalla, 137
value, average, 151, 303
Vatican, 43
Verne, Jules, 37, 222
Vesuvius, 329
Virgil, 30
Voltaire, 41, 49, 141, 189, 291
   a few last quotes, 331
   a mathematician?, xi
   a poet, 117
   an axiom of, xiv

   and comets, 108
   and religion, 237
   and the Berlin Academy, 191
   as a mufti, 10
   exile, 11, 95
   height, xiii, 4
   in the Bastille, 71
   life principle, 10
   life summary, 279
   love life, 160
   major talent, 13
   meeting Swift, 44
   name change, 72
   pen-names, 175
   smile of, 12
   social commentary of, 32
   storyteller, 3
   wit of, 13
   with Swift, 258
volume, 16
   of solid of revolution, 130
vortex, 38, 121, 122

water closet, 32
Way Things Are, The (*De Rerum Natura*), 217
Wedell Sea, 270
Wells, H. G., 222, 296
Westminster Abbey, 95
Whitehead, Alfred North, 298, 330
Wilkins, John, 221
winter, 103
word wizard, 331

York, 321

zodiac, 139, 142
Zuckermann, 244, 247

About the Author

Andrew J. Simoson hails from Minnesota, finished high school in the upper peninsula of Michigan, earned a B.S. in mathematics in Oklahoma, and then a Ph.D. in mathematics from Wyoming, found his wife in New York, and raised two sons in Tennessee. He has twice been a Fulbright professor, one year at the University of Botswana in Gaborone, located at the edge of the Kalahari Desert, and another year at the University of Dar es Salaam in Tanzania near the equator on the Indian Ocean.

For thirty years, he has been the chairman of the mathematics and physics department at King College, located in the foothills of the Appalachian Mountains. By providence and good luck, he has won the Chauvenet Prize in 2007 and the George Pólya Award in 2008, the latter prize for the article "Pursuit curves for *the Man in the Moone*" appearing in *The College Mathematics Journal*, a version of which is Chapter IX of this volume. Besides trying to be a mathematician, he and his wife teach ballroom dance as a college activity class each spring term, and go scuba diving for at least a week each summer, usually in the Caribbean. He currently is a member of the MAA and is on the editorial board of *Primus*, a journal on undergraduate mathematics teaching.